T0140260

# Current Topics in Microbiology and Immunology

## Volume 417

More information about this series at http://www.springer.com/series/82

Guido Silvestri · Mathias Lichterfeld
Editors

# HIV-1 Latency

Responsible series editor: Michael B. A. Oldstone

 Springer

*Editors*
Guido Silvestri
School of Medicine
Emory University
Atlanta, GA, USA

Mathias Lichterfeld
The Brigham and Women's Hospital and the
  Ragon Institute of MGH, MIT and
  Harvard
Boston, MA, USA

ISSN 0070-217X          ISSN 2196-9965   (electronic)
Current Topics in Microbiology and Immunology
ISBN 978-3-030-13214-9          ISBN 978-3-030-02816-9   (eBook)
https://doi.org/10.1007/978-3-030-02816-9

This Springer imprint is published by the registered company Springer Nature Switzerland AG
The registered company address is: Gewerbestrasse 11, 6330 Cham, Switzerland

# Preface

The history of medicine includes only few examples of individual patients who changed an entire scientific paradigm. Approximately 10 years ago, Tim Brown, an HIV-infected individual with leukemia, received a bone marrow transplant with hematopoietic stem cells expressing a defective version of the viral co-receptor CCR5, conferring cell-intrinsic resistance to HIV. As the first and only patient so far, he subsequently developed what appears to be a sterilizing cure of HIV infection, with no residual detectable HIV-1 infected cells using the most sensitive detection technologies. This episode appeared to violate long-established concepts of HIV-1 disease pathogenesis: HIV integrates into host chromosomes, becomes part of the human DNA, remains transcriptionally silent, and stays inside the human body for the lifetime of a patient. If at all, a cure of HIV-1 infection could only be expected in this scenario in patients undergoing extremely long periods of continuous, completely suppressive antiretroviral therapy (>70 years), due to the remarkably long half-life of viral reservoir cells. Tim Brown's story initiated nothing short of a scientific revolution: A cure for HIV infection, previously considered elusive, was suddenly within the range of what was thinkable, and could be regarded as an increasingly realistic objective, at least for some infected patients. The search for strategies to induce a cure has since then triggered large investments by the NIH and philanthropic organizations, which allowed to initiate global research efforts to identify and develop ways to eliminate residual viral reservoirs, enhance host immunity to HIV and/or combine these efforts to induce a long-term drug-free remission of HIV-1 infection.

This book is published approximately 10 years after the transformative first description of a cure of HIV-1 infection and in many ways reflects the remarkable progress that has been made since then in defining the mechanisms of HIV-1 long-term persistence in the human body, and in understanding the fundamental challenges and scientific problems that would have to be overcome to achieve a cure. Central to these efforts is the concept of HIV-1 latency, which is operationally defined as a replication-competent HIV-1 provirus that is integrated in host DNA, but not actively expressed, due to a variety of host factors that actively repress or insufficiently support viral transcription. Increasingly, it is recognized that this transcriptional silence offers unique advantages to HIV-1: During latency, HIV-1 infected cells remain unrecognizable and undetectable by immune cells, which represents a highly effective strategy to escape from antiviral host immunity. In addition, viral latency reduces cytopathic effects associated with HIV-1 replication, which arguably enhances the survival and persistence of cells harboring chromosomally integrated HIV-1. Finally, the fact that HIV-1 remains transcriptionally silent does not mean that it cannot be amplified and expanded; indeed, substantial evidence from a number of studies now demonstrates that whenever the infected host cell divides and proliferates, the HIV-1 genome is automatically duplicated—a highly elegant mechanism by which the number of virally infected cells can be exponentially expanded, as a passive bystander of host cell proliferation. From a clinical perspective, viral latency is now frequently regarded as the main barrier to viral eradication and cure, and a variety of pharmacological agents that disrupt the transcriptional silence of latently infected cells and sensitize cells to host immune recognition are now in clinical development.

This book includes multiple chapters approaching the fascinating area of HIV latency and persistence from multiple perspectives and scientific directions: Van Lint et. al provide a detailed summary of the current understanding of molecular pathways and mechanisms that govern HIV transcription and latency, and identify key targets for pharmacological interventions designed to manipulate viral latency. Drs. Siliciano & Siliciano contributed an update on technologies used for viral reservoir quantification—an area that they have pioneered from the very beginning when viral latency was first recognized. Interactions between viral reservoirs and antiviral immune responses are discussed in a detailed review by Blankson et al., and the possible role of HIV-1-associated immune activation and pro-inflammatory stimuli for maintaining and supporting viral persistence are described by Sereti et al. Two dedicated chapters focus on interventions to reduce or eliminate persisting viral reservoirs: Anaworanich et al. discuss immune-based interventions, while Kiem et al. review cell- and gene-therapy-oriented approaches. Dr. Hill has kindly contributed a comprehensive summary of mathematical models and computational approaches that can support and inform the understanding of viral reservoir persistence. Finally, Lifson et al. summarize non-human primate and animal models for HIV-1 cure research, and Clemens et al. specifically focus on the possible role of infected macrophages for viral persistence. Together, these manuscripts provide a diverse, in-depth analysis of current concepts, paradigms, and ideas that drive the HIV-1 cure agenda, and identify areas that require specific emphasis in future studies. We sincerely thank all authors for their time and effort in writing these manuscripts and hope that this book will become an interesting and informative resource for readers committed to finding a cure for HIV-1 infection.

Atlanta, USA                                                    Guido Silvestri, M.D.
Boston, USA                                          Mathias Lichterfeld, M.D., Ph.D.
July 2018

# Contents

# Molecular Control of HIV and SIV Latency

**Gilles Darcis, Benoit Van Driessche, Sophie Bouchat, Frank Kirchhoff and Carine Van Lint**

**Abstract** The HIV latent reservoirs are considered as the main hurdle to viral eradication. Numerous mechanisms lead to the establishment of HIV latency and act at the transcriptional and post-transcriptional levels. A better understanding of latency is needed in order to ultimately achieve a cure for HIV. The mechanisms underlying latency vary between patients, tissues, anatomical compartments, and cell types. From this point of view, simian immunodeficiency virus (SIV) infection and the use of nonhuman primate (NHP) models that recapitulate many aspects of HIV-associated latency establishment and disease progression are essential tools since they allow extensive tissue sampling as well as a control of infection parameters (virus type, dose, route, and time).

Gilles Darcis, Benoit Van Driessche—Equal contribution.

G. Darcis · B. Van Driessche · S. Bouchat · C. Van Lint (✉)
Service of Molecular Virology, Département de Biologie Moléculaire (DBM), Université Libre de Bruxelles (ULB), Rue des Professeurs Jeener et Brachet 12, 6041 Gosselies, Belgium
e-mail: cvlint@ulb.ac.be

G. Darcis
Service des Maladies Infectieuses, Université de Liège, CHU de Liège, Domaine Universitaire du Sart-Tilman, B35, 4000 Liège, Belgium

G. Darcis
Laboratory of Experimental Virology, Department of Medical Microbiology, Academic Medical Center of the University of Amsterdam, Meibergdreef 15, 1105, AZ Amsterdam, The Netherlands

F. Kirchhoff
Institute of Molecular Virology, Ulm University Medical Center, Meyerhofstraße 1, 89081 Ulm, Germany

Current Topics in Microbiology and Immunology (2018) 417:1–22
DOI 10.1007/82_2017_74
© Springer International Publishing AG 2017
Published Online: 26 October 2017

**Contents**

# 1 Introduction

HIV latency is a key hurdle to curing HIV. The HIV latent reservoirs are defined as a cell type or anatomical site where a replication-competent form of the virus persists for a longer time than in the main pool of actively replicating virus (Van Lint et al. 2013). This definition mainly restricts the viral reservoirs to latently infected resting CD4$^+$ memory T cells carrying stably integrated, transcriptionally silent but replication-competent proviruses. These cells do not produce virus particles while in resting state, but can give rise to infectious virus following activation by several stimuli, leading to viral rebound when antiretroviral therapy (ART) is stopped (Chun et al. 1995, 1997, 2000; Finzi et al. 1999; Siliciano et al. 2003; Davey et al. 1999). A less conventional, wider definition of HIV reservoirs has also been proposed: all infected cells and tissues containing all forms of HIV persistence that can participate in HIV pathogenesis (Avettand-Fenoel et al. 2016). This definition includes defective proviruses which participate to HIV pathogenesis through viral transcription and synthesis of viral proteins without new virion production. These proteins can induce and maintain immune activation, thus participating in the vicious circle of HIV pathogenesis (Avettand-Fenoel et al. 2016).

The mechanisms conducting to the establishment of HIV latency but also to its maintenance probably vary from one patient to the other, from one tissue or one anatomical compartment to the other, and also from one cell type to the other (Darcis et al. 2017). Therefore, a cure for HIV is unlikely achievable without considering all latent cellular and anatomical reservoirs such as the brain (Kumar et al. 2014).

HIV is divided into HIV type 1 (HIV-1) and HIV type 2 (HIV-2). HIV-1 is responsible for the HIV pandemic and is related to viruses found in chimpanzees and gorillas, while HIV-2 is related to viruses found in primate sooty mangabey. HIV-1 may be further divided into groups (M, N, O, and P) and subtypes within the M group. Simian immunodeficiency virus (SIV) infection and the use of nonhuman primate (NHP) models that recapitulate HIV-associated disease progression are essential tools. Indeed, NHP models of ART-treated macaques infected with the simian immunodeficiency virus of macaques (SIVmac), which is more closely related to HIV-2 in comparison with HIV-1, have been validated and help to characterize the type, establishment, maintenance, and activation of latent viral reservoirs (Deleage et al. 2016). Importantly, unlike nonpathogenic infection in its African natural host, SIVmac induces an AIDS-like disease in Asian rhesus macaque monkeys with similar symptoms and immunological consequences seen in HIV-infected humans. The use of SIV latter in this chapter specifically refers to SIVmac.

NHP infected with SIV provides several significant advantages, including the possibility to perform extensive tissue sampling in animal and elective necropsy. Since the huge majority of viruses persists under ART resides in tissues that are difficult to access in human clinical settings, this is undoubtedly the main benefit.

During the past few years, important progress has been made to characterize the viral reservoirs, to understand the molecular mechanisms underlying HIV/SIV latency, and to better investigate and address the crucial questions of the complexity, diversity, and dynamics of these mechanisms.

In this chapter, we consider our present knowledge of the molecular mechanisms involved in HIV-1 and SIV latency. To begin, we present a brief description of the HIV-1 and SIVmac promoters, which will be of great importance for the subsequent discussion.

## 2   The HIV-1 and SIVmac Promoters

Most of the HIV-1 and SIVmac transcripts are initiated at the main viral promoter located in the 5' long terminal repeat (5'LTR) region. The 5'LTR has been divided into three regions [U3 (unique in 3'), R (repeated), and U5 (unique in 5')] and into four functional domains (from the 5'end to the 3'end: the modulatory region, the enhancer composed of a distal region and a proximal region, the core promoter and the leader region that extends until the first codon of the gag gene) (Fig. 1a). Importantly, this latter region encodes the trans-activating response (TAR) element whose RNA forms a stable stem-loop structure (Fig. 1b). The TAR hairpin is present at the 5'end of each transcript and allows the recruitment of the viral transactivator protein Tat.

The strength of the HIV-1 promoter is modulated by cellular factors and its chromatin environment (see below). Indeed, the 5'LTR of HIV-1 contains several DNA-binding sites for various cellular transcription factors (TFs), including Sp1 and NF-κB, that are important for HIV-1 replication, whereas other sites, such as NF-AT, LEF-1, COUP-TF, Ets1, USF, and AP-1 binding sites, enhance transcription without being indispensable (Colin and Van Lint 2009).

In the absence of Tat, critical TFs, such as NF-κB and Sp1, are required for the formation of the pre-initiation complex leading only to the production of short transcripts, while in the presence of Tat, transcription is enhanced and full-length

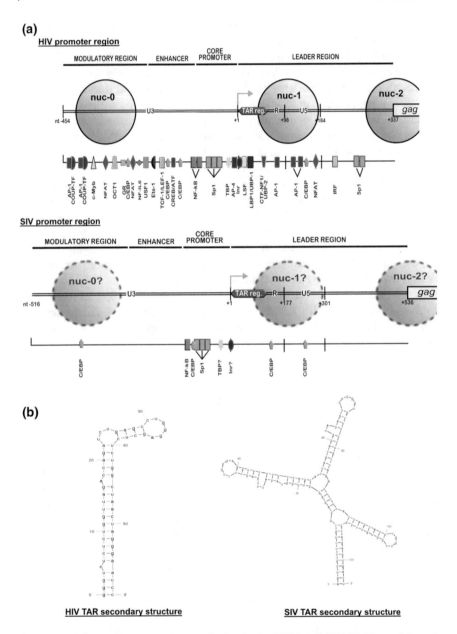

**Fig. 1** Comparison of the molecular organization in the HIV-1 and SIV 5'LTRs. **a** Schematic representation of the main transcription factor binding sites located in the 5'LTR and in the leader region of HIV- 1 (upper panel) and of SIV (lower panel). Nucleotide +1 (nt +1) is the transcriptional start site for both viruses. The U3, R, U5, and leader regions as well as the different functional regions involved in transcriptional regulation are indicated. Moreover, nucleosomal organization of the 5'LTR is shown for HIV-1. Putative nucleosome positions on the SIV 5'LTR are also shown with dashed lines. **b** TAR secondary structures for HIV-1 (HXB2 isolate; GenBank: K03455.1) and for SIV (SIVmac239 isolate; GenBank: M33262.1) were determined using the Mfold web server. While HIV-1 TAR exhibits a hairpin structure, most of the SIV TARs present a three-loop structure (Berkhout 1992)

viral transcripts are synthetized. However, in addition to its classically recognized role in the induction of transcriptional elongation and chromatin remodeling, Tat may also influence transcriptional initiation by facilitating the assembly of the pre-initiation complex requiring the Sp1 and NF-κB binding sites (Brady and Kashanchi 2005). Interestingly, in this context, recent studies from Ben Berkhout's laboratory demonstrate that Tat(HIV) and Tat(SIV) also stimulate HIV-1 or SIV gene expression, respectively, independent of the TAR hairpin, via Sp1 sequence elements in the U3 promoter region (van der Velden et al. 2012; Das et al. 2011).

The three Sp1 binding sites present on the core promoter play a role on HIV-1 transcription recruiting the pre-initiation complex and the transcriptional factor Sp1 that serves as a recruitment platform for modifying chromatin complexes. The TF Sp1 is a ubiquitous factor that can lead to a positive or negative transcriptional effect depending on additional recruited factors. Sp1 bound to the U3 sites can have a negative effect by recruiting histone deacetylases (HDAC1 and HDAC2) to promote histone H3 and H4 deacetylations (Marban et al. 2005, 2007). In microglial cells, the CNS-resident macrophages, this recruitment requires the cofactor CTIP-2 (COUP-TF interacting protein 2). Indeed, the group of Rohr, in collaboration with our laboratory, has demonstrated that Sp1 recruits a multi-enzymatic chromatin-modifying complex including HDAC1, HDAC2, and SUV39H1 to the viral promoter, where CTIP-2 allows deacetylation of the ninth lysine of the N-terminal tail of histone H3 (H3K9), which is a prerequisite for H3K9 trimethylation by SUV39H1 (Marban et al. 2005). This last histone modification allows heterochromatin protein 1 (HP1) binding and polymerization. Interestingly, the Rohr's group reported displacement of CTIP-2 and subsequent recruitment of CREB-binding protein (CBP) through Sp1 following HIV-1 activation with phorbol esters (Marban et al. 2007). In CD4+ T lymphocytes, another study demonstrated that c-Myc is recruited to the HIV-1 5'LTR by Sp1 and in turn recruits HDAC1 in order to blunt HIV-1 promoter expression (Jiang et al. 2007). Following activation, cellular histone acetyltransferases (HATs), including p300/CBP, PCAF, and Gcn5, are recruited to the promoter region, leading to the acetylation of both H3 and H4 histones via several TFs such as Sp1 (Marsili et al. 2004). Interestingly, HMBA causes the release of P-TEFb from HEXIM1 and triggers CDK9 recruitment to the HIV-1 5'LTR via an unexpected interaction with the transcription factor Sp1 (Choudhary et al. 2008).

Otherwise, NF-κB binding sites are found in the enhancer region of all primate lentiviral LTRs, although their numbers may vary between different groups of SIV and HIV-1. Most subtypes of pandemic HIV-1 group M strains (A, B, D, F, G, H, J, and K) and some SIVs contain two NF-κB binding sites located −104 to −80 bp upstream of the transcriptional start site (Fig. 1a). However, HIV-1 group M subtype C strains, which account for almost 50% of HIV-1 infections worldwide, typically contain three binding sites for NF-κB in their enhancer region (Heusinger and Kirchhoff 2017). In contrast, subtype A/E recombinants of HIV-1 group M, the human immunodeficiency virus type 2 (HIV-2), and several SIV lineages contain just a single NF-κB binding site. Typically, mutations in the NF-κB binding sites of HIV-1 LTRs prevent efficient proviral transcription.

Another well-characterized cellular TF is the C/EBP (CCAAT/enhancer-binding protein) family for which three binding sites have been identified in the HIV-1 LTR and four binding sites in the SIVmac LTR (Ravimohan et al. 2010; Hogan et al. 2003). Functionally, these sites are involved in activation of HIV-1 transcription and are important for viral replication in the monocyte–macrophage lineage, but not in T cell lines. The regulation of HIV-1 transcription and replication in macrophages is mediated primarily by the two isoforms of C/EBPβ, the liver-enriched transcriptional activator protein (LAP) and liver-enriched transcriptional inhibitory protein (LIP) translated from the second and third in-frame AUG, respectively, and in these cells at least one functional C/EBP binding site within the HIV-1 LTR is necessary for basal level transcription and replication (Ravimohan et al. 2012). In the context of SIV, three of the four sites have been shown acting as negative regulators of SIV basal transcription, while the last binding site is associated with positive regulation of basal viral transcription [reviewed in (Liu et al. 2009)]. These differences could be explained by the differential recruitment to the SIV LTR of the C/EBPβ2 isoform (LAP) or the C/EBPβ3 isoform (LIP), which present an activator or repressor activity, respectively (Barber et al. 2006).

HIV-1 and SIV transcriptions are consequently coupled with the cellular activation status and by the abundance of cellular transcription factors that can either induce or repress viral promoter activity depending on the cell types. Interestingly, besides the presence of DNA-binding sites in the HIV-1 promoter region, several ubiquitous and cell-specific TFs have also been shown to be recruited to part of the *pol* gene coding for the integrase and to have an important impact on viral infectivity [(Goffin et al. 2005), reviewed in (Van Lint et al. 2013)].

Moreover, nucleosome positioning in the HIV-1 promoter appears to be specific and dynamic, supporting a major implication during latency and transcriptional activation. In latent conditions, two nucleosomes (named nuc-0 and nuc-1) are precisely situated at the proviral promoter (Fig. 1a). Nuc-0 is located immediately upstream of the modulatory region and nuc-1 immediately downstream of the viral transcription start site (TSS). The position of those nucleosomes in the 5'LTR appears to be an intrinsic property of the LTR. Indeed, the same positions were observed independently of the integration sites in different cell lines (Van Lint et al. 2013). Notably, during HIV-1 transcriptional activation, the organization of nuc-1 but not of others nucleosomes present on the HIV-1 genome is disrupted (Verdin et al. 1993). To our knowledge, such a precise nucleosome organization of the SIV promoter has not been described yet.

# 3 Regulation of HIV-1/SIV Transcription

Latency is established and maintained through multiples mechanisms acting in concert and operating mostly at the transcriptional level but also at several post-transcriptional steps. Regarding transcriptional regulation, HIV-1/SIV latency results in a complex and variable combination of multiple elements acting at the initiation and/or at the elongation phases of transcription. This heterogeneous and dynamic combination of transcriptional repression mechanisms impedes the synthesis of the viral trans-activating factor Tat, a viral protein indispensable for profound activation of HIV-1 and SIV transcription.

## 3.1  Nuclear Topography

Besides the organization of the genetic information itself, the cellular factors associated with transcription, replication, and genomic architecture are structured in sophisticated patterns within the nucleus (Lamond and Sleeman 2003). TFs, chromatin-associated proteins, and RNA-processing factors are confined to precise nuclear areas corresponding to distinctive tasks. In addition, transcription as well as replication occurs at spatially definite nuclear sites (Misteli 2007). Therefore, the nuclear topography of HIV-1/SIV integration may drastically influence its transcriptional level.

The HIV-1 pre-integration complex targets regions of chromatin that are close to the nuclear pore (NP). In contrast, it excludes the internal regions in the nucleus and the peripheral regions associated with the nuclear lamina (Marini et al. 2015). This integration near the NP corresponds to the first open chromatin regions that HIV-1 meets after its entrance into the nucleus (Marini et al. 2015).

The nuclear pore complex (NPC) interacts with specific chromosomal areas, called nucleoporin-associated regions, and contributes to the organization of the three-dimensional nuclear architecture (Capelson et al. 2010). Therefore, the NPC provides a chromatin topology and a nuclear environment favoring HIV-1 transcription. Indeed, the roles of the nucleoporins Tpr and Nup153 have been well demonstrated since the silencing of Tpr and Nup153 leads to a reduction of HIV-1 transcription in infected CD4+ T cells (Marini et al. 2015). Nup153 and Tpr play distinct but complementary roles in the HIV-1 integration process (Lelek et al. 2015). Nup153 is required for HIV-1 nuclear import. Tpr remodels chromatin regions proximal to NPC in a state encouraging HIV-1 transcription (Lelek et al. 2015). Those NPC components are consequently needed for proper HIV-1 genome integration. Following the nucleoporin knockdown, the structure of the chromatin is reformed. This remodeling leads to HIV-1 integration into nuclear regions that are less favorable to an efficient viral gene transcription (Wong et al. 2015).

Transcriptionally silenced but replication-competent HIV-1 proviruses might therefore reside in areas refractory to viral transcription, for instance, in close proximity to promyelocytic leukemia nuclear bodies (PML NB). HIV-1 gene expression inhibition in those nuclear regions is dependent on epigenetic mechanisms. Proviruses typically exhibit transcriptionally inactive heterochromatic marks such G9a-mediated H3K9 dimethylation (Lusic et al. 2013). Other studies suggest that PMLs impede transcriptional elongation through the sequestration of cyclin T1, a subunit of the transcription elongation factor P-TEFb (Marcello et al. 2003;

Doucas et al. 1999). The HIV-1 nuclear topography is therefore directly linked to various mechanisms implicated in HIV-1 transcriptional regulation.

## 3.2 Viral Integration Site

After reverse transcription in the cytoplasm, both viral and cellular proteins associated with the viral cDNA form the pre-integration complex that next migrates into the nucleus. Upon nuclear entry, the viral DNA is integrated into chromatin (see above). This process of HIV-1 integration is a nonrandom process. Indeed, the cellular lens epithelium-derived growth factor (LEDGF/p75) that binds both cellular chromosomal DNA and HIV integrase directs integration preferentially to introns of actively transcribed genes (Wagner et al. 2014). Interestingly, in the absence of LEDGF/p75, integration is still not a random process. Residual integration is then largely facilitated by the Hepatoma-derived growth factor-related protein 2 (HRP-2), another unique cellular protein containing an integrase-binding domain (Schrijvers et al. 2012).

While nothing is known about the SIV nuclear topography, it has been shown that SIVmac integration is also predominant in introns of actively transcribed genes (Crise et al. 2005; Hematti et al. 2004). However, SIV integration sites were determined after in vitro infection of a human lymphoid cell line (Crise et al. 2005) or in rhesus monkeys transplanted for at least 6 months with autologous SIV-transduced CD34$^+$ cells (Hematti et al. 2004). Therefore, additional analyses of the SIV integration sites during natural infection are needed to ensure that SIV and HIV-1 present similar integration site preferences.

How can we explain HIV-1 transcriptional repression when integration occurs in highly expressed genes? Several mechanisms impeding promoter activity including steric hindrance, enhancer trapping, and promoter occlusion could occur depending on the orientation of the HIV-1 genome within the cellular transcriptional unit (Van Lint et al. 2013; Shan et al. 2011):

- Steric hindrance is a phenomenon that may occur when the provirus integrates downstream and in the same transcriptional orientation as the cellular host gene. The "read-through" RNA polymerase transcription from the upstream cellular promoter displaces key transcription factors from the HIV-1 promoter and prevents assembly of the pre-initiation complex on the viral promoter.
- Enhancer trapping may occur when the enhancer located in the HIV-1 5'LTR is placed near the promoter of a cellular gene and acts on the transcriptional activity of this cellular promoter, thereby preventing the enhancer action on the viral promoter.
- Promoter occlusion occurs when a provirus integrates into the opposite orientation compared to the host gene. This may lead to collisions between the RNA polymerase complexes elongating from the viral and cellular promoters, resulting in a premature termination of transcription from the weaker or from both promoters.

Latently infected transformed cell lines provide good examples of the influence of HIV-1 integration site on basal transcriptional rate. For instance, J-Lat cell lines carry a unique provirus and can be distinguished by the integration site of this provirus in the cellular genome. Moreover, the HIV-1 genome integrated into these cells contains the green fluorescent protein (GFP) gene (Jordan et al. 2003). It is therefore easy to evaluate the HIV-1 transcriptional level. Intriguingly, there is a 75-fold difference in basal expression level between the highest and lowest expressing clones. Those differences in expression levels are due to diversity of integration sites. Differential levels of GFP expression correlate with integration in (i) gene deserts, (ii) centromeric heterochromatin, and (iii) very highly expressed cellular genes, suggesting that viral integration site, along with cellular environment, influences the balance between latency and proviral expression (Lewinski et al. 2005).

In this context, Chen et al. developed a method called barcoded where HIV ensembles to map the chromosomal locations of thousands of proviruses while tracking their transcriptional activities in an infected cell population (Chen et al. 2017). They showed that HIV-1 expression is strongest close to endogenous enhancers and that the insertion site also affects the response of latent proviruses to reactivation.

Undoubtedly, the insertion context of HIV-1 is a critical determinant of latency and viral response to reactivation therapies (see chapter "LATENCY REACTIVATION AS A STRATEGY TO CURE HIV"). The site of integration may also have another important effect on HIV-1 persistence: it may impact the survival of latently infected cells if HIV-1 integration occurs into genes associated with cell cycle regulation, such as MKL2 or BACH2. In this precise situation, HIV-1 integration thus confers a survival advantage that allows these cells to proliferate and expand despite a potent and suppressive ART (Wagner et al. 2014), and hence propagating HIV-1 without viral replication. The presence of such a clonal extension in the context of SIV infection has never been reported so far.

## 3.3   Epigenetic Modifications of HIV-1/SIV Promoter

In eukaryotic cells, DNA is wrapped around a nucleosome composed of a histone octamer. The histone tails are subject to multiple post-translational modifications including acetylation, phosphorylation, sumoylation, ubiquitination, and methylation. These reversible epigenetic marks modify gene expression by changing chromatin condensation which dictates the accessibility of DNA to TFs and transcription machineries. Epigenetic modifications are catalyzed by several chromatin-modifying enzymes such as histone acetyltransferases (HAT), histone deacetylases (HDAC), DNA methyltransferases (DNMT), or histone methyltransferases (HMT).

The chromatin structure and the epigenetic control of the HIV-1 promoter (5'LTR) are key mechanisms underlying transcriptional regulation and thus latency.

As stated earlier in this chapter, two nucleosomes named nuc-0 and nuc-1 are localized in the HIV-1 5'LTR in latently infected cell lines (Verdin et al. 1993). Nuc-1 is situated immediately downstream of TSS and contributes to the blockage of transcriptional elongation. The two nucleosomes on the promoter of latent proviruses are characterized by epigenetic modifications, described below, that contribute to transcriptional repression. Nevertheless, multiple stimuli can change these epigenetic modifications to induce nuc-1 remodeling and therefore favor transcriptional initiation and elongation (Fig. 2) (Van Lint et al. 1996).

HATs and HDACs influence transcription by selectively acetylating or deacetylating the $\epsilon$-amino group of lysine residues in histone tails, respectively. HATs favor chromatin opening and thus increase the accessibility of TFs to their binding sites. Moreover, histone acetylation marks enable the recruitment of bromodomain-containing proteins, such as chromatin remodeling complexes and TFs, which in turn regulate gene expression. In contrast, deacetylation by HDACs promotes a repressive heterochromatin environment (Yang et al. 2007). HDAC1, HDAC2, and HDAC3 are recruited by transcriptional repressors to the HIV-1 LTR5' that typically displays deacetylated histones in latent condition. Therefore, deacetylation of the HIV-1 promoter chromatin by these enzymes plays a role in the establishment and maintenance of HIV-1 latency (Colin and Van Lint 2009). Indeed, treatment of infected cells with HDAC inhibitors allowing a global increase of histone acetylation and the remodeling of nuc-1 is coinciding with activation of HIV-1 gene expression (Verdin et al. 1993). Mechanistically, in microglial cells which constitute an important reservoir in the brain (Alexaki et al. 2008), the corepressor CTIP2 (COUP-TF Interacting Protein 2) acts as a recruitment platform for HDAC1 and HDAC2 on the HIV-1 promoter, leading to a heterochromatin environment [reviewed in (Le Douce et al. 2014)].

While histone hypoacetylation is generally associated with transcriptional repression, histone methylation can be either associated with transcriptional repression or activation, depending on the site of modification. H3K9 trimethylation (H3K9me3) and H3K27 trimethylation (H3K27me3) are patterns associated with transcriptional repression and have been shown to be associated with HIV-1 transcriptional silencing in different postintegration latency models (Imai et al. 2010; Friedman et al. 2011).

The HMT enhancer of Zeste homolog 2 (EZH2) is required for H3K27me3. This HMT is present at high levels in the LTR region of silenced HIV-1 proviruses and is rapidly displaced following proviral reactivation (Friedman et al. 2011). EZH2 seems to be of particular importance since the knockdown of this enzyme strongly induced HIV-1 expression compared to the knockdown of SUV39H1, an HMT required for H3K9me3 that has been shown to be recruited by CTIP2 to the HIV-1 LTR in microglial cells (Nguyen et al. 2017). Notably, EZH2 interacts—within the context of the Polycomb repressive complexes 2 and 3 (PRC2/3)—with DNA methyltransferases (DNMTs) and associates itself with DNMT activity in vivo, highlighting a direct connection between two key epigenetic repression systems (Vire et al. 2006). Additionally, the euchromatic histone-lysine N-methyltransferase 2 (EHMT2 also called G9a) is implicated in the control of

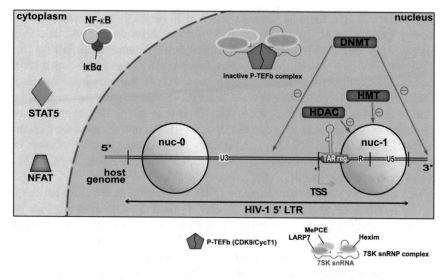

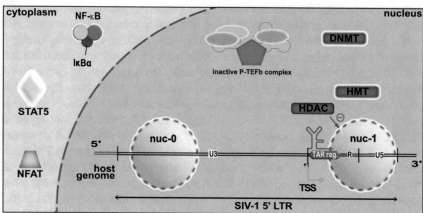

**Fig. 2** Comparison of the molecular mechanisms of transcriptional repression in HIV-1 and SIV. During latency, HIV-1 transcription (upper panel) is repressed (i) by the presence of a nucleosome (nuc-1) located immediately downstream of the transcription start site (TSS), (ii) by a repressive epigenetic environment (DNA methylation, and histone deacetylation and methylation through the action of different classes of epigenetic writers), (iii) by sequestration of important inducible cellular transcription factors (such as NF-κB, STAT5, or NF-AT) in the cytoplasm, and (iv) by the sequestration of P-TEFb in the inactive 7SK snRNP complex. In the case of SIV (lower panel), tight regulation of the NF-κB pathway and importance of histone acetylation are involved in viral latency. However, the other mechanisms have not been yet studied

latency. While both EZH2 and EHMT2 are required to silence HIV-1 proviruses and are recruited to the LTR in latently infected Jurkat T cells, PRC2 is distinctive because it controls the major rate-limiting step restricting proviral reactivation. In contrast, in both the primary cell models and in cells isolated from HIV[+]-treated patients, PRC2 and EHMT2 are both required to establish and maintain HIV-1 latency (Nguyen et al. 2017). Boehm et al. further performed a systematic small hairpin RNA (shRNA) knockdown of cellular HMT (Boehm et al. 2017). They identified SET and MYND domain-containing protein 2 (SMYD2), a member of the SMYD family of methyltransferases, as a HIV-1 transcriptional repressor. SMYD2 has been previously shown to regulate transcription by methylating H3K36 and H3K4 (Brown et al. 2006; Abu-Farha et al. 2008). Boehm et al. further demonstrated the recruitment of SMYD2 to HIV-1 latent promoter and the presence of H4K20me1 at the 5'LTR. Interestingly, this epigenetic mark allows the recruitment of an MBT (malignant brain tumor) family member, L3MBTL1, a reader protein linked to PRC1 (Boehm et al. 2017).

In addition to histone modifications, DNA methylation at cytosines located in CpG dinucleotides also participates in HIV-1 transcriptional silencing. During latency, the HIV-1 promoter is hypermethylated at two CpG islands surrounding the HIV-1 transcription start site. Methylation of the promoter region is generally associated with gene silencing, either by directly blocking binding of transcription factors to their recognition sequences or indirectly through the recruitment of methyl-CpG-binding domain proteins (MBDs) which in turn interact with HMTs and with HDACs, leading to a repressive chromatin structure (Suzuki and Bird 2008). This link between DNA methylation and histone epigenetic marks is important for our understanding of the establishment of a latent infection (Blazkova et al. 2009; Kauder et al. 2009). In patients' cells, DNA methylation of the HIV-1 promoter increases progressively during ART treatment, suggesting that this epigenetic mark could participate more to viral persistence than to latency establishment (Trejbalova et al. 2016). Indeed, Trejbalova et al. detected low levels of 5'LTR DNA methylation in resting CD4+ T cells of patients who were ART-treated for up to 3 years. But, after long-term suppressive ART, they observed an accumulation of 5'LTR DNA methylation in the latent reservoir (Trejbalova et al. 2016). The exact mechanism of this DNA methylation accumulation in the latent reservoir of HIV-1-infected individuals remains unclear but has a potential impact on HIV-1 reactivation from latency, one of the most explored cure strategies.

Regarding HIV-1 transcriptional regulation, it has been well demonstrated that a great number of epigenetic modifications participate in the establishment or the maintenance of HIV-1 latency [reviewed in (Van Lint et al. 2013)]. However, much less is known in the context of SIV infection (Fig. 2). It has been shown that acetylation of histone H4 is detected during active/acute SIV replication in the brain (Barber et al. 2006). In contrast, during the asymptomatic infection, when full-length viral transcripts become undetectable, acetylation of histone H4 is lost (Barber et al. 2006). This reduction of acetylation seems linked to the recruitment of LIP to the SIV LTR. Indeed, in contrast to LAP, LIP is unable to interact and recruit to the LTR histone acetyltransferases activity (Barber et al. 2006). In addition, administration of the HDAC inhibitor SAHA (vorinostat) induced an increase in viral expression in ex vivo culture of CD4+ T cells (Ling et al. 2014) or in virally suppressed macaques (Gama et al. 2017), respectively. These data reinforce the importance of histone acetylation and epigenetic modifications in the control of SIV expression and are consistent with data obtained in HIV-1 infection studies.

## 3.4 Regulation of HIV-1 Transcription by Tat/P-TEFb

The enzyme that transcribes messenger RNA (mRNA) from protein-encoding genes is the RNA polymerase II (Pol II). This protein executes a series of distinct steps: it binds to promoters, initiates RNA synthesis, and then pauses in early transcriptional elongation to allow capping of the neo-synthetized mRNA before resumption of the transcription. In the case of HIV-1 transcription, due to the presence of the nuc-1 nucleosome, the paused Pol II is not able to resume transcription directly, therefore leaving the nascent RNA TAR. Further signals are needed to elicit the transition from the paused Pol II to a productive elongation complex (Adelman and Lis 2012). The switch from promoter-proximal pausing to productive elongation is mediated by the couple Tat/P-TEFb, an essential elongation transcription factor constituted of two subunits: cyclin T1 (CycT1) and the cyclin-dependent kinase 9 (CDK9).

Resting CD4$^+$ T cells are characterized by extremely low levels of cyclin T1 due to actions of specific miRNAs (see below section "POST-TRANSCRIPTIONAL REGULATION OF HIV-1 EXPRESSION") and of the cellular factor NF-90, which blocks translation of CycT1 mRNA (Budhiraja et al. 2013; Chiang and Rice 2012; Chiang et al. 2012). Moreover, in the absence of Tat, the elongation block is reinforced by the sequestration of P-TEFb within the 7SK small nuclear ribonu-cleoprotein (snRNP) repressive complex including the 7SK snRNA, the hexam-ethylene bisacetamide inducible protein 1 (HEXIM1), the 5'methylphosphate capping enzyme (MePCE), and the La-related protein (LARP7), as well as the combined inhibition by the negative elongation factor (NELF) and the 5,6-Dichloro-1-β-D-ribofuranosylbenzimidazole (DRB) sensitivity-inducing factor (DSIF) (Fig. 2).

Upon cellular activation (stress signals) and when Tat is not produced yet, P-TEFb is released from the HEXIM-1/7SK snRNA complex, and associated with the BET bromodomain protein 4 (BRD4), thereby forming the active P-TEFb complex. P-TEFb is then recruited to the HIV-1 LTR via interactions of the BRD4 bromodomains with acetylated histones (Darcis et al. 2015).

Once Tat has been synthesized, on one hand, Tat competes with BRD4 for binding to P-TEFb and on the other hand, Tat is also able to directly disrupt the inactive P-TEFb and to form a stable complex with P-TEFb. Tat then recruits P-TEFb to the HIV-1 promoter through TAR and increases transcription elongation. Tat can also recruit, in addition to P-TEFb, other elongation factors (such as ELL2, AFF4, ENL, and AF9), thereby forming the superelongation complex (SEC). P-TEFb is thereby positioned to phosphorylate the C-terminal domain (CTD) of Pol II, resulting in efficient elongation of viral transcription.

Thus, P-TEFb is present in two forms: a free active form and a 7SK-associated inactive form in which the kinase activity of the CDK9 is repressed. The balance between these two forms controls the activity of P-TEFb and the subsequent viral transcriptional reactivation processes. More clearly, since HIV-1 gene expression critically depends on P-TEFb function, factors that contribute to P-TEFb inactivation will also favor the persistence of latently infected cells.

We have previously observed that the CTIP2 is a corepressor of HIV-1 transcription in microglial cells since it acts as a recruitment platform for HDAC and HMT (Marban et al. 2007). In a separate complex, CTIP2 also associates with the P-TEFb inactive complex and represses P-TEFb functions by inhibiting CDK9 activity (Cherrier et al. 2013). Knocking down CTIP2 increases Tat-dependent transcriptional activity of the HIV-1 promoter. In contrast, overexpression of CTIP2 increases the recruitment of the inactive P-TEFb complex to the HIV-1 core promoter (Cherrier et al. 2013). Interestingly, HMGA1 (High Mobility Group A1), a nonhistone chromatin protein, also participates in the recruitment of the CTIP2-repressed P-TEFb to the HIV-1 core promoter through interaction with the 7SKsnRNA. Thus, HMGA1 and CTIP2 cooperatively repress HIV-1 gene expression by a HMGA1-mediated recruitment of CTIP2-inactivated P-TEFb to the HIV-1 promoter (Eilebrecht et al. 2014).

Another pathway has recently been implicated in HIV-1 latency, or rather in latency reversal, through P-TEFb activity. Besnard et al. (Besnard et al. 2016) showed that knockdown of mammalian target of rapamycin (mTOR) complex subunits or pharmacological inhibition of mTOR activity suppresses reversal of latency in various HIV-1 latency models and HIV-infected patient cells; mTOR inhibitors suppress HIV-1 transcription both through the viral transactivator Tat and via Tat-independent mechanisms. This inhibition occurs at least in part via blocking the phosphorylation of CDK9 (Besnard et al. 2016), further supporting the essential function of the P-TEFb complex in HIV-1 transcription.

## 3.5   The Role of Cellular Transcription Factors

The NF-κB transcription factor plays a complex role during the replication of primate lentiviruses. On one hand, NF-κB is crucial for induction of efficient proviral gene expression. On the other hand, NF-κB activation also induces expression of genes involved in the innate immune response and the cellular antiviral response.

NF-κB is sequestered in the cytoplasm of unstimulated cells in an inactive form through its interaction with an inhibitory protein from the family of inhibitors of NF-κB (IκB). Following cellular activation, the phosphorylation of IκB by IKK (IκB kinase) leads to its ubiquitination and proteosomal degradation, allowing the translocation of NF-κB into the nucleus and the transcriptional trans-activation of NF-κB-dependent genes. Notably, NF-κB also stimulates HIV-1 transcriptional elongation by interacting with P-TEFb and directs the recruitment of a co-activator complex of HATs to the HIV-1 LTR (Barboric et al. 2001; Perkins et al. 1997).

HIV-1 and SIV have been shown to modulate NF-κB activation through their regulatory and accessory proteins. Indeed, Tat and Nef proteins are able to activate or enhance the NF-κB activation. Later, during the infection, the viral protein Vpu that is only expressed by HIV-1 and its simian precursors suppress NF-κB activation [reviewed in (Heusinger and Kirchhoff 2017)]. Notably, only the Nef proteins of these Vpu-containing viruses are unable to down-modulate the TCR–CD3 complex from the cell surface and render virally infected T cells hyper-responsive to stimulation and increase the induction of NF-κB and NF-AT (Fortin et al. 2004; Schindler et al. 2006). In contrast, efficient down-modulation of TCR–CD3 by the Nef proteins of most SIVs and HIV-2 blocks the responsiveness of CD4+ T cells to stimulation and is associated with low levels of NF-κB and NF-AT activation (Schindler et al. 2008; Khalid et al. 2012). Thus, regulation of T cell activation and the NF-κB and NF-AT pathways by the accessory viral proteins Vpu and Nef may also have implications in viral latency. The importance of the NF-κB pathway in HIV-1 and SIV latency is further demonstrated by the reactivation of viral production upon treatment with NF-κB inducers such as PKC agonists (Gama et al. 2017; Darcis et al. 2015).

Additionally, the capacity of HIV-1 to establish latent infection is partially controlled by a four-nucleotide AP-1 element just upstream of the NF-κB element in the HIV-1 5'LTR (Duverger et al. 2013). Indeed, deletion of this AP-1 site mostly deprived HIV-1 of its ability to establish latent infection. This observation supports the idea that HIV-1 latency is a transcription factor restriction phenomenon (Duverger et al. 2013). The lack of active forms of other key cellular transcription factors (such as NF-AT and STAT5) is another element involved in repression of initiation and elongation of viral transcription in resting CD4+ T cells [reviewed in (Van Lint et al. 2013)]. Abdel-Mohsen et al. also showed that human Galectin-9 (Gal-9) is a potent mediator of HIV-1 transcription and reactivation (Abdel-Mohsen et al. 2016). Recombinant Gal-9 potently reverses HIV-1 latency in vitro and ex vivo through the induction of several HIV-1 transcription factors (NF-κB, AP-1, and NF-AT) expression and the inhibition of several chromatin modification and remodeling factors (including HDAC1, 2, and 3, EZH2, DNMT1) gene expression (Abdel-Mohsen et al. 2016). This pathway well illustrates the impact of different but dynamically linked molecular mechanisms on HIV-1 latency.

## 4 Post-transcriptional Regulation of HIV-1 Expression

### 4.1 HIV-1 mRNA Processing and Latency

Transcriptional regulation, described in the previous section, takes place and is influenced by nuclear co-transcriptional processes, including pre-mRNA capping, splicing, and polyadenylation that occur mostly co-transcriptionally (Karn and Stoltzfus 2012). First, pre-mRNAs are capped at their 5'end by capping enzymes:

RNA triphosphatase, RNA guanylyltransferase, and RNA (guanine-N7) methyltransferase. Further processing of viral pre-mRNA by the host splicing machinery produces various transcripts and therefore various proteins. Nascent transcripts indeed result in over 40 differently spliced mRNAs (Purcell and Martin 1993), which can be divided into three classes: (i) fully spliced RNA expressing Tat exon1 +2, Rev and Nef; (ii) singly spliced RNA encoding Tat exon1, Vif, Vpu-Env, and Vpr; or (iii) unspliced RNA serving as genomic RNA or to produce the Gag and Gag-Pol precursors.

P-TEFb, in addition to its crucial role in transcriptional regulation, also links the co-transcriptional processes of pre-mRNA capping and alternative splicing to transcriptional elongation (Lenasi et al. 2011). P-TEFb therefore facilitates the generation and processing of protein-coding mRNA.

Before the transport to the cytoplasm, the processing of the nascent transcript is completed by polyadenylation. The 3' processing and polyadenylation of pre-mRNAs involve recognition of the upstream AAUAAA and downstream GU-rich motifs surrounding the cleavage and poly(A) addition site. For this, host cellular proteins such as the cleavage/polyadenylation specificity factor (CPSF), cleavage stimulation factor (CstF), CF1m, CF2m, and poly(A) polymerase are required for endonucleolytic cleavage and polyadenylation of viral pre-mRNA.

The viral Rev protein is involved in the transport of unspliced and partially spliced mRNAs from the nucleus to the cytoplasm, following its interaction with the Rev-responsive element (RRE). Nuclear export occurs upon association of Rev with the nuclear export factor Exportin 1 (Crm-1) and translocation of the Rev/RNA complex to the cytoplasm where it is either translated or packaged into virions [reviewed in (Kula and Marcello 2012)]. Therefore, defects in viral RNA export, which could be due to insufficient levels of either Rev (Huang et al. 2007) or the HIV-1 RNA-binding factors Matrin 3 and PTB (polypyrimidine tract-binding protein)-associated factor PSF (Zolotukhin et al. 2003; Yedavalli and Jeang 2011; Kula et al. 2013) or inhibition of HIV-1 mRNA translation, are also implicated in HIV-1 latency.

## 4.2 Noncoding mRNAs and HIV-1 Latency

MicroRNAs (miRNAs) are short single-stranded noncoding RNAs of 19–25 nucleotides that mediate post-transcriptional gene silencing. In general, following RNA Pol II transcription, primary miRNA transcripts are sequentially processed via the nuclear RNases III Drosha and Dicer to generate mature miRNAs which interact with a complementary sequence in the 3' untranslated region of target mRNAs by partial sequence matching, resulting in degradation of the mRNA and/or translational repression. The level of specific mRNA translation can therefore be modulated by miRNAs.

Interestingly, modifications of the miRNA profile have been observed in HIV-1 infected patients (Houzet et al. 2008; Witwer et al. 2012; Bignami et al. 2012).

Mechanistically, Tat and Vpr are known to function as RNA silencing suppressors by modulating miRNA expression levels in infected cells (Qian et al. 2009; Coley et al. 2010).

Cellular or viral miRNAs can target either cellular or virally expressed mRNAs. For example, PCAF, a HAT that is involved in chromatin remodeling, is targeted by miR-17/92 and miR-20a, both of which are downregulated in HIV-1 infection (Hayes et al. 2011; Triboulet et al. 2007). Cyclin T1 is repressed by various miRNAs in resting CD4+ T cells (miR-27b, miR-29b, miR-150, and miR-223) (Chiang and Rice 2012; Sung and Rice 2009). Moreover, cellular miRNAs, miR-28, miR-125b, miR-150, miR-223, and miR-382, known to be upregulated in resting CD4+ T cells, recognize the 3'end of HIV-1 mRNAs (Huang et al. 2007) and thus participate in the repression of HIV-1 gene expression.

Several viral miRNAs (vmiRNAs) have also been identified, including TAR-derived miRNA-TAR5p/3p (Klase et al. 2007; Ouellet et al. 2008) and the Nef-derived miR-N367 (Omoto et al. 2004). In contrast, some miRNAs can also have a positive effect on HIV-1 expression, for instance, when targeting HDAC involved in both the regulation of NF-κB and Tat. Indeed, the acetylation of these key factors is needed to allow for their proper action (Darcis et al. 2015).

The role of long noncoding RNAs (lncRNAs) has recently been observed in gene expression regulation, from transcriptional initiation to protein translation and degradation. For instance, the 7SK RNA is a lncRNA involved in the regulation of active P-TEFb levels (see the previous section). Nuclear-enriched abundant transcript 1 (NEAT1) is another example of a lncRNA that is involved in HIV-1 gene expression. This lncRNA is associated with the pathway of HIV mRNA export dependent on Rev and other cellular cofactors (Zhang et al. 2013) and play a crucial role in the post-translational regulation of HIV-1 expression. In addition, expression levels of noncoding repressor of NF-AT (NRON), a lncRNA involved in the HIV-1 latency establishment by targeting Tat for degradation (Imam et al. 2015), was observed to be inversely correlated with levels of HIV mRNA in resting CD4+ T cells (Li et al. 2016).

# 5 Concluding Remarks

For several years now, intensive efforts have been made by the scientific community to better characterize the HIV-1 latent reservoir and to investigate the molecular mechanisms regulating latency in infected cells. Improved knowledge of these mechanisms of persistence has paved the way for innovative strategies to attempt to eradicate latent HIV-1 but have also highlighted hurdles that should be overcome to reach this goal. One of them is the heterogeneity of latency, resulting from the multiplicity of the molecular mechanisms of HIV-1 transcriptional repression.

Numerous advances in our understanding of viral transmission, pathogenesis, and latency can be attributed to the use of SIV and NHP models. Multiple studies investigated whether these models recapitulate known and newly discovered features of HIV persistence in humans, including molecular mechanisms underlying latency. By many aspects, SIV and NHP models well reflect HIV infection of the human body. However, molecular mechanisms underlying SIV latency have been barely studied. Therefore, it is not excluded that some differences exist between HIV-1 and SIV latency, imposing a prudent analysis of the data obtained from the SIV/NHP models.

**Funding** This project has received funding from the Belgian Fund for Scientific Research (FRS-FNRS, Belgium), the *European Union's Horizon 2020 research and innovation programme* (grant agreement N° 691119 EU4HIVCURE H2020-MSCA-RISE-2015), the ANRS (France Recherche Nord & Sud Sida-HIV Hépatites), the "Fondation Roi Baudouin", the NEAT Program, the Walloon Region (the Excellence Program "Cibles" and the "Fond de maturation" program), the ARC program (ULB) and the Internationale Brachet Stiftung (IBS). BVD and SB are postdoctoral fellows (ARC program and PDR project from the FRS-FNRS, respectively). CVL is "Directeur de Recherches" of the FRS-FNRS (Belgium).

# References

Abdel-Mohsen M et al (2016) Human galectin-9 is a potent mediator of HIV transcription and reactivation. PLoS Pathog 12:e1005677

Abu-Farha M et al (2008) The tale of two domains: proteomics and genomics analysis of SMYD2, a new histone methyltransferase. Mol Cell Proteomics 7:560–572

Adelman K, Lis JT (2012) Promoter-proximal pausing of RNA polymerase II: emerging roles in metazoans. Nat Rev Genet 13:720–731

Boehm D et al (2017) SMYD2-mediated histone methylation contributes to HIV-1 latency. Cell Host Microbe 21:569–579 e566

Alexaki A, Liu Y, Wigdahl B (2008) Cellular reservoirs of HIV-1 and their role in viral persistence. Curr HIV Res 6:388–400

Avettand-Fenoel V et al (2016) Total HIV-1 DNA, a marker of viral reservoir dynamics with clinical implications. Clin Microbiol Rev 29:859–880

Barber SA et al (2006) Mechanism for the establishment of transcriptional HIV latency in the brain in a simian immunodeficiency virus-macaque model. J Infect Dis 193:963–970

Barboric M, Nissen RM, Kanazawa S, Jabrane-Ferrat N, Peterlin BM (2001) NF-kappaB binds P-TEFb to stimulate transcriptional elongation by RNA polymerase II. Mol Cell 8:327–337

Berkhout B (1992) Structural features in TAR RNA of human and simian immunodeficiency viruses: a phylogenetic analysis. Nucleic Acids Res 20:27–31

Besnard E et al (2016) The mTOR complex controls HIV latency. Cell Host Microbe 20:785–797

Bignami F et al (2012) Stable changes in CD4+ T lymphocyte miRNA expression after exposure to HIV-1. Blood 119:6259–6267

Blazkova J et al (2009) CpG methylation controls reactivation of HIV from latency. PLoS Pathog 5:e1000554

Brady J, Kashanchi F (2005) Tat gets the "green" light on transcription initiation. Retrovirology 2:69

Brown MA, Sims RJ 3rd, Gottlieb PD, Tucker PW (2006) Identification and characterization of Smyd2: a split SET/MYND domain-containing histone H3 lysine 36-specific methyltransferase that interacts with the Sin3 histone deacetylase complex. Mol Cancer 5:26

Budhiraja S, Famiglietti M, Bosque A, Planelles V, Rice AP (2013) Cyclin T1 and CDK9 T-loop phosphorylation are downregulated during establishment of HIV-1 latency in primary resting memory CD4+ T cells. J Virol 87:1211–1220

Capelson M, Doucet C, Hetzer MW (2010) Nuclear pore complexes: guardians of the nuclear genome. Cold Spring Harb Symp Quant Biol 75:585–597

Chen HC, Martinez JP, Zorita E, Meyerhans A, Filion GJ (2017) Position effects influence HIV latency reversal. Nat Struct Mol Biol 24:47–54

Cherrier T et al (2013) CTIP2 is a negative regulator of P-TEFb. Proc Natl Acad Sci USA 110:12655–12660

Chiang K, Rice AP (2012) MicroRNA-mediated restriction of HIV-1 in resting CD4+ T cells and monocytes. Viruses 4:1390–1409

Chiang K, Sung TL, Rice AP (2012) Regulation of cyclin T1 and HIV-1 Replication by microRNAs in resting CD4+ T lymphocytes. J Virol 86:3244–3252

Choudhary SK, Archin NM, Margolis DM (2008) Hexamethylbisacetamide and disruption of human immunodeficiency virus type 1 latency in CD4(+) T cells. J Infect Dis 197:1162–1170

Chun TW et al (1995) In vivo fate of HIV-1-infected T cells: quantitative analysis of the transition to stable latency. Nat Med 1:1284–1290

Chun TW et al (1997) Quantification of latent tissue reservoirs and total body viral load in HIV-1 infection. Nature 387:183–188

Chun TW et al (2000) Relationship between pre-existing viral reservoirs and the re-emergence of plasma viremia after discontinuation of highly active anti-retroviral therapy. Nat Med 6:757–761

Coley W et al (2010) Absence of DICER in monocytes and its regulation by HIV-1. J Biol Chem 285:31930–31943

Colin L, Van Lint C (2009) Molecular control of HIV-1 postintegration latency: implications for the development of new therapeutic strategies. Retrovirology 6:111

Crise B et al (2005) Simian immunodeficiency virus integration preference is similar to that of human immunodeficiency virus type 1. J Virol 79:12199–12204

Darcis G et al (2015) An in-depth comparison of latency-reversing agent combinations in various in vitro and ex vivo HIV-1 latency models identified bryostatin-1+ JQ1 and ingenol-B+ JQ1 to potently reactivate viral gene expression. PLoS Pathog 11:e1005063

Darcis G et al (2017) Reactivation capacity by latency-reversing agents ex vivo correlates with the size of the HIV-1 reservoir. AIDS 31:181–189

Das AT, Harwig A, Berkhout B (2011) The HIV-1 Tat protein has a versatile role in activating viral transcription. J Virol 85:9506–9516

Davey RT Jr et al (1999) HIV-1 and T cell dynamics after interruption of highly active antiretroviral therapy (HAART) in patients with a history of sustained viral suppression. Proc Natl Acad Sci USA 96:15109–15114

Deleage C, Turkbey B, Estes JD (2016) Imaging lymphoid tissues in nonhuman primates to understand SIV pathogenesis and persistence. Curr Opin Virol 19:77–84

Doucas V, Tini M, Egan DA, Evans RM (1999) Modulation of CREB binding protein function by the promyelocytic (PML) oncoprotein suggests a role for nuclear bodies in hormone signaling. Proc Natl Acad Sci U S A 96:2627–2632

Duverger A et al (2013) An AP-1 binding site in the enhancer/core element of the HIV-1 promoter controls the ability of HIV-1 to establish latent infection. J Virol 87:2264–2277

Eilebrecht S et al (2014) HMGA1 recruits CTIP2-repressed P-TEFb to the HIV-1 and cellular target promoters. Nucleic Acids Res 42:4962–4971

Finzi D et al (1999) Latent infection of CD4+ T cells provides a mechanism for lifelong persistence of HIV-1, even in patients on effective combination therapy. Nat Med 5:512–517

Fortin JF, Barat C, Beausejour Y, Barbeau B, Tremblay MJ (2004) Hyper-responsiveness to stimulation of human immunodeficiency virus-infected CD4+ T cells requires Nef and Tat virus gene products and results from higher NFAT, NF-kappaB, and AP-1 induction. J Biol Chem 279:39520–39531

Friedman J et al (2011) Epigenetic silencing of HIV-1 by the histone H3 lysine 27 methyltransferase enhancer of Zeste 2. J Virol 85:9078–9089

Gama L et al (2017) Reactivation of simian immunodeficiency virus reservoirs in the brain of virally suppressed macaques. AIDS 31:5–14

Goffin V et al (2005) Transcription factor binding sites in the pol gene intragenic regulatory region of HIV-1 are important for virus infectivity. Nucleic Acids Res 33:4285–4310

Hayes AM, Qian S, Yu L, Boris-Lawrie K (2011) Tat RNA silencing suppressor activity contributes to perturbation of lymphocyte miRNA by HIV-1. Retrovirology 8:36

Hematti P et al (2004) Distinct genomic integration of MLV and SIV vectors in primate hematopoietic stem and progenitor cells. PLoS Biol 2:e423

Heusinger E, Kirchhoff F (2017) Primate Lentiviruses Modulate NF-kappaB Activity by Multiple Mechanisms to Fine-Tune Viral and Cellular Gene Expression. Front Microbiol 8:198

Hogan TH et al (2003) Structural and functional evolution of human immunodeficiency virus type 1 long terminal repeat CCAAT/enhancer binding protein sites and their use as molecular markers for central nervous system disease progression. J Neurovirol 9:55–68

Houzet L et al (2008) MicroRNA profile changes in human immunodeficiency virus type 1 (HIV-1) seropositive individuals. Retrovirology 5:118

Huang J et al (2007) Cellular microRNAs contribute to HIV-1 latency in resting primary CD4+ T lymphocytes. Nat Med 13:1241–1247

Imai K, Togami H, Okamoto T (2010) Involvement of histone H3 lysine 9 (H3K9) methyltransferase G9a in the maintenance of HIV-1 latency and its reactivation by BIX01294. J Biol Chem 285:16538–16545

Imam H, Bano AS, Patel P, Holla P, Jameel S (2015) The lncRNA NRON modulates HIV-1 replication in a NFAT-dependent manner and is differentially regulated by early and late viral proteins. Sci Rep 5:8639

Jiang G, Espeseth A, Hazuda DJ, Margolis DM (2007) c-Myc and Sp1 contribute to proviral latency by recruiting histone deacetylase 1 to the human immunodeficiency virus type 1 promoter. J Virol 81:10914–10923

Jordan A, Bisgrove D, Verdin E (2003) HIV reproducibly establishes a latent infection after acute infection of T cells in vitro. EMBO J 22:1868–1877

Karn J, Stoltzfus CM (2012) Transcriptional and posttranscriptional regulation of HIV-1 gene expression. Cold Spring Harb Perspect Med 2:a006916

Kauder SE, Bosque A, Lindqvist A, Planelles V, Verdin E (2009) Epigenetic regulation of HIV-1 latency by cytosine methylation. PLoS Pathog 5:e1000495

Khalid M et al (2012) Efficient Nef-mediated downmodulation of TCR-CD3 and CD28 is associated with high CD4+ T cell counts in viremic HIV-2 infection. J Virol 86:4906–4920

Klase Z et al (2007) HIV-1 TAR element is processed by Dicer to yield a viral micro-RNA involved in chromatin remodeling of the viral LTR. BMC Mol Biol 8:63

Kula A, Marcello A (2012) Dynamic post-transcriptional regulation of HIV-1 gene expression. Biology (Basel) 1:116–133

Kula A, Gharu L, Marcello A (2013) HIV-1 pre-mRNA commitment to Rev mediated export through PSF and Matrin 3. Virology 435:329–340

Kumar A, Abbas W, Herbein G (2014) HIV-1 latency in monocytes/macrophages. Viruses 6:1837–1860

Lamond AI, Sleeman JE (2003) Nuclear substructure and dynamics. Curr Biol 13:R825–828

Le Douce V, Cherrier T, Riclet R, Rohr O, Schwartz C (2014) The many lives of CTIP2: from AIDS to cancer and cardiac hypertrophy. J Cell Physiol 229:533–537

Lelek M et al (2015) Chromatin organization at the nuclear pore favours HIV replication. Nat Commun 6:6483

Lenasi T, Peterlin BM, Barboric M (2011) Cap-binding protein complex links pre-mRNA capping to transcription elongation and alternative splicing through positive transcription elongation factor b (P-TEFb). J Biol Chem 286:22758–22768

Lewinski MK et al (2005) Genome-wide analysis of chromosomal features repressing human immunodeficiency virus transcription. J Virol 79:6610–6619

Li J et al (2016) Long noncoding RNA NRON contributes to HIV-1 latency by specifically inducing tat protein degradation. Nat Commun 7:11730

Ling B et al (2014) Effects of treatment with suppressive combination antiretroviral drug therapy and the histone deacetylase inhibitor suberoylanilide hydroxamic acid; (SAHA) on SIV-infected Chinese rhesus macaques. PLoS ONE 9:e102795

Liu Y, Nonnemacher MR, Wigdahl B (2009) CCAAT/enhancer-binding proteins and the pathogenesis of retrovirus infection. Future Microbiol 4:299–321

Lusic M et al (2013) Proximity to PML nuclear bodies regulates HIV-1 latency in CD4+ T cells. Cell Host Microbe 13:665–677

Marban C et al (2005) COUP-TF interacting protein 2 represses the initial phase of HIV-1 gene transcription in human microglial cells. Nucleic Acids Res 33:2318–2331

Marban C et al (2007) Recruitment of chromatin-modifying enzymes by CTIP2 promotes HIV-1 transcriptional silencing. EMBO J 26:412–423

Marcello A et al (2003) Recruitment of human cyclin T1 to nuclear bodies through direct interaction with the PML protein. EMBO J 22:2156–2166

Marini B et al (2015) Nuclear architecture dictates HIV-1 integration site selection. Nature 521:227–231

Marsili G, Remoli AL, Sgarbanti M, Battistini A (2004) Role of acetylases and deacetylase inhibitors in IRF-1-mediated HIV-1 long terminal repeat transcription. Ann N Y Acad Sci 1030:636–643

Misteli T (2007) Beyond the sequence: cellular organization of genome function. Cell 128:787–800

Nguyen K, Das B, Dobrowolski C, Karn J (2017) Multiple histone lysine methyltransferases are required for the establishment and maintenance of HIV-1 latency. MBio 8

Omoto S et al (2004) HIV-1 nef suppression by virally encoded microRNA. Retrovirology 1:44

Ouellet DL et al (2008) Identification of functional microRNAs released through asymmetrical processing of HIV-1 TAR element. Nucleic Acids Res 36:2353–2365

Perkins ND et al (1997) Regulation of NF-kappaB by cyclin-dependent kinases associated with the p300 coactivator. Science 275:523–527

Purcell DF, Martin MA (1993) Alternative splicing of human immunodeficiency virus type 1 mRNA modulates viral protein expression, replication, and infectivity. J Virol 67:6365–6378

Qian S et al (2009) HIV-1 Tat RNA silencing suppressor activity is conserved across kingdoms and counteracts translational repression of HIV-1. Proc Natl Acad Sci U S A 106:605–610

Ravimohan S, Gama L, Barber SA, Clements JE (2010) Regulation of SIV mac 239 basal long terminal repeat activity and viral replication in macrophages: functional roles of two CCAAT/enhancer-binding protein beta sites in activation and interferon beta-mediated suppression. J Biol Chem 285:2258–2273

Ravimohan S, Gama L, Engle EL, Zink MC, Clements JE (2012) Early emergence and selection of a SIV-LTR C/EBP site variant in SIV-infected macaques that increases virus infectivity. PLoS ONE 7:e42801

Schindler M et al (2006) Nef-mediated suppression of T cell activation was lost in a lentiviral lineage that gave rise to HIV-1. Cell 125:1055–1067

Schindler M et al (2008) Inefficient Nef-mediated downmodulation of CD3 and MHC-I correlates with loss of CD4+ T cells in natural SIV infection. PLoS Pathog 4:e1000107

Schrijvers R et al (2012) LEDGF/p75-independent HIV-1 replication demonstrates a role for HRP-2 and remains sensitive to inhibition by LEDGINs. PLoS Pathog 8:e1002558

Shan L et al (2011) Influence of host gene transcription level and orientation on HIV-1 latency in a primary-cell model. J Virol 85:5384–5393

Siliciano JD et al (2003) Long-term follow-up studies confirm the stability of the latent reservoir for HIV-1 in resting CD4+ T cells. Nat Med 9:727–728

Sung TL, Rice AP (2009) miR-198 inhibits HIV-1 gene expression and replication in monocytes and its mechanism of action appears to involve repression of cyclin T1. PLoS Pathog 5:e1000263

Suzuki MM, Bird A (2008) DNA methylation landscapes: provocative insights from epigenomics. Nat Rev Genet 9:465–476

Trejbalova K et al (2016) Development of 5' LTR DNA methylation of latent HIV-1 provirus in cell line models and in long-term-infected individuals. Clin Epigenetics 8:19

Triboulet R et al (2007) Suppression of microRNA-silencing pathway by HIV-1 during virus replication. Science 315:1579–1582

van der Velden GJ, Vink MA, Berkhout B, Das AT (2012) Tat has a dual role in simian immunodeficiency virus transcription. J Gen Virol 93:2279–2289

Van Lint C, Emiliani S, Ott M, Verdin E (1996) Transcriptional activation and chromatin remodeling of the HIV-1 promoter in response to histone acetylation. EMBO J 15:1112–1120

Van Lint C, Bouchat S, Marcello A (2013) HIV-1 transcription and latency: an update. Retrovirology 10:67

Verdin E, Paras P Jr, Van Lint C (1993) Chromatin disruption in the promoter of human immunodeficiency virus type 1 during transcriptional activation. EMBO J 12:3249–3259

Vire E et al (2006) The Polycomb group protein EZH2 directly controls DNA methylation. Nature 439:871–874

Wagner TA et al (2014) HIV latency. Proliferation of cells with HIV integrated into cancer genes contributes to persistent infection. Science 345:570–573

Witwer KW, Watson AK, Blankson JN, Clements JE (2012) Relationships of PBMC microRNA expression, plasma viral load, and CD4+ T-cell count in HIV-1-infected elite suppressors and viremic patients. Retrovirology 9:5

Wong RW, Mamede JI, Hope TJ (2015) The impact of nucleoporin mediated chromatin localization and nuclear architecture on HIV integration site selection. J Virol

Yang G, Thompson MA, Brandt SJ, Hiebert SW (2007) Histone deacetylase inhibitors induce the degradation of the t(8;21) fusion oncoprotein. Oncogene 26:91–101

Yedavalli VS, Jeang KT (2011) Matrin 3 is a co-factor for HIV-1 Rev in regulating post-transcriptional viral gene expression. Retrovirology 8:61

Zhang Q, Chen CY, Yedavalli VS, Jeang KT (2013) NEAT1 long noncoding RNA and paraspeckle bodies modulate HIV-1 posttranscriptional expression. MBio 4:e00596–00512

Zolotukhin AS et al (2003) PSF acts through the human immunodeficiency virus type 1 mRNA instability elements to regulate virus expression. Mol Cell Biol 23:6618–6630

# Assays to Measure Latency, Reservoirs, and Reactivation

**Janet D. Siliciano and Robert F. Siliciano**

**Abstract** HIV-1 persists even in patients who are successfully treated with combination antiretroviral therapy. The major barrier to cure is a small pool of latently infected resting CD4$^+$ T cells carrying an integrated copy of the viral genome that is not expressed while the cells remain in a resting state. Targeting this latent reservoir is a major focus of HIV-1 cure research, and the development of a rapid and scalable assay for the reservoir is a rate-limiting step in the search for a cure. The most commonly used assays are standard PCR assays targeting conserved regions of the HIV-1 genome. However, because the vast majority of HIV-1 proviruses are defective, such assays may not accurately capture changes in the minor subset of proviruses that are replication-competent and that pose a barrier to cure. On the other hand, the viral outgrowth assay that was used to initially define the latent reservoir may underestimate reservoir size because not all replication-competent proviruses are induced by a single round of T cell activation in this assay. Therefore, this assay is best regarded as a definitive minimal estimate of reservoir size. The best approach may be to measure all of the proviruses with the potential to cause viral rebound. A variety of novel assays have recently been described. Ultimately, the assay that best predicts time to viral rebound will be the most useful to the cure effort.

## Contents

J. D. Siliciano
Johns Hopkins University School of Medicine, Baltimore, MD 21205, USA

R. F. Siliciano (✉)
Johns Hopkins University School of Medicine, Howard Hughes Medical Institute,
Baltimore, MD 21205, USA
e-mail: rsiliciano@jhmi.edu

Current Topics in Microbiology and Immunology (2018) 417:23–42
DOI 10.1007/82_2017_75
© Springer International Publishing AG 2017
Published Online: 26 October 2017

# 1   Introduction

The major barrier to curing HIV-1 infection is a small pool of latently infected resting $CD4^+$ T cells (Chun et al. 1995; Chun et al. 1997a) that persist even in patients on optimal antiretroviral therapy (Finzi et al. 1997; Chun et al. 1997b, Wong et al. 1997). The cells contain an integrated copy of the viral genome that is not expressed while the cells remain in a resting state (Chun et al. 1995; Chun et al. 1997a; Hermankova et al. 2003). Targeting this latent reservoir is a major focus of HIV-1 cure research, and the availability of a rapid and scalable assay for the reservoir is considered to be essential for assessing the efficacy of cure strategies and engaging the pharmaceutical industry in the search for a cure (Deeks et al. 2016). The development of an assay for HIV-1 virions in the plasma (Piatak et al. 1993) greatly accelerated the development of effective combination antiretroviral therapy (ART). Antiretroviral drugs belonging to the protease inhibitor and non-nucleoside reverse transcriptase inhibitors classes caused a rapid reduction in plasma HIV-1 RNA levels (Ho et al. 1995; Wei et al. 1995), and three drug combinations rapidly reduced viremia to below the limit of detection of clinical assays (Gulick et al. 1997; Hammer et al. 1997; Perelson et al. 1997). Thus drug efficacy could be assessed without the need to wait for clinical endpoints. Similarly, an accurate and scalable assay is needed to monitor interventions targeting the latent reservoir. Here, we discuss the various approaches currently being used to measure the reservoir and the advantages and drawbacks of each approach.

# 2   Definitions

In the HIV-1 cure field, the terms latency and reservoir are frequently used but seldom defined, and it is helpful at the outset to define precisely what we mean by these and other key terms (Box 1). **Latency** is a reversibly nonproductive state of infection of individual cells. While cells are in a latent state of infection, the viral genome persists in some form within the cells, but viral gene expression is limited.

For some viruses, particularly those of the herpesvirus family, latency is an essential mechanism for viral persistence and immune evasion (Speck and Ganem 2010; Perng and Jones 2010). It is somewhat surprising that HIV-1 can establish latent infection, given that viral replication continues throughout the course of untreated HIV-1 infection (Piatak et al. 1993), with rapid viral evolution providing that main mechanism for escaping immune pressure (Borrow et al. 1997; Richman et al. 2003; Wei et al. 2003). Recent work suggests that HIV-1 latency may be an accidental consequence of infection of CD4$^+$ T cells in a narrow time window after activation when they are permissive for viral entry and reverse transcription, but only slightly and transiently permissive for viral gene expression (Shan et al. 2017). In any event, it is clear that latently infected cells are present in all HIV-1-infected individuals and that these cells constitute a reservoir that prevents cure with antiretroviral therapy (Finzi et al. 1999; Siliciano et al. 2003; Strain et al. 2003).

A **reservoir** can be defined as a cell type or anatomical site in which replication-competent forms of HIV-1 persist on a timescale of years in patients on optimal ART (Eisele and Siliciano 2012). This is a practical definition that reflects the fact that curative interventions will be only be attempted in patients who have had effective suppression of active viral replication on ART for long enough to allow labile populations of infected cells to decay. Based on the measured decay rates of labile infected cell populations, this is likely to be a period of at least 6 months (Zack et al. 1990; Wei et al. 1995; Ho et al. 1995; Perelson et al. 1997; Blankson et al. 2000). To date, the latent reservoir in resting CD4$^+$ T cells is the only cell population yet convincingly shown to meet this definition. There is great current interest in the controversial issue of whether other infected cell populations, including tissue macrophages, represent stable reservoirs for HIV-1 (Calantone et al. 2014; Honeycutt et al. 2016; Gama et al. 2017). Addressing this issue is difficult because it requires sampling of tissues. This can be done in the SIV model of HIV-1 infection. SIV establishes a latent reservoir in resting CD4$^+$ T cells (Shen et al. 2003; Dinoso et al. 2009b), and investigations of SIV persistence in tissues sites such as the central nervous system are ongoing (Gama et al. 2017). Importantly, such studies must be carried out in animals on long-term suppressive ART in order to establish that a given cell type meets the definition of a reservoir given above.

There are some common uses of the terms latent and reservoir that are clearly ill-advised. A stable population of productively infected cells could in principle serve as a reservoir, and therefore the term "latent reservoir" should only be used if the relevant cell population is shown to be in a reversibly nonproductive state of infection. As is discussed below, PCR-based assays for proviral DNA are often used to measure persistent HIV-1, and the term "DNA reservoir" is frequently used. This term makes little sense and should be avoided, especially since the vast majority of proviral DNA detected by standard PCR assays is defective (see below). Similarly, the term "active reservoir" is often used in situations where viral RNA is detected. However, infected CD4$^+$ T cells expressing HIV-1 RNA are likely to be in an activated state and to have a much shorter half-life than the resting CD4$^+$ T cells that comprise the latent reservoir. Therefore, it is not clear that cells expressing

HIV-1 RNA will survive long enough to be considered a reservoir by the above definition.

The term reactivation is used to indicate situations which latency has been reversed, allowing viral gene expression and virus production. This term can cause confusion, however, because reactivation of the T cell can reactivate a latent provirus. It is preferable to the terms induction or latency reversal to refer to situations in which viral gene expression from a latent provirus is induced.

---

Box 1

Latency—a reversibly nonproductive state of infection of individual cells

Reservoir—a cell type or anatomical site in which replication-competent forms of HIV can persist on a timescale of years in patients on optimal ART

Latency reversing agent: a pharmacologic or biologic agent that induces HIV-1 gene expression in cells with latent proviruses

---

## 3 The Viral Outgrowth Assay

### 3.1 Assay Design and Rationale

The presence of a latent reservoir for HIV-1 in vivo was originally demonstrated using viral outgrowth experiments in which resting CD4$^+$ T cells, which do not produce virus, were purified and activated with the mitogen phytohemagglutinin (PHA) and irradiated allogeneic peripheral blood mononuclear cells to reverse latency. In these experiments, CD4$^+$ T lymphoblasts were then added to the culture so that virus released from infected cells could be expanded and eventually detected by an ELISA assay for HIV-1 p24 antigen in the supernatant. As shown in Fig. 1, experiments of this kind can be converted into a quantitative viral outgrowth assay (QVOA) for latently infected cells by plating the resting CD4$^+$ T cells in limiting dilution *before* adding the activating stimulus. Virus release from a single latently infected cell can be expanded to give a readily detectable ELISA signal in 2–3 weeks, and the frequency of latently infected cells can be determined using Poisson statistics (Siliciano and Siliciano 2005; Rosenbloom et al. 2015; Laird et al. 2016). This assay is positive in almost all patients with HIV-1 infection provided that a sufficient number of patient cells are plated. Because the frequency of latently infected cells is low, 0.1–10 infectious units per million (IUPM) resting CD4$^+$ T cells, a minimum of $20 \times 10^6$ purified resting CD4$^+$ cells is usually required.

The QVOA is based on a model of HIV-1 latency that takes into account the relationship between viral gene expression and the state of cellular activation. Pioneering studies by Nabel and Baltimore demonstrated that HIV-1 gene

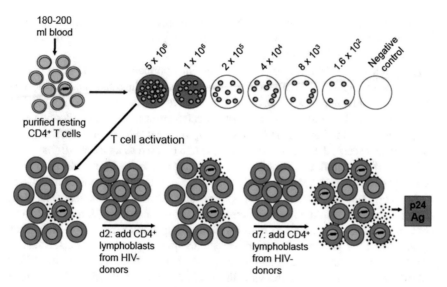

**Fig. 1** The quantitative viral outgrowth assay. Resting CD4⁺ T cells from a large volume blood sample are plated in serial dilution and activated with PHA and irradiated allogeneic PBMC causing some of the latently infected cells to produce virus. This virus then replicates in CD4+ T lymphoblasts from normal donors that are added to the culture. Over the course of 2–3 weeks, virus from a single cell replicates to the point where it can be detected by an ELISA for HIV-1 p24 antigen in the supernatant. The frequency of latently infected cells can then be calculated using limiting dilution statistics ($\sim 1/10^6$ cells in the example shown). More sensitive assays can be used to detect outgrowth, but it is essential to demonstrate exponential increases in the quantity measured. As discussed in the text, this assay provides a definitive minimal estimate of the frequency of latently infected cells. Detailed features of the assay protocol (Siliciano and Siliciano 2005; Laird et al. 2016) and statistical analysis (Rosenbloom et al. 2015) have been published

expression is dependent upon the host transcription factor NFκB, which is translocated to the nucleus in activated T cells (Nabel and Baltimore 1987). Work from many groups has shown that NFκB and other host transcription factors required for HIV-1 gene expression, including NFAT and pTEFb, are also mobilized in activated T cells (Bohnlein et al. 1988; Duh et al. 1989; Adams et al. 1994; Zhu et al. 1997; Kinoshita et al. 1998; Lin et al. 2003; Rice and Herrmann 2003). The sequestration of these factors as activated T cells return to a resting memory state like contributes to the establishment of latent infection. Additional epigenetic modifications may then enforce the state of latency (Van Lint et al. 1996; Pearson et al. 2008). The molecular events that lead to reactivation of latent HIV-1 are discussed in detail in other chapters in this issue.

The QVOA is regarded as the definitive assay for the latent reservoir. It was initially used to show that the population of latently infected resting CD4⁺ T cells is extremely stable, with a high life of 3.7 years in patients on suppressive ART (Finzi et al. 1999; Siliciano et al. 2003). Recently, David Margolis and colleagues have repeated this analysis in patients on newer regimens and reported a very similar half-life of 3.6 years (Crooks et al. 2015), indicating that all the improvement in ART that has occurred in the interval have not changed the fundamental problem of a stable, nonreplicating form of the virus.

Several aspects of the QVOA deserved special comment. One important issue is the choice of input cell population. In our original studies, the input cells were highly purified resting $CD4^+$ T cells. Without activation, these cells do not produce virus (Chun et al. 1995). Thus any virus isolated from these cells following activation can be said to have come from a latently infected cell by the definition given above. In this sense, the viral outgrowth assay can be used as a measure of the *latent* reservoir. If the input cell population is not first shown to be unable to spontaneously produce virus, then the assay cannot be said to measure latent infection.

Another important issue is the state of viral suppression in the patient. The assay gives meaningful results only in patients who have had suppression of detectable viral replication for at least 6 months. Pioneering studies by Mario Stevenson demonstrated the presence in untreated patients of many recently infected resting $CD4^+$ T cells carrying unintegrated HIV-1 DNA (Bukrinsky et al. 1991). Following T cell activation, the viral life cycle can be completed, and these cells can produce virus and thus be detected in the QVOA even though they are not part of the stable latent reservoir. Similarly, there are likely to be resting $CD4^+$ T cells with integrated HIV-1 DNA that are not destined to become part of the stable reservoir. As a result of these two populations, the frequency of infected resting $CD4^+$ T cells that produce virus in the QVOA is very high ($1000/10^6$) in samples from viremic patients (Blankson et al. 2000) and does not fall to the stable average level of $1/10^6$ level until 6–8 months after initiation of suppressive ART. Thus, it is common practice to carry out the QVOA only in patients who have had suppression of detectable viremia for at least that long.

## 3.2  Recent Improvements and Potential Problems with the QVOA

Over the years, some alternative versions of the QVOA have been described. In some versions of the assay, the activation of resting $CD4^+$ T cells is accomplished with anti-CD3/anti-CD28 antibodies instead of PHA and irradiated allogeneic PBMC (Chun et al. 1997b; Beliakova-Bethell et al. 2017). A considerable simplification of the assay can also be realized by using a transformed $CD4^+$ T cell line expressing CCR5 in place of normal donor $CD4^+$ T cells blasts for expanding virus released from latently infected cells (Laird et al. 2013).

In place of the final ELISA assay for virus in the culture supernatant, more sensitive assays can be used including RT-PCR for HIV-1 RNA in cells or in the supernatant virus (Shan et al. 2013; Laird et al. 2013), novel ultrasensitive assays for p24 protein (Wu et al. 2017; Passaes et al. 2017), or transfer of infection to a reporter cell line (Sanyal et al. 2017). This may shorten the time needed to detect outgrowth. However, a major problem with the use of more sensitive assays is the potential detection of virus release from defective proviruses. As is discussed below, the landscape of HIV-1 proviruses that persist in treated patients is dominated by defective proviruses (Ho et al. 2013; Bruner et al. 2016), and some of these are capable of giving rise to viral RNA and protein (Imamichi et al. 2016; Pollack et al. 2017). The value of the original version of the assay, which utilizes a relatively insensitive ELISA assay to detect viral outgrowth, is that exponential viral growth in vitro is required before virus released from a single infected cell becomes detectable. Thus, the assay detects viruses capable of producing the kind of exponential growth that is seen during viral rebound following ART interruption (Davey et al. 1999). When more sensitive assays are used to detect outgrowth, it is essential that the exponential increases in the measured parameter over time be demonstrated. Otherwise, the reported values may reflect defective proviruses irrelevant to cure efforts.

Although the standard QVOA reliably detects the latent proviruses that are capable of causing viral rebound, some problems with the assay remain. First, it requires large blood samples, typically 100–200 ml. This is not a problem of assay sensitivity, but rather of the low frequency of latently infected cells. Second, the assay is labor-intensive, requiring careful tissue culture work in a BSL3 laboratory. Third, it has a slow turnaround time, requiring 1–3 weeks for detection of outgrowth, depending on whether RT-PCR or ELISA is used to detect outgrowth. These are technical issues that limit the scalability of the assay. However, a more fundamental problem has become apparent through the analysis of the landscape of proviruses that persist in treated patients.

# 4 The Proviral Landscape: Defective and Non-induced Proviruses

Due to the difficulties associated with the QVOA, many investigators have resorted to simple PCR assays to detect proviral DNA in latently infected cells. Early studies demonstrated that the frequency of infected cells detected by PCR is considerably greater than the frequency measured by QVOA (Chun et al. 1997a), and this has been confirmed in a recent study that directly compared the QVOA with various PCR assays on a shared set of samples from well-characterized patients (Eriksson et al. 2013). PCR assays gave infected cell frequencies that were ∼300-fold higher than, and poorly correlated with the QVOA. One implication of this finding is that QVOA cultures negative for viral outgrowth must contain proviruses that are not induced to produce replication-competent virus in this assay. To understand the nature of these non-induced proviruses, Ho et al. carried out full-length, single-genome analysis of proviruses in negative QVOA cultures (Ho et al. 2013), and Bruner et al. used the same approach to analyze proviruses in freshly isolated resting CD4$^+$ T cells (Bruner et al. 2016). The results were striking. In patients who start ART during chronic infection, 98% of proviruses have major defects precluding replication, including large internal deletions and/or APOBEC3G-mediated lethal hypermutation. Many of these proviruses would be detected by standard PCR assays or Alu-PCR assays that detect integrated HIV-1 DNA. These defects accumulate rapidly and are readily apparent even in patients who start ART in the first few months of infection (Bruner et al. 2016). Thus standard PCR assay vastly overestimates latent reservoir size in all patients and should not be used to evaluate cure strategies, particularly since cells with defective proviruses may be affected differently than cells with intact proviruses in patients receiving interventions targeting the reservoir. The analysis of the proviral landscape by full genome sequencing may prove useful in understanding the distribution of latent HIV-1 in different T cells subsets (Lee et al. 2017) and the role of defective proviruses in other pathological processes (Imamichi et al. 2016).

Although the vast majority of proviruses are defective, some proviruses lacking deletions, hypermutation, and premature stop codons can be detected by full-length, single-genome sequencing (Ho et al. 2013; Bruner et al. 2016). The frequency of cells carrying these intact proviruses still exceeds the frequency of latently infected cells measured in the QVOA by 30–60-fold (Fig. 2). Thus there are a large number

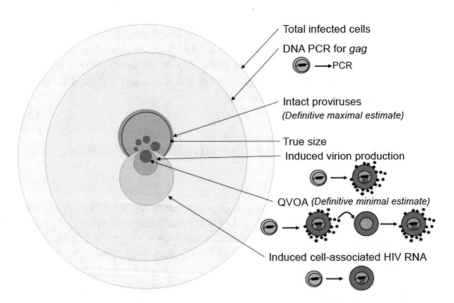

**Fig. 2** Venn diagram representation of the proviral landscape as detected by various assays. The total number of infected cells (light yellow) is approximated by standard PCR assays (dark yellow) but these assays miss proviruses with deletions spanning the relevant amplicons. The vast majority of proviruses detected by standard PCR assays are defective. The QVOA (larger dark pink circle) provides a definitive minimal estimate of reservoir size but underestimates the true size of the reservoir (light pink) because additional rounds of T cell activation induce additional intact proviruses (other dark pink circles). Other induction assays that measure viral RNA (gray) levels or virion production (green) also miss proviruses that are not induced by a single round of T cell activation. Therefore, the number of intact proviruses (blue) may provide the best measure of the true size of the reservoir

of intact proviruses that are not induced to produce replication-competent virus in the standard QVOA. To determine whether these intact, non-induced proviruses are replication-competent, Ho et al. reconstructed these proviruses by gene synthesis, transfected them into virus-producing cells to generate viral stocks, and compared their replication kinetics to those of reference isolates and replication-competent QVOA isolates from the same patient. All of these viruses replicated normally (Ho et al. 2013). This result indicates that most of the defects that render proviruses replication-incompetent are readily apparent defects such as large deletions or hypermutation.

Although the intact non-induced proviruses were capable of producing infectious virus in transfection experiments, it remained unclear whether they could be induced in vivo. Further studies showed that for the most part they had functional, non-methylated LTRs and were integrated into host genes that were actively expressed in resting and activated CD4$^+$ T cells (Ho et al. 2013). These features suggested that they were potentially inducible. This was formally demonstrated by recovering cells from QVOA wells negative for viral outgrowth and subjecting the cells to additional rounds of T cell activation. In these experiments, each round of T cell activation induced additional proviruses to release replication-competent virus (Ho et al. 2013; Hosmane et al. 2017). These experiments indicate that while standard PCR assays vastly overestimate reservoir size, the QVOA may underestimate reservoir size since some of the intact, non-induced proviruses can be induced with additional stimulation. It is possible that some intact, non-induced proviruses may be permanently silenced by epigenetic modifications or integrated into chromosomal locations that preclude viral gene expression. Nevertheless, the QVOA should be regarded as a definitive *minimal* estimate of reservoir size, while the direct measurement of cells with genetically intact proviruses may provide the best measure of the true frequency of latently infected cells.

# 5 Other Reservoir Assays

## 5.1 PCR and Hybridization-Based Assays

The most common approach to reservoir measurement involves quantation of proviral DNA by PCR (for a review, see (Massanella and Richman 2016). PCR assays generally amplify short conserved regions of the provirus. Recent approaches utilize digital droplet PCR which provides better quantitation when the number of proviruses in the sample is low (Strain and Richman 2013). Alu-PCR assays amplify the regions between an Alu element and the integrated provirus and provide discrimination between integrated and unintegrated proviruses (O'Doherty et al. 2002; Yu et al. 2008; Liszewski et al. 2009). This is important in some situations, such as in samples from viremic patients that will contain large numbers of recently infected cells with unintegrated HIV-1 DNA (Bukrinsky et al. 1991). However,

Alu-PCR assays and all other subgenomic PCR assays suffer from the major drawback that they do not accurately distinguish between intact and defective proviruses. Subgenomic PCR assays will not amplify proviruses with deletions that overlap the primer binding sites but do amplify proviruses with defects outside of the region amplified.

Recent studies have described hybridization-based assays for viral DNA and RNA (Deleage et al. 2016; Baxter et al. 2016; Grau-Exposito et al. 2017). These studies allow visualization of infected cells in tissues and will be useful in understanding the anatomical distribution of infected cells in vivo. However, the same caveats regarding the detection of defective proviruses apply. As is discussed below, some defective proviruses can give rise to viral RNA. Therefore, the detection of viral RNA$^+$ cells does not necessarily indicate that these cells can produce infectious virus.

One argument for the use of PCR-based assays is that they will provide a surrogate measure for reservoir size that will be correlated with measurements that detect replication-competent virus. Plasma virus levels certainly vary widely among untreated patients, and it is likely various measures of the extent of infection within individuals will be correlated. However, a problem arises when there is a need to measure reservoir reductions in response to cure strategies. PCR assays may give incorrect answers because cells containing defective proviruses may not be affected by the interventions in the same manner as cells carrying replication-competent proviruses.

## 5.2  Induction Assays

Another general approach to reservoir measurement involves assays in which latently infected cells are treated in vitro with a latency reversing agent and the induction of viral gene expression is then quantified (Cillo et al. 2014; Bullen et al. 2014; Procopio et al. 2015). These assays can be termed "induction assays" because they depend on a stimulus that induces the latent proviruses. The QVOA belongs in this class, but most inductions assays measure viral RNA in cells or in culture supernatants rather than release of infectious virus. Thus they are faster and easier than the QVOA. If the cell population to be assayed is plated in limiting dilution before the addition of the latency reversing agent, then the assay can be used to measure the frequency of cells that respond to the inducing stimulus. However, if the cells are stimulated and cultured before plating at limiting dilution, then cell proliferation, viral spread, and other complications may preclude the accurate measurement of infected cell frequency (Sanyal et al. 2017).

The induction assays suffer from two major drawbacks. First, as is the case with the QVOA, they fail to detect intact proviruses that are not induced by a single round of in vitro stimulation (Fig. 2). Second, with the exception of the QVOA, these assays can detect viral gene expression from defective proviruses. Recent studies have shown that some defective proviruses can be transcribed and that some can even give rise to proteins (Ho et al. 2013; Imamichi et al. 2016). Defective proviruses with small deletions in the packaging signal can even give rise to virus-like particles (Ho et al. 2013). Therefore, depending on the output parameter measured, induction assays other than the QVOA may detect defective as well as intact proviruses.

The induction assays are particularly useful in measuring the response to LRAs (Bullen et al. 2014; Wei et al. 2014; Cillo et al. 2014). Although LRA efficacy is often evaluated using cell line or primary cell models of latency, recent studies have shown that many agents that work well in this model systems fail to induce latent proviruses from patient cells (Bullen et al. 2014). Induction assays using cells from patients on ART have proven to be very useful in identifying LRAs that can effectively reverse latency in vivo.

# 6 Other Approaches to Reservoir Measurement

## 6.1 Residual Viremia

In patients on ART, plasma virus levels fall to below the limit of detection of clinical assays but level off at levels around one copy of HIV-1 RNA/ml of plasma (Dornadula et al. 1999; Palmer et al. 2003; Maldarelli et al. 2007). Sequencing of this residual viremia indicates that it is composed of archival drug-sensitive virus that is continually released over long period of time without evolutionary change (Hermankova et al. 2001; Persaud et al. 2004; Kieffer et al. 2004; Nettles et al. 2005; Bailey et al. 2006). In addition, intensification of ART by addition of another drug from a different class does not further reduce residual viremia (Dinoso et al. 2009a; Gandhi et al. 2010). All of these findings are consistent with the hypothesis that residual viremia represents virus release from a small number of latently infected cells that become activated every day. The data are not consistent with the alternative hypothesis that the residual viremia represents ongoing cycles of replication that continue despite ART.

Given that residual viremia represents virus release from stable viral reservoirs, it is reasonable to ask whether the measurement of residual viremia could provide an alternative approach to reservoir measurement. It is likely that the level of residual viremia is generally related to reservoir size, but the precise nature of this relationship is not yet clear. There are two additional problems with the use of residual viremia as a reservoir measure. First, the level of residual viremia is very low and is frequently below the level of detection of sensitive research assays that have limits of detection of one copy of HIV-1 RNA/ml of plasma. Thus for some patients, pelleting of large volumes of plasma would be required. Second, the residual viremia is often dominated by a clonal population of viruses (Tobin et al. 2005; Bailey et al. 2006). Recent studies have shown that the latent reservoir is also dominated by clonal populations of latently infected cells (Bui et al. 2017; Lorenzi et al. 2016;

Hosmane et al. 2017). This finding appears to reflect the fact that $CD4^+$ T cells can proliferate after infection, thereby copying unmodified forms of the viral genome into progeny cells. Although the frequency of latently infected cells is remarkably stable (Finzi et al. 1999; Siliciano et al. 2003; Crooks et al. 2015), the reservoir is not static and is likely composed of many clones of infected $CD4^+$ T cells that expand and contract over time in response to stimuli that are not yet fully understood. At any given time, the residual viremia may be dominated by a single clone, and it is not clear that the level of residual viremia will accurately indicate the size of the entire reservoir.

## 6.2 Biomarkers

Given the difficulties associated with reservoir measurement, there has been great interest in the potential use of nonviral biomarkers as surrogates for reservoir size. No such biomarker has yet been identified. A recent report suggesting that CD32a, an Fc receptor not normally expressed on $CD4^+$ T cells, is specifically upregulated by latent HIV-1 infection and can therefore uniquely identify latently infected cells (Descours et al. 2017). No mechanism was provided for this remarkable observation, and it awaits confirmation.

## 6.3 Time to Rebound

The goal of curative interventions is to allow patients to interrupt ART without immediate rebound of viremia. Normally, viremia rebounds to detectable levels approximately 2 weeks after interruption of ART (Davey et al. 1999; Rothenberger et al. 2015). Reductions in the reservoir should cause a delay in rebound, and this has been observed in several "near cure" cases in which reservoir size has been dramatically reduced by allogeneic hematopoietic stem cell transplantation (Henrich et al. 2013; Henrich et al. 2014) or limited by very early treatment (Persaud et al. 2013; Luzuriaga et al. 2015). The relationship between the size of the latent reservoir and time until rebound of viremia after interruption of ART has been described in a mathematical model that accurately predicts rebound in the near cure cases (Hill et al. 2014).

Because of this relationship, time to viral rebound is used as an outcome measure in some clinical studies. However, there are several problems with this approach. There are potential risks to the patient, including the evolution of drug resistance (Henrich et al. 2014). In addition, the model mentioned above predicts that reservoir reductions of >2 logs will be required to produce meaningful delays in viral rebound. The frequency of latently infected cells varies among patients over a 2-log range (Finzi et al. 1999; Siliciano et al. 2003), and this variation, together with the stochastic nature of the reactivation of latently infected cells, introduces a large degree of variability in rebound time for a given degree of reservoir reduction. Therefore, very large clinical studies would be needed to detect small reductions in the latent reservoir by measuring time to rebound.

# 7 Conclusions

Currently, over 100 clinical trials of curative strategies are in progress or under development (Treatment Action Group June 12, 2017). An accurate and scalable assay for the latent reservoir is urgently needed to allow assessment of the efficacy of curative interventions targeting the reservoir. The most commonly used assays are standard PCR assays targeting conserved regions of the HIV-1 genome. However, because the vast majority of HIV-1 proviruses are defective, such assays may not accurately capture changes in the minor subset of proviruses that are replication-competent and that pose a barrier to cure. The QVOA, on the other hand, may underestimate reservoir size because not all replication-competent proviruses are induced by a single round of T cell activation. This assay is best regarded as a definitive minimal estimate of reservoir size. The best approach may be to measure all of the proviruses with the potential to cause viral rebound. A variety of novel assays have recently been described. Ultimately, the assay that best predicts time to viral rebound will be the most useful to the cure effort.

**Acknowledgements** This work was supported by the NIH Martin Delaney I4C, Beat-HIV and DARE Collaboratories, by the Johns Hopkins Center for AIDS Research (P30AI094189), by NIH grant 43222, and by the Howard Hughes Medical Institute and the Bill and Melinda Gates Foundation.

# References

Adams M, Sharmeen L, Kimpton J, Romeo JM, Garcia JV, Peterlin BM, Groudine M, Emerman M (1994) Cellular latency in human immunodeficiency virus-infected individuals with high CD4 levels can be detected by the presence of promoter-proximal transcripts. Proc Natl Acad Sci USA 91(9):3862–3866

Bailey JR, Sedaghat AR, Kieffer T, Brennan T, Lee PK, Wind-Rotolo M, Haggerty CM, Kamireddi AR, Liu Y, Lee J, Persaud D, Gallant JE, Cofrancesco J Jr, Quinn TC, Wilke CO, Ray SC, Siliciano JD, Nettles RE, Siliciano RF (2006) Residual human immunodeficiency virus type 1 viremia in some patients on antiretroviral therapy is dominated by a small number of invariant clones rarely found in circulating CD4+ T cells. J Virol 80(13):6441–6457

Baxter AE, Niessl J, Fromentin R, Richard J, Porichis F, Charlebois R, Massanella M, Brassard N, Alsahafi N, Delgado GG, Routy JP, Walker BD, Finzi A, Chomont N, Kaufmann DE (2016) Single-cell characterization of viral translation-competent reservoirs in HIV-infected individuals. Cell Host Microbe 20(3):368–380

Beliakova-Bethell N, Hezareh M, Wong JK, Strain MC, Lewinski MK, Richman DD, Spina CA (2017) Relative efficacy of T cell stimuli as inducers of productive HIV-1 replication in latently infected CD4 lymphocytes from patients on suppressive cART. Virology 508:127–133

Blankson JN, Finzi D, Pierson TC, Sabundayo BP, Chadwick K, Margolick JB, Quinn TC, Siliciano RF (2000) Biphasic decay of latently infected CD4+ T cells in acute human immunodeficiency virus type 1 infection. J Infect Dis 182(6):1636–1642

Bohnlein E, Lowenthal JW, Siekevitz M, Ballard DW, Franza BR, Greene WC (1988) The same inducible nuclear proteins regulates mitogen activation of both the interleukin-2 receptor-alpha gene and type 1 HIV. Cell 53(5):827–836

Borrow P, Lewicki H, Wei X, Horwitz MS, Peffer N, Meyers H, Nelson JA, Gairin JE, Hahn BH, Oldstone MB, Shaw GM (1997) Antiviral pressure exerted by HIV-1-specific cytotoxic T lymphocytes (CTLs) during primary infection demonstrated by rapid selection of CTL escape virus. Nat Med 3(2):205–211

Bruner KM, Murray AJ, Pollack RA, Soliman MG, Laskey SB, Capoferri AA, Lai J, Strain MC, Lada SM, Hoh R, Ho YC, Richman DD, Deeks SG, Siliciano JD, Siliciano RF (2016) Defective proviruses rapidly accumulate during acute HIV-1 infection. Nat Med 22(9):1043–1049

Bui JK, Sobolewski MD, Keele BF, Spindler J, Musick A, Wiegand A, Luke BT, Shao W, Hughes SH, Coffin JM, Kearney MF, Mellors JW (2017) Proviruses with identical sequences comprise a large fraction of the replication-competent HIV reservoir. PLoS Pathog 13(3): E1006283

Bukrinsky MI, Stanwick TL, Dempsey MP, Stevenson M (1991) Quiescent T lymphocytes as an inducible virus reservoir in HIV-1 infection. Science (New York, N.Y.) 254(5030):423–427

Bullen CK, Laird GM, Durand CM, Siliciano, JD, Siliciano RF (2014) New ex vivo approaches distinguish effective and ineffective single agents for reversing HIV-1 latency in vivo. Nat Med

Calantone N, Wu F, Klase Z, Deleage C, Perkins M, Matsuda K, Thompson EA, Ortiz AM, Vinton CL, Ourmanov I, Lore K, Douek DC, Estes JD, Hirsch VM, Brenchley JM (2014) Tissue myeloid cells in SIV-infected primates acquire viral DNA through phagocytosis of infected T cells. Immunity 41(3):493–502

Chun TW, Finzi D, Margolick J, Chadwick K, Schwartz D, Siliciano RF (1995) In vivo fate of HIV-1-infected T cells: quantitative analysis of the transition to stable latency. Nat Med 1 (12):1284–1290

Chun TW, Carruth L, Finzi D, Shen X, Digiuseppe JA, Taylor H, Hermankova M, Chadwick K, Margolick J, Quinn TC, Kuo YH, Brookmeyer R, Zeiger MA, Barditch-Crovo P, Siliciano RF (1997a) Quantification of latent tissue reservoirs and total body viral load in HIV-1 infection. Nature 387(6629):183–188

Chun TW, Stuyver L, Mizell SB, Ehler LA, Mican JA, Baseler M, Lloyd AL, Nowak MA, Fauci AS (1997b) Presence of an inducible HIV-1 latent reservoir during highly active antiretroviral therapy. Proc Natl Acad Sci USA 94(24):13193–13197

Cillo AR, Sobolewski MD, Bosch RJ, Fyne E, Piatak M Jr, Coffin JM, Mellors JW (2014) Quantification of HIV-1 latency reversal in resting CD4+ T cells from patients on suppressive antiretroviral therapy. Proc Natl Acad Sci USA 111(19):7078–7083

Crooks AM, Bateson R, Cope AB, Dahl NP, Griggs MK, Kuruc JD, Gay CL, Eron JJ, Margolis DM, Bosch RJ, Archin NM (2015) Precise quantitation of the latent HIV-1 reservoir: implications for eradication strategies. J Infect Dis

Davey RT Jr, Bhat N, Yoder C, Chun TW, Metcalf JA, Dewar R, Natarajan V, Lempicki RA, Adelsberger JW, Miller KD, Kovacs JA, Polis MA, Walker RE, Falloon J, Masur H, Gee D, Baseler M, Dimitrov DS, Fauci AS, Lane HC (1999) HIV-1 and T cell dynamics after interruption of highly active antiretroviral therapy (HAART) in patients with a history of sustained viral suppression. Proc Natl Acad Sci USA 96(26):15109–15114

Deeks SG, Lewin SR, Ross AL, Ananworanich J, Benkirane M, Cannon P, Chomont N, Douek D, Lifson JD, Lo YR, Kuritzkes D, Margolis D, Mellors J, Persaud D, Tucker JD, Barre-Sinoussi F, International Aids Society Towards A Cure Working Group, Alter G, Auerbach J, Autran B, Barouch DH, Behrens G, Cavazzana M, Chen Z, Cohen EA, Corbelli GM, Eholie S, Eyal N, Fidler S, Garcia L, Grossman C, Henderson G, Henrich TJ, Jefferys R, Kiem HP, Mccune J, Moodley K, Newman PA, Nijhuis M, Nsubuga MS, Ott M, Palmer S, Richman D, Saez-Cirion

A, Sharp M, Siliciano J, Silvestri G, Singh J, Spire B, Taylor J, Tolstrup M, Valente S, Van Lunzen J, Walensky R, Wilson I, Zack J (2016) International AIDS society global scientific strategy: towards an HIV cure 2016. Nat Med 22(8):839–850

Deleage C, Wietgrefe SW, Del Prete G, Morcock DR, Hao XP, Piatak M Jr, Bess J, Anderson JL, Perkey KE, Reilly C, Mccune JM, Haase AT, Lifson JD, Schacker TW, Estes JD (2016) Defining HIV and SIV reservoirs in lymphoid tissues. Pathog Immun 1(1):68–106

Descours B, Petitjean G, Lopez-Zaragoza JL, Bruel T, Raffel R, Psomas C, Reynes J, Lacabaratz C, Levy Y, Schwartz O, Lelievre JD, Benkirane M (2017) CD32a is a marker of a CD4 T-cell HIV reservoir harbouring replication-competent proviruses. Nature 543(7646):564–567

Dinoso JB, Kim SY, Wiegand AM, Palmer SE, Gange SJ, Cranmer L, O'shea A, Callender M, Spivak A, Brennan T, Kearney MF, Proschan MA, Mican JM, Rehm CA, Coffin JM, Mellors JW, Siliciano RF, Maldarelli F (2009a) Treatment intensification does not reduce residual HIV-1 viremia in patients on highly active antiretroviral therapy. Proc Natl Acad Sci USA 106(23):9403–9408

Dinoso JB, Rabi SA, Blankson JN, Gama L, Mankowski JL, Siliciano RF, Zink MC, Clements JE (2009b) A simian immunodeficiency virus-infected macaque model to study viral reservoirs that persist during highly active antiretroviral therapy. J Virol 83(18):9247–9257

Dornadula G, Zhang H, Vanuitert B, Stern J, Livornese L Jr, Ingerman MJ, Witek J, Kedanis RJ, Natkin J, Desimone J, Pomerantz RJ (1999) Residual HIV-1 RNA in blood plasma of patients taking suppressive highly active antiretroviral therapy. JAMA, J Am Med Assoc 282 (17):1627–1632

Duh EJ, Maury WJ, Folks TM, Fauci AS, Rabson AB (1989) Tumor necrosis factor alpha activates human immunodeficiency virus type 1 through induction of nuclear factor binding to the NF-kappa B sites in the long terminal repeat. Proc Natl Acad Sci USA 86(15):5974–5978

Eisele E, Siliciano RF (2012) Redefining the viral reservoirs that prevent HIV-1 eradication. Immunity 37(3):377–388

Eriksson S, Graf EH, Dahl V, Strain MC, Yukl SA, Lysenko ES, Bosch RJ, Lai J, Chioma S, Emad F, Abdel-Mohsen M, Hoh R, Hecht F, Hunt P, Somsouk M, Wong J, Johnston R, Siliciano RF, Richman DD, O'doherty U, Palmer S, Deeks SG, Siliciano JD (2013) Comparative analysis of measures of viral reservoirs in HIV-1 eradication studies. PLoS Pathog 9(2):E1003174

Finzi D, Hermankova M, Pierson T, Carruth LM, Buck C, Chaisson RE, Quinn TC, Chadwick K, Margolick J, Brookmeyer R, Gallant J, Markowitz M, Ho DD, Richman DD, Siliciano RF (1997) Identification of a reservoir for HIV-1 in patients on highly active antiretroviral therapy. Science (New York, N.Y.) 278(5341):1295–1300

Finzi D, Blankson J, Siliciano JD, Margolick JB, Chadwick K, Pierson T, Smith K, Lisziewicz J, Lori F, Flexner C, Quinn TC, Chaisson RE, Rosenberg E, Walker B, Gange S, Gallant J, Siliciano RF (1999) Latent infection of CD4+ T cells provides a mechanism for lifelong persistence of HIV-1, even in patients on effective combination therapy. Nat Med 5(5):512–517

Gama L, Abreu CM, Shirk EN, Price SL, Li M, Laird GM, Pate KA, Wietgrefe SW, O'connor SL, Pianowski L, Haase AT, Van Lint C, Siliciano RF, Clements JE, Lra-Siv Study Group (2017) Reactivation of simian immunodeficiency virus reservoirs in the brain of virally suppressed macaques. AIDS (London, England) 31(1):5–14

Gandhi RT, Zheng L, Bosch RJ, Chan ES, Margolis DM, Read S, Kallungal B, Palmer S, Medvik K, Lederman MM, Alatrakchi N, Jacobson JM, Wiegand A, Kearney M, Coffin JM, Mellors JW, Eron JJ, Aids Clinical Trials Group A5244 Team (2010) The effect of raltegravir intensification on low-level residual viremia in HIV-infected patients on antiretroviral therapy: a randomized controlled trial. PLOS Med 7(8):E1000321

Grau-Exposito J, Serra-Peinado C, Miguel L, Navarro J, Curran A, Burgos J, Ocana I, Ribera E, Torrella A, Planas B, Badia R, Castellvi J, Falco V, Crespo M, Buzon MJ (2017) A novel single-cell fish-flow assay identifies effector memory Cd4+ T Cells as a major niche for HIV-1 transcription in HIV-infected patients. Mbio 8(4). doi:10.1128/Mbio.00876-17

Gulick RM, Mellors JW, Havlir D, Eron JJ, Gonzalez C, Mcmahon D, Richman DD, Valentine FT, Jonas L, Meibohm A, Emini EA, Chodakewitz JA (1997) Treatment with indinavir, zidovudine, and lamivudine in adults with human immunodeficiency virus infection and prior antiretroviral therapy. N Engl J Med 337(11):734–739

Hammer SM, Squires KE, Hughes MD, Grimes JM, Demeter LM, Currier JS, Eron JJ Jr, Feinberg JE, Balfour HH Jr, Deyton LR, Chodakewitz JA, Fischl MA (1997) A controlled trial of two nucleoside analogues plus indinavir in persons with human immunodeficiency virus infection and CD4 cell counts of 200 per cubic millimeter or less. AIDS Clinical Trials Group 320 Study Team. N Engl J Med 337(11):725–733

Henrich TJ, Hu Z, Li JZ, Sciaranghella G, Busch MP, Keating SM, Gallien S, Lin NH, Giguel FF, Lavoie L, Ho VT, Armand P, Soiffer RJ, Sagar M, Lacasce AS, Kuritzkes DR (2013) Long-term reduction in peripheral blood HIV type 1 reservoirs following reduced-intensity conditioning allogeneic stem cell transplantation. J Infect Dis 207(11):1694–1702

Henrich TJ, Hanhauser E, Marty FM, Sirignano MN, Keating S, Lee TH, Robles YP, Davis BT, Li JZ, Heisey A, Hill AL, Busch MP, Armand P, Soiffer RJ, Altfeld M, Kuritzkes DR (2014) Antiretroviral-free HIV-1 remission and viral rebound after allogeneic stem cell transplantation: report of 2 cases. Ann Intern Med 161(5):319–327

Hermankova M, Ray SC, Ruff C, Powell-Davis M, Ingersoll R, D'aquila RT, Quinn TC, Siliciano JD, Siliciano RF, Persaud D (2001) HIV-1 drug resistance profiles in children and adults with viral load of< 50 copies/ml receiving combination therapy. JAMA, J Am Med Assoc 286(2):196–207

Hermankova M, Siliciano JD, Zhou Y, Monie D, Chadwick K, Margolick JB, Quinn TC, Siliciano RF (2003) Analysis of human immunodeficiency virus type 1 gene expression in latently infected resting CD4+ T lymphocytes in vivo. J Virol 77(13):7383–7392

Hill AL, Rosenbloom DI, Fu F, Nowak MA, Siliciano RF (2014) Predicting the outcomes of treatment to eradicate the latent reservoir for HIV-1. Proc Natl Acad Sci USA

Ho DD, Neumann AU, Perelson AS, Chen W, Leonard JM, Markowitz M (1995) Rapid turnover of plasma virions and CD4 lymphocytes in HIV-1 infection. Nature 373(6510):123–126

Ho YC, Shan L, Hosmane NN, Wang J, Laskey SB, Rosenbloom DI, Lai J, Blankson JN, Siliciano JD, Siliciano RF (2013) Replication-competent noninduced proviruses in the latent reservoir increase barrier to HIV-1 cure. Cell 155(3):540–551

Honeycutt JB, Wahl A, Baker C, Spagnuolo RA, Foster J, Zakharova O, Wietgrefe S, Caro-Vegas C, Madden V, Sharpe G, Haase AT, Eron JJ, Garcia JV (2016) Macrophages sustain HIV replication in vivo independently of T cells. J Clin Invest

Hosmane NN, Kwon KJ, Bruner KM, Capoferri AA, Beg S, Rosenbloom DI, Keele BF, Ho YC, Siliciano JD, Siliciano RF (2017) Proliferation of latently infected CD4+ T cells carrying replication-competent HIV-1: Potential role in latent reservoir dynamics. J Exp Med 214 (4):959–972

Imamichi H, Dewar RL, Adelsberger JW, Rehm CA, O'doherty U, Paxinos EE, Fauci AS, Lane HC (2016) Defective HIV-1 proviruses produce novel protein-coding RNA species in HIV-infected patients on combination antiretroviral therapy. Proc Natl Acad Sci USA

Kieffer TL, Finucane MM, Nettles RE, Quinn TC, Broman KW, Ray SC, Persaud D, Siliciano RF (2004) Genotypic analysis of HIV-1 drug resistance at the limit of detection: virus production without evolution in treated adults with undetectable HIV loads. J Infect Dis 189(8):1452–1465

Kinoshita S, Chen BK, Kaneshima H, Nolan GP (1998) Host control of HIV-1 parasitism in T cells by the nuclear factor of activated T cells. Cell 95(5):595–604

Laird GM, Eisele EE, Rabi SA, Lai J, Chioma S, Blankson JN, Siliciano JD, Siliciano RF (2013) Rapid quantification of the latent reservoir for HIV-1 using a viral outgrowth assay. PLoS Pathog 9(5):E1003398

Laird GM, Rosenbloom DI, Lai J, Siliciano RF, Siliciano JD (2016) Measuring the frequency of latent HIV-1 in resting Cd4(+) T cells using a limiting dilution coculture assay. Methods Mol Biol (Clifton, N.J.) 1354:239–253

Lee GQ, Orlova-Fink N, Einkauf K, Chowdhury FZ, Sun X, Harrington S, Kuo HH, Hua S, Chen HR, Ouyang Z, Reddy K, Dong K, Ndung'u T, Walker BD, Rosenberg ES, Yu XG, Lichterfeld M (2017) Clonal expansion of genome-intact HIV-1 in functionally polarized Th1 CD4+ T cells. J Clin Investig 127(7):2689–2696

Lin X, Irwin D, Kanazawa S, Huang L, Romeo J, Yen TS, Peterlin BM (2003) Transcriptional profiles of latent human immunodeficiency virus in infected individuals: effects of Tat on the host and reservoir. J Virol 77(15):8227–8236

Liszewski MK, Yu JJ, O'doherty U (2009) Detecting HIV-1 integration by repetitive-sampling Alu-gag PCR. Methods (San Diego, Calif.) 47(4):254–260

Lorenzi JC, Cohen YZ, Cohn LB, Kreider EF, Barton JP, Learn GH, Oliveira T, Lavine CL, Horwitz JA, Settler A, Jankovic M, Seaman MS, Chakraborty AK, Hahn BH, Caskey M, Nussenzweig MC (2016) Paired quantitative and qualitative assessment of the replication-competent HIV-1 reservoir and comparison with integrated proviral DNA. Proc Natl Acad Sci USA 113(49):E7908–E7916

Luzuriaga K, Gay H, Ziemniak C, Sanborn KB, Somasundaran M, Rainwater-Lovett K, Mellors JW, Rosenbloom D, Persaud D (2015) Viremic relapse after HIV-1 remission in a perinatally infected child. N Engl J Med 372(8):786–788

Maldarelli F, Palmer S, King MS, Wiegand A, Polis MA, Mican J, Kovacs JA, Davey RT, Rock-Kress D, Dewar R, Liu S, Metcalf JA, Rehm C, Brun SC, Hanna GJ, Kempf DJ, Coffin JM, Mellors JW (2007) ART suppresses plasma HIV-1 RNA to a stable set point predicted by pretherapy viremia. PLoS Pathog 3(4):E46

Massanella M, Richman DD (2016) Measuring the latent reservoir in vivo. J Clin Investig 126 (2):464–472

Nabel G, Baltimore D (1987) An inducible transcription factor activates expression of human immunodeficiency virus in T cells. Nature 326(6114):711–713

Nettles RE, Kieffer TL, Kwon P, Monie D, Han Y, Parsons T, Cofrancesco J Jr, Gallant JE, Quinn TC, Jackson B, Flexner C, Carson K, Ray S, Persaud D, Siliciano RF (2005) Intermittent HIV-1 viremia (Blips) and drug resistance in patients receiving HAART. JAMA, J Am Med Assoc 293(7):817–829

O'doherty U, Swiggard WJ, Jeyakumar D, Mcgain D, Malim MH (2002) A sensitive, quantitative assay for human immunodeficiency virus type 1 integration. J Virol 76(21):10942–10950

Palmer S, Wiegand AP, Maldarelli F, Bazmi H, Mican JM, Polis M, Dewar RL, Planta A, Liu S, Metcalf JA, Mellors JW, Coffin JM (2003) New real-time reverse transcriptase-initiated PCR assay with single-copy sensitivity for human immunodeficiency virus type 1 RNA in plasma. J Clin Microbiol 41(10):4531–4536

Passaes CP, Bruel T, Decalf J, David A, Angin M, Monceaux V, Muller-Trutwin M, Noel N, Bourdic K, Lambotte O, Albert ML, Duffy D, Schwartz O, Saez-Cirion A, Anrs Rhiviera Consortium (2017) Ultrasensitive HIV-1 P24 assay detects single infected cells and differences in reservoir induction by latency reversal agents. J Virol 91(6). doi:10.1128/Jvi.02296-16. Print Mar 15 2017

Pearson R, Kim YK, Hokello J, Lassen K, Friedman J, Tyagi M, Karn J (2008) Epigenetic silencing of human immunodeficiency virus (HIV) transcription by formation of restrictive chromatin structures at the viral long terminal repeat drives the progressive entry of HIV into latency. J Virol 82(24):12291–12303

Perelson AS, Essunger P, Cao Y, Vesanen M, Hurley A, Saksela K, Markowitz M, Ho DD (1997) Decay characteristics of HIV-1-infected compartments during combination therapy. Nature 387 (6629):188–191

Perng GC, Jones C (2010) Towards an understanding of the herpes simplex virus type 1 latency-reactivation cycle. Interdisc Perspect Infect Dis 2010:262415

Persaud D, Siberry GK, Ahonkhai A, Kajdas J, Monie D, Hutton N, Watson DC, Quinn TC, Ray SC, Siliciano RF (2004) Continued production of drug-sensitive human immunodeficiency virus type 1 in children on combination antiretroviral therapy who have undetectable viral loads. J Virol 78(2):968–979

Persaud D, Gay H, Ziemniak C, Chen YH, Piatak M Jr, Chun TW, Strain M, Richman D, Luzuriaga K (2013) Absence of detectable HIV-1 viremia after treatment cessation in an infant. N Engl J Med 369(19):1828–1835

Piatak M Jr, Saag MS, Yang LC, Clark SJ, Kappes JC, Luk KC, Hahn BH, Shaw GM, Lifson JD (1993) High levels of HIV-1 in plasma during all stages of infection determined by competitive PCR. Science (New York, N.Y.) 259(5102):1749–1754

Pollack RA, Jones RB, Pertea M, Bruner KM, Martin AR, Thomas AS, Capoferri AA, Beg SA, Huang SH, Karandish S, Hao H, Halper-Stromberg E, Yong PC, Kovacs C, Benko E, Siliciano RF, Ho YC (2017) Defective HIV-1 Proviruses Are Expressed and Can Be Recognized by Cytotoxic T Lymphocytes, which Shape the Proviral Landscape. Cell Host Microbe 21(4):494–506. E4

Procopio FA, Fromentin R, Kulpa DA, Brehm JH, Bebin AG, Strain MC, Richman DD, O'doherty U, Palmer S, Hecht FM, Hoh R, Barnard RJ, Miller MD, Hazuda DJ, Deeks SG, Sekaly RP, Chomont N (2015) A novel assay to measure the magnitude of the inducible viral reservoir in HIV-infected individuals. Ebiomedicine 2(8):872–881

Rice AP, Herrmann CH (2003) Regulation of TAK/P-TEFb in CD4+ T lymphocytes and macrophages. Curr HIV Res 1(4):395–404

Richman DD, Wrin T, Little SJ, Petropoulos CJ (2003) Rapid evolution of the neutralizing antibody response to HIV type 1 infection. Proc Natl Acad Sci USA 100(7):4144–4149

Rosenbloom DI, Elliott O, Hill AL, Henrich TJ, Siliciano JM, Siliciano RF (2015) Designing and interpreting limiting dilution assays: general principles and applications to the latent reservoir for human immunodeficiency virus-1. Open Forum Infect Dis 2(4):Ofv123

Rothenberger MK, Keele BF, Wietgrefe SW, Fletcher CV, Beilman GJ, Chipman JG, Khoruts A, Estes JD, Anderson J, Callisto SP, Schmidt TE, Thorkelson A, Reilly C, Perkey K, Reimann TG, Utay NS, Nganou Makamdop K, Stevenson M, Douek DC, Haase AT, Schacker TW (2015) Large number of rebounding/founder HIV variants emerge from multifocal infection in lymphatic tissues after treatment interruption. Proc Natl Acad Sci USA 112(10):E1126–E1134

Sanyal A, Mailliard RB, Rinaldo CR, Ratner D, Ding M, Chen Y, Zerbato JM, Giacobbi NS, Venkatachari NJ, Patterson BK, Chargin A, Sluis-Cremer N, Gupta P (2017) Novel assay reveals a large, inducible, replication-competent HIV-1 reservoir in resting CD4+ T cells. Nat Med 23(7):885–889

Shan L, Rabi SA, Laird GM, Eisele E, Zhang H, Margolick JB, Siliciano RF (2013) A novel PCR assay for quantification of HIV-1 RNA. J Virol

Shan L, Deng K, Xing S, Capoferri A, Gao H, Durand CM, Rabi SA, Laird GM, Kim M, Hosmane NN, Yang HC, Zhang H, Margolick JB, Ke R, Siliciano JD, Siliciano RF (2017) Unique features of effector to memory transition render CD4$^+$ T cells permissive for latent HIV-1 infection. Immunity. In Press

Shen A, Zink MC, Mankowski JL, Chadwick K, Margolick JB, Carruth LM, Li M, Clements JE, Siliciano RF (2003) Resting CD4+ T lymphocytes but not thymocytes provide a latent viral reservoir in a simian immunodeficiency virus-Macaca nemestrina model of human immunodeficiency virus type 1-infected patients on highly active antiretroviral therapy. J Virol 77 (8):4938–4949

Siliciano JD, Kajdas J, Finzi D, Quinn TC, Chadwick K, Margolick JB, Kovacs C, Gange SJ, Siliciano RF (2003) Long-term follow-up studies confirm the stability of the latent reservoir for HIV-1 in resting CD4+ T cells. Nat Med 9(6):727–728

Siliciano JD, Siliciano RF (2005) Enhanced culture assay for detection and quantitation of latently infected, resting CD4+ T-cells carrying replication-competent virus in HIV-1-infected individuals. Methods Mol Biol (Clifton, N.J.) 304:3–15

Speck SH, Ganem D (2010) Viral latency and its regulation: lessons from the gamma-herpesviruses. Cell Host Microbe 8(1):100–115

Strain MC, Gunthard HF, Havlir DV, Ignacio CC, Smith DM, Leigh-Brown AJ, Macaranas TR, Lam RY, Daly OA, Fischer M, Opravil M, Levine H, Bacheler L, Spina CA, Richman DD, Wong JK (2003) Heterogeneous clearance rates of long-lived lymphocytes infected with HIV: intrinsic stability predicts lifelong persistence. Proc Natl Acad Sci USA 100(8):4819–4824

Strain MC, Richman DD (2013) New assays for monitoring residual HIV burden in effectively treated individuals. Curr Opin HIV AIDS 8(2):106–110

Tobin NH, Learn GH, Holte SE, Wang Y, Melvin AJ, Mckernan JL, Pawluk DM, Mohan KM, Lewis PF, Mullins JI, Frenkel LM (2005) Evidence that low-level viremias during effective highly active antiretroviral therapy result from two processes: expression of archival virus and replication of virus. J Virol 79(15):9625–9634

Treatment Action Group (2017, June 12), 2017-Last Update, Research toward a cure

Van Lint C, Emiliani S, Ott M, Verdin E (1996) Transcriptional activation and chromatin remodeling of the HIV-1 promoter in response to histone acetylation. Embo J 15(5):1112–1120

Wei DG, Chiang V, Fyne E, Balakrishnan M, Barnes T, Graupe M, Hesselgesser J, Irrinki A, Murry JP, Stepan G, Stray KM, Tsai A, Yu H, Spindler J, Kearney M, Spina CA, Mcmahon D, Lalezari J, Sloan D, Mellors J, Geleziunas R, Cihlar T (2014) Histone deacetylase inhibitor romidepsin induces HIV expression in CD4 T cells from patients on suppressive antiretroviral therapy at concentrations achieved by clinical dosing. PLoS Pathog 10(4):E1004071

Wei X, Ghosh SK, Taylor ME, Johnson VA, Emini EA, Deutsch P, Lifson JD, Bonhoeffer S, Nowak MA, Hahn BH (1995) Viral dynamics in human immunodeficiency virus type 1 infection. Nature 373(6510):117–122

Wei X, Decker JM, Wang S, Hui H, Kappes JC, Wu X, Salazar-Gonzalez JF, Salazar MG, Kilby JM, Saag MS, Komarova NL, Nowak MA, Hahn BH, Kwong PD, Shaw GM (2003) Antibody neutralization and escape by HIV-1. Nature 422(6929):307–312

Wong JK, Hezareh M, Gunthard HF, Havlir DV, Ignacio CC, Spina CA, Richman DD (1997) Recovery of replication-competent HIV despite prolonged suppression of plasma viremia. Science (New York, N.Y.) 278(5341):1291–1295

Wu G, Swanson M, Talla A, Graham D, Strizki J, Gorman D, Barnard RJ, Blair W, Sogaard OS, Tolstrup M, Ostergaard L, Rasmussen TA, Sekaly RP, Archin NM, Margolis DM, Hazuda DJ, Howell BJ (2017) HDAC inhibition induces HIV-1 protein and enables immune-based clearance following latency reversal. JCI Insight 2(16). doi:10.1172/Jci.Insight.92901

Yu JJ, Wu TL, Liszewski MK, Dai J, Swiggard WJ, Baytop C, Frank I, Levine BL, Yang W, Theodosopoulos T, O'doherty U (2008) A more precise HIV integration assay designed to detect small differences finds lower levels of integrated DNA in HAART treated patients. Virology 379(1):78–86

Zack JA, Arrigo SJ, Weitsman SR, Go AS, Haislip A, Chen IS (1990) HIV-1 entry into quiescent primary lymphocytes: molecular analysis reveals a labile, latent viral structure. Cell 61(2):213–222

Zhu Y, Pe'ery T, Peng J, Ramanathan Y, Marshall N, Marshall T, Amendt B, Mathews MB, Price DH (1997) Transcription elongation factor P-TEFb is required for HIV-1 tat transactivation in vitro. Genes Dev 11(20):2622–2632

# The Antiviral Immune Response and Its Impact on the HIV-1 Reservoir

**Rebecca T. Veenhuis and Joel N. Blankson**

**Abstract** Latently infected resting memory CD4$^+$ T cells represent a major barrier to HIV-1 eradication. Studies have shown that it will not be possible to cure HIV-1 infection unless these cells are eliminated. Latently infected cells probably do not express viral antigens and thus may not be susceptible to the HIV-1 specific immune response, nevertheless the size and composition of the reservoir is influenced by the immune system. In this chapter, we review the different components of the HIV-1 specific immune response and discuss how the immune system can be harnessed to eradicate the virus.

## Contents

This work was supported by NIH grants (P30AI094189), 2R56AI080328-05A1 and 1R01AI120024-01.

R. T. Veenhuis · J. N. Blankson (✉)
Center for AIDS Research, Department of Medicine, Johns Hopkins University
School of Medicine, Baltimore, MD, USA
e-mail: jblanks@jhmi.edu

Current Topics in Microbiology and Immunology (2018) 417:43–68
DOI 10.1007/82_2017_72
Published Online: 26 October 2017

# 1   Introduction

The best characterized HIV-1 reservoir is the pool of long-lived resting memory CD4$^+$ T cells that have proviral DNA integrated into their genome (Blankson 2006). It is clear that this latent reservoir represents a significant barrier to the eradication of HIV-1 by the host immune system (Siliciano et al. 2003). One of the fundamental characteristics of HIV-1 pathogenesis is the failure of the immune system to recognize, control, and eliminate the virus. While both innate and adaptive immune responses are raised, they appear to be insufficient or too late to eliminate the virus prior seeding of the reservoir. The innate immune system is the first line of defense against viral infection and has evolved to rapidly sense and nonspecifically eliminate pathogens. During acute infection, the innate system senses HIV-1 using pattern recognition receptors (PRRs), which triggers a signaling cascade that initiates innate intracellular antiviral defenses aimed at restricting replication and spread of virus (Altfeld and Gale 2015). This initial signal leads to the production of cytokines and chemokines that inform the surrounding environment of the invading pathogen, activating, and attracting innate immune cells to the site of infection and to the lymph nodes. Antiviral innate effector cells can subsequently contribute to the control of viremia and modulate the quality of the adaptive immune response to HIV-1 as it develops (Altfeld and Gale 2015). The adaptive immune response has evolved to provide a broader and more finely tuned response and is heavily influenced by the innate response to viral infections. This highly specific response takes several weeks to develop and elicits a direct and potent attack that can eliminate most invading pathogens (Bonilla and Oettgen 2010). The adaptive arm consists of the humoral response, virus specific antibodies, and cell-mediated response, virus specific T-cells. Although there have been no cases where the host immune response has been able to eliminate HIV-1 infection on its own, there are examples of robust responses leading to control of viral replication and impact on the size of the latent reservoir, suggesting that with manipulation via preventative or therapeutic treatments or vaccinations the host immune system could be empowered to eliminate HIV-1 infection. Here, we discuss ways both the innate and adaptive immune systems impact and shape the HIV-1 reservoir and the barriers that exist for eradication of the virus by the immune system.

# 2   Innate Immune Response

## 2.1   Type I Interferon and Restriction Factors

One of the initial functions of the innate immune system is the recognition of viral pathogen-associated molecular patterns (PAMPs) by pattern recognition receptors (PPRs), which leads to type I interferon (IFN) signaling, and the release of cytokines. Type I IFN induces antiviral restriction factors called interferon simulated genes (ISGs) and has been shown to suppress HIV-1 replication in vitro (Doyle et al. 2015; Hardy et al. 2013). In addition, IFN has been shown to inhibit early HIV-1 infection in humanized mice and SIV infection in rhesus macaques (Lavender et al. 2016; Sandler et al. 2014). These observations suggest that an initial robust IFN response can help to limit HIV-1 or SIV infection and seeding of the reservoir.

The induction of viral restriction factors has been shown to have a large impact on the types of proviruses that accumulate in the HIV-1 reservoir. APOBEC3G, one of the most widely characterized ISG encoded proteins, plays a substantial role in development of the reservoir. This protein is a cytidine deaminase, which causes G to A hypermutation in retroviral genomes (Goff 2003; Harris et al. 2003; Lecossier et al. 2003; Mangeat et al. 2003; Sheehy et al. 2002; Yu et al. 2004; Zhang et al. 2003). APOBEC3G is incorporated into assembling virions where it deaminates cytidines on the single-stranded viral cDNA that is synthesized by reverse transcriptase (RT) upon entry of the virus into a new host cell (Harris et al. 2003; Lecossier et al. 2003; Mangeat et al. 2003; Yu et al. 2004; Zhang et al. 2003). This induces many mutations in the HIV-1 genome and can render the virus nonfunctional. The hypermutation of the virus leads to a massive accumulation of defective proviruses in the reservoir that are incapable of producing functional virus when reactivated (Kieffer et al. 2005; Bruner et al. 2016; Ho et al. 2013). Additional IFN-induced restriction factors have been shown to have a substantial effect on the infectivity of HIV-1. MX2 is known to localize to the nuclear envelope and has been shown to inhibit divergent strains of HIV-1, however the protein's mechanism of function is not yet fully understood. TRIM5α is known to inhibit RT and prevent viral cDNA synthesis in SIV infection and SAMHD1 depletes dNTP levels in nondividing cells, thereby depriving RT of the substrates it requires for effective cDNA generation in HIV-1 infection. Tetherin prevents the release of budded HIV-1 virions from infected cells (Doyle et al. 2015). Despite the fact that these ISGs have not been directly linked to an effect on the latent reservoir, the general prevention of further CD4$^+$ T cell infection does have an effect on the development of the reservoir.

Type I IFN responses during acute HIV-1 infection have been shown to be very effective and essential for the initial control of HIV-1 infection (Sandler et al. 2014). However, it is still widely debated whether the innate response only contributes to viral control or can also contribute to chronic activation and act as a mediator of disease progression (Chehimi et al. 2010; Tomescu et al. 2007; Boasso and Shearer

2008). The results of several clinical trials support a predominately antiviral activity when pegylated-IFN (peg-IFN) is administered in HIV-1 infected persons in the absence of ART (Lane et al. 1988; Boue et al. 2011; Asmuth et al. 2010; Dianzani et al. 2008). However, it was noted that IFN in this setting was not entirely suppressive and was suggested that the immune system may have deteriorated too far in the presence of ongoing replication to show the full effect of IFN treatment. To test this further, an additional clinical trial was conducted in which HIV-1 infected individuals were treated with peg-IFN post ART interruption. This trial showed that peg-IFN not only suppressed viral load but also decreased cell associated HIV-1 DNA and extended rebound time following ART interruption from 2 weeks in controls to 12–24 weeks in treated individuals. These results suggest that treatment with exogenous IFN may have a significant effect on both HIV-1 replication and the latent reservoir (Azzoni et al. 2013). However, interestingly another study showed that the blockade of endogenous interferon signaling in chronic infection was found to lead to smaller reservoirs and delayed viral rebound following ART interruption in a humanized mouse model (Cheng et al. 2017). These data underscore the complex roles IFNs have on viral replication and the viral reservoir.

## 2.2   Natural Killer Cells

Natural killer (NK) cells occupy a unique niche in the immune response, bridging the innate and adaptive immune systems. They are the critical antiviral effectors of the innate immune system with the potential to directly respond to viruses, they are able to develop memory-like responses after initial infection and are essential in shaping the adaptive immune response (Scully and Alter 2016). Population-level genetic associations between NK cell receptor expression, HIV-1 outcomes and evolution revealed the impact of NK cells can have on HIV-1 disease progression (Scully and Alter 2016). NK cell inhibitory receptors including the killer immunoglobulin-like receptors (KIRs) heavily influence cellular activation. Interactions between KIRs and their cognate HLA ligands set a threshold of NK activity and have been shown to critically influence the course of viral infection (Khakoo et al. 2004). In HIV-1 infection, HLA and KIR combinations have been associated with the pace of disease progression (Martin et al. 2002, 2007), protection against disease acquisition (Boulet et al. 2008a, b) and in the natural control of HIV-1 infection (O'Connell et al. 2009; Marras et al. 2013; Walker-Sperling et al. 2017). The mechanisms conferring this protection may include both NK cell education through inhibitory receptor engagement and the direct interactions of KIRs with HIV-1 specific peptides presented on HLAs (Scully and Alter 2016).

Alternative mechanisms in which HIV-1 can activate NK cells involve the virus' ability to evade detection by the immune system. HIV-1-mediated downregulation of HLA molecules prevents detection by CD8[+] T cells (Collins et al. 1998), but can induce NK cell activation, offering the "missing self" trigger for NKs. However, this is limited because the downregulation of HLA A and B by Nef is coupled to the preservation HLA C and E (Cohen et al. 1999). The presence of HLA C and E maintains self-signals and preventing mass activation of NK cells (Specht et al. 2008; Cohen et al. 1999). In addition, infection by HIV-1 naturally upregulates stress signals, such as NKG2D, which serve to activate NK cells. However, HIV-1 is able to limit the expression of the ligands and others regulating their expression via the virus' accessory proteins (Norman et al. 2011; Richard et al. 2010; Shah et al. 2010). Overall, NK cell activation by HIV-1 infection is a fine balance and each pathway that leads to activation could be important in overall control of HIV-1 infection. An additional means of NK cell control in HIV-1 infection is antibody-dependent cellular cytotoxicity (ADCC). ADCC is mediated predomi-nately by NKs and involves the engagement of the Fc gamma receptor 3A (CD16) by antibody immune complexes. CD16 engagement is a strong activator of NK cell function and allows for antigen specific recruitment of NK responses (Scully and Alter 2016). Most importantly, ADCC activity and or polyfunctional were asso-ciated with a modest protective effect in one HIV-1 vaccine trial (Haynes et al. 2012). The role of ADCC responses in the natural control of HIV-1 infection is controversial with some studies showing a correlation with control (Lambotte et al. 2009; Wren et al. 2013; Ackerman et al. 2016) whereas another study showed no correlation between ADCC and viral loads or CD4 counts in untreated patients (Smalls-Mantey et al. 2012). Additionally, one study found an inverse correlation between ADCC activity and viral loads in CPs but not in patients with slowly progressive disease (Isitman et al. 2016), and was followed up by a recent study that suggested while ADCC alone is not increased in patients who control HIV-1 infection naturally, these patients are more likely to have polyfunctional antibody responses which may control HIV-1 replication through NK cell, monocyte, and neutrophil effector function (Ackerman et al. 2016). However, as with other effector functions, ADCC is also limited by viral evasion. The viral accessory protein Vpu antagonizes the antiviral factor tetherin, altering the release of virus aggregates and disabling ADCC mediate recognition (Alvarez et al. 2014; Li et al. 2014; Pham et al. 2014). These data highlight the importance of NK cells in HIV-1 disease. Although, there is no current literature to support the direct effect NK cells have on the latent reservoir, it is likely that NK cell recognition of generic stress signals induced early in HIV-1 infection would have a substantial effect of the development of the reservoir. Additionally, the ability of ADCC to prevent further infection may alter maintenance of the reservoir.

# 3 Adaptive Immune Response: Humoral Immunity

## 3.1 HIV-1-Specific Antibodies

Antibodies (Abs) have the potential to block HIV-1 infection through multiple pathways, exerting immune pressure on the virus that often leads to viral escape. Neutralizing antibodies (nAbs) bind cell-free virus and prevent virions from entering host target cells, thereby disrupting subsequent rounds of infection (Overbaugh and Morris 2012). HIV-1 specific Abs can also bind HIV-1 antigens expressed on the surface of infected cells. When complexed to Fc gamma receptors (FcγR) on effector cells this can lead to antibody-dependent cellular cytotoxicity (ADCC) or antibody-dependent cellular phagocytosis (ADCP) and the destruction of the infected cell. These two independent processes have the potential to contain cell–cell spread of the virus (Overbaugh and Morris 2012) and allow Abs to direct the cytotoxic and antiviral activity of the innate immune system. This immunologic activity has been widely exploited by the advanced engineering and use of Abs as therapeutic agents against cancer and autoimmune disease (Goede et al. 2014; Shibata-Koyama et al. 2009; Chan and Carter 2010). The successes in these fields have given new life to the concept that Abs could also be effective at targeting the HIV-1 reservoir. In an ideal situation, a latently infected cell that becomes activated, whether naturally or through reactivation by latency reversal agents (LRAs), would express HIV-1 antigens on its surface. The expression of these antigens would allow the infected cell to be bound by HIV-1 specific Abs and therefore targeted for destruction by the immune system. Furthermore, it is likely that latently infected cells reside in multiple compartments, including blood and tissues, where there is limited T cell access. Compared T cells Abs would most likely be able to diffuse more freely and gain access to these hard to reach sites potentially eliminating these areas of the HIV-1 reservoir.

## 3.2   The Development of Broadly Neutralizing Abs

Ab responses to HIV-1 infection develop within a week of detectable viremia (Tomaras et al. 2008). However, this initial Ab response has been shown not to have an effect on viremia or exert any selective immune pressure on the envelope (Tomaras et al. 2008; Keele et al. 2008). It is not until neutralizing Abs (NAbs) develop that selective immune pressure is exerted on circulating virus (Overbaugh and Morris 2012). Though, once an Ab that can suppress infection develops, the virus quickly escapes and the Ab is no longer able to control the virus (Overbaugh and Morris 2012). These escape mutants are likely to be arcHIV-1ed into the viral reservoir and this may have implications for strategies that rely on autologous antibodies to control viral rebound. The delay in development of an effective Ab response leaves the humoral immune system at a disadvantage.

One of the major goals in developing an HIV-1 vaccine has been to elicit a strong Ab response, specifically broadly neutralizing Ab (bNAbs) that could neutralize a wide spectrum of HIV-1 variants (Caskey et al. 2016). This concept would be beneficial in both a preventative strategy as well as a therapeutic strategy if the bNAbs developed prevent new infection and deplete the HIV-1 reservoir. However, the development of naturally occurring bNAbs is not well understood. Only a fraction of infected individuals tend to develop bNAbs and in contrast to a typical Ab response, bNAbs can take years to develop (Caskey et al. 2016). The extensive amount of time required is thought to be due to several unusual characteristics that

HIV-1 bNAbs feature. The most prominent of these features is an unusually high level of somatic hypermutation which is required to accommodate binding to the highly glycosylated viral envelope (Caskey et al. 2016). It is most likely because of this unusual Ab characteristic that it has not yet been possible to elicit bNAbs via traditional immunization techniques. Therefore, it may be necessary to use laboratory-engineered bNAbs as opposed to those developed through a vaccination strategy to help control the HIV-1 reservoir.

## 3.3  Abs as a Preventative or Therapeutic Treatment

Engineered bNAbs are being widely investigated as an option for therapeutic and preventative strategies. There are two distinct domains of an Ab, the antigen binding domain (Fab) and the constant domain (Fc). The Fab is responsible for antigen specificity and the Fc is responsible for delivering instructions to the innate system on how to destroy what the Ab is bound to. The Fc domain can elicit a variety of effector functions, such as ADCC largely mediated by NK cells and ADCP largely mediated by macrophages (Euler and Alter 2015). While a great emphasis has been placed on the Fab function of the Ab to neutralize virus it is likely that the effector functions of the Fc domain are equally important. The ability of the Ab to not only bind free virus but also recruit the proper immune cells to destroy the virus is essential. These concepts have recently been put to the test. Animal studies have shown that bNAbs are effective therapeutically (Barouch et al. 2013; Klein et al. 2012; Horwitz et al. 2013; Shingai et al. 2013). In a non-human primate study, 3 of 18 monkeys exhibited prolonged virological control after the animals were treated with a bNAb cocktail and virus was cleared from the blood. However, while the bNAbs cleared systemic virus transiently, the Abs were unable to eradicate the reservoir (Barouch et al. 2013). The lack of reservoir reduction seen in this study suggests that the effector function elicited by the bNAb is equally important to its recognition of free virus or infected $CD4^+$ T cells. Techniques to enhance Ab effector functions are being widely studied. One such strategy has been to employ immunotoxins as potential therapies against HIV-1. These interventions have an effector domain from a plant or bacterial toxin and a targeting domain with affinity for the viral HIV-1 envelope. Work completed in a humanized mouse model illustrated that immunotoxins can be very effective at reducing the size of the latent reservoir (Denton et al. 2014). Animals treated with both ART and the immunotoxin 3B3-PE38, a combination of the 3B3 HIV-1-specific Ab and the pseudomonas exotoxin A, showed a dramatic drop in the size of the latent reservoir as measured by cell associated HIV-1 RNA in a variety of tissues (Denton et al. 2014). Mice treated with both ART and 3B3-PE338 had significantly less cell associated RNA than mice treated with ART alone. An additional strategy has been to genetically modify the Fc domain of Abs to enhance their binding affinities for particular FcγRs. One study in humanized mice demonstrated that enhanced in vivo potency of bnAbs was associated with preferential engagement of activating FcγRs

(Halper-Stromberg et al. 2014). The bnAbs engineered to have this selective binding capacity for activating receptors, but not inhibitory, displayed enhanced protection upon HIV-1 challenge (Halper-Stromberg et al. 2014). The importance of Fc domains was also demonstrated in an additional humanized mouse study that utilized LRAs in addition to bNAb treatment. A substantial increase in time to viral rebound following ART interruption was reported for mice that were treated with multiple LRAs and bNAbs compared to those treated with LRAs alone. The therapeutic nature of these Abs was heavily dependent on their Fc effector function, as the delivery of bNAbs with nonfunctional Fc domains was not as effective at prolonging viral rebound (Halper-Stromberg et al. 2014).

The advances made by testing bNAb treatments in HIV-1 animal models are quickly being translated into human trials. Most recently, four independent clinical trials utilizing bNAbs have been reported (Scheid et al. 2016; Bar et al. 2016; Lynch et al. 2015). One trial testing ART interruption followed by treatment with 3BNC117, a bNAb against the CD4 binding site of the HIV-1 envelope, resulted in a delay in viral rebound of 5–9 weeks with two infusions of the Ab and 19 weeks with four infusions of the Ab, compared to a delay of 2.6 weeks seen in historical controls (Scheid et al. 2016). An additional report of two clinical trials that tested VRC01 administration post ART interruption, a bNAb also against the CD4 binding site on the HIV-1 envelope, reported a slight delay in viral rebound of 4 and 5.6 weeks compared to 2.6 week as seen with historical controls (Bar et al. 2016). An earlier clinical trial that also tested VRC01 found similar results, a single infusion of the Ab had very little effect on viral rebound, but should a significant drop in plasma viral load in individuals not on ART (Lynch et al. 2015). These reports indicate that bNAbs are capable of successfully targeting the HIV-1 reservoir in vivo, however it was reported in these studies that the predominant viral clone to rebound had resistance mutations to the bNAb used in each study. This suggests that the use of a single bNAb selects for preexisting or emerging viral clones that are resistant to the therapeutic bNAb, therefore using multiple bNAbs may be more effective. These studies suggest that the next generation of bNAbs must successfully bind many HIV-1 variants as well as elicit innate cell effector functions and that use of combinations of bNAbs may be necessary to consider this a successful therapeutic option.

## 3.4   Bispecific Abs

Abs are limited by the fact that they are unable to kill infected cells without the help of effector cells. In an effort to overcome this limitation, bispecific Abs or Dual-Affinity Re-Targeting (DARTs) proteins have been developed. These proteins recognize both the HIV-1 envelope and the CD3 molecule present on T cells. This dual recognition activates T cells through the engagement of CD3 and redirects them to kill infected cells by binding to the viral envelope expressed on the cell surface (Sung et al. 2015; Pegu et al. 2015; Sloan et al. 2015). This allows for killing by polyclonal CD8$^+$ T cells and does not require the specificity of CTL clones. These molecules have been effective at targeting latently infected cells in vitro following latency reversal and may have great potential in cure strategies.

# 4 Adaptive Immune Response: Cellular Immunity

## 4.1 HIV-1-Specific CD8$^+$ T Cells

CD8$^+$ T cells are a critical component of the cellular immune response against viral infections. During infection, CD8$^+$ T cells recognize HIV-1 through an HLA class I dependent mechanism and are able to lyse cells harboring the virus by the secretion of perforin and granzymes. These cytotoxic T-lymphocytes (CTL) can also elimi-nate virally infected cells through the engagement of death inducing ligands on the target cells as well as the secretion of other soluble factors that suppress viral budding and transcription (Gulzar and Copeland 2004). CTLs place a tremendous amount of pressure on HIV-1 and in order to survive the immune system's attack the virus has adopted numerous strategies to evade the CD8$^+$ T cell response. The high mutation rate of HIV-1 has allowed the virus to escape the CTL response. Escape mutations develop in CD8$^+$ T cell targeted epitopes shortly after infection (Goonetilleke et al. 2009) and these escape mutations are archived into the latent reservoir unless viremia is controlled during primary infection by either the immune system (Bailey et al. 2006) or by ART (Deng et al. 2015). The CTL response is also evaded by the virus' ability to downregulate surface HLA class I expression in infected cells (Collins et al. 1998). Additionally, since CD4$^+$ T cells are the primary target of HIV-1 infection, the virus is able to disrupt proper cytokine signaling and maturation of CD8$^+$ T cells leading to the development of exhaustion and aberrant function of these cells (Gougeon 2003; Gulzar and Copeland 2004). Finally, because latency is established very early in the CD4$^+$ T cell memory population and the CTL response takes time to develop, the virus is given a great temporal advantage over this arm of the immune system (Blankson 2006).

However, despite the advantage the virus receives, CD8$^+$ T cell responses play a very significant role in controlling HIV-1 pathogenesis. Sequencing of both cir-culating and provirus virus from infected individuals reveals evidence of immune selection pressure mediated by the CD8$^+$ T cell response and an association with the initial decline in peak viremia during acute infection (Borrow et al. 1994; Koup et al. 1994; Phillips et al. 1991). There is now also evidence that CD8$^+$ T cells can recognize antigens expressed by some defective provirus, and this shapes the proviral landscape (Pollack et al. 2017). One of the strongest associations discov-ered with disease outcome was the expression of certain HLA class I alleles, which implicated class I restricted CTLs as the major modulator of disease progression (Altman et al. 2011; Migueles et al. 2000). Additionally, the relationship between CD8$^+$ T cells responses and viral control was shown by experimental depletion of CD8$^+$ T cells from animal models of HIV-1 infection (Jin et al. 1999; Schmitz et al. 1999); a quick rebound in viremia is seen in these animals even when they are on suppressive ART regimens (Cartwright et al. 2016). There is accumulating evidence that CTLs play an important role in at least partial containment of HIV-1 replication in chronic infection but in the majority of infected individuals robust $CD8^+$ T cell responses, as measured by interferon-$\gamma$ secretion, do not correlate with protection from infection. It is likely that the many defects in $CD8^+$ T cell function the virus induces prevent these responses from clearing the infection and in most individuals leading to the eventual development of AIDS.

## 4.2   $CD8^+$ T Cell Responses in Patients with Natural Control of HIV-1 Infection

In contrast to the majority of HIV-1 infected individuals, there are some individuals who have both functional and effective $CD8^+$ T cell responses. These individuals are referred to as long-term nonprogressors (LNTPs) and elite controllers or suppressors (ES). LTNPs are a subset of HIV-1 infected individuals who maintain stable $CD4^+$ T cell counts greater than 500 cells/uL for several years, in the absence of ART. LTNPs are a phenotypically diverse population compromised of individuals with varying HIV-1 plasma RNA levels. In contrast, ES maintain HIV-1 RNA levels below the limit of detection of standard assays (<50 copies/mL) in the absence of ART and are not defined by their $CD4^+$ T cell counts. The ES population of HIV-1 infected individuals represent less than 1% of the total HIV-1 infected population (Okulicz and Lambotte 2011; Buckheit et al. 2013a). Patients who exhibit partial control of viral replication in the absence of ART are called viremic controllers and these patients maintain viral loads of between 50 and 2000 copies/mL (Pereyra et al. 2008). The ability of these select populations to control HIV-1 has been widely studied. It is now generally accepted that host factors play a substantial role in elite control. The HLA-B*57 and B*27 alleles are over represented in ES (Migueles et al. 2000, 2008; Lambotte et al. 2005; Emu et al. 2008; Pereyra et al. 2008; Han et al. 2008), these two alleles along with a polymorphism in the promoter of HLA-C have been associated with slow progression in multiple GWAS studies (Fellay et al. 2007; International HIV-1 Controllers Study et al. 2010; Catano et al. 2008; Dalmasso et al. 2008; Limou et al. 2009; van Manen et al. 2009). HLA class I proteins are involved in presentation of peptides to $CD8^+$ T cells and the presence of these specific genetic characteristics may explain why HIV-1-specific $CD8^+$ T cells from ES are more effective at controlling HIV-1 replication than $CD8^+$ T cells from chronic progressors (CP) (Migueles et al. 2008, 2002; Hersperger et al. 2010; Saez-Cirion et al. 2009, 2007). Additionally, ES maintain polyfunctional $CD8^+$ responses and have an enhanced ability to suppress viral replication without additional stimulation when compared to CPs (Betts et al. 2006; Almeida et al. 2007; Ferre et al. 2009). The best evidence of an effective $CD8^+$ T cell response in these individuals is seen in animal models of elite control where the depletion of $CD8^+$ T cells results in immediate loss of viral control (Friedrich et al. 2007; Pandrea et al. 2011).

In addition to understanding how these individuals control ongoing viral replication, there has been extensive work completed to elucidate the latent reservoir in ES. Interestingly, in addition to having undetectable circulating viremia, ES also have a reduced number of latently infected cells compared to ART treated individuals (Lambotte et al. 2005; Graf et al. 2011; Blankson et al. 2007; Julg et al. 2010). In fact, the frequency of latently infected cells in the blood of ES is 10–50 fold lower than the frequency observed for ART treated CPs (Blankson et al. 2007). ES also have significantly lower levels of total (Lambotte et al. 2005; Julg et al. 2010) and integrated (Graf et al. 2011) proviral DNA compared to CPs. The lower level of peak viremia during acute infection observed in ES may partially explain the lower frequency of latently infected cells in these individuals (Altfeld et al. 2003; Goujard et al. 2009). In a primary model system of latency, $CD8^+$ T cells from ES were observed to be more effective at targeting reactivated latently infected cells than $CD8^+$ T cells from individuals on ART (Shan et al. 2012). Additional studies have shown that $CD8^+$ T cells from ES are also more capable eliminating nonproductively infected $CD4^+$ T cells which may represent a subpopulation of cells that are precursors to latently infected cells (Buckheit et al. 2013b; Graf et al. 2013). The elimination of these cells may help to explain the smaller reservoir in ES compared to CP.

## 4.3   CD8+ T Cell Based Vaccines

Overall, the development of qualitatively superior $CD8^+$ T cell responses as observed with ES would be the goal for the development of a successful vaccine. Thus, far T cell based vaccines have not been very successful. However, there have been a few recent studies with promising results that may translate well into HIV-1 infected individuals. A recently developed therapeutic dendritic cell (DC) based vaccine tested in small double-blinded controlled study of untreated CPs showed a modest yet significant decrease in plasma HIV-1 levels in vaccinated individuals compared to controls. The decrease in plasma RNA was correlated with an increase in HIV-1-specific T cell responses (GarcÃ-a et al. 2011). Further analysis of the vaccine elicited T cell responses revealed that they had a profound effect on the viral reservoir and significantly delayed replenishment of integrated HIV-1 DNA after ART interruption (Andres et al. 2015). Suggesting that $CD8^+$ T cells from a CP could be educated to attack the latent reservoir. An additional study in an animal model of SIV infection tested a therapeutic vaccine that included an Ad26/MVA prime and boost in addition to stimulation by a Toll-like receptor 7 (TLR7) agonist. This vaccination technique resulted in decreased levels of viral DNA in lymph nodes and peripheral blood, improved virologic control, and delayed viral rebound following ART interruption. The breadth of cellular immune responses correlated inversely with viral set point and correlated directly with time to viral rebound (Borducchi et al. 2016). Suggesting that therapeutic vaccination in combination with innate immune stimulation may be a useful strategy for controlling the reservoir.

In addition to therapeutic vaccines, preventative vaccines are also being investigated. The most notable of which is a monkey model of HIV-1 infection that utilizes rhesus cytomegalovirus (RhCMV) vectors to establish persistent, high-frequency, SIV-specific T cell responses. This CMV-based strategy was shown to provide impressive control of SIV replication in a subset of vaccinated animals through unconventional CD8+ T responses where the CD8+ T cells are restricted by MHC-II (Hansen et al. 2013) or the nonclassical MHC class Ib molecule MHC-E (Hansen et al. 2016). The vaccine resulted in long-term protection characterized by controlled viremia, undetectable cell associated viral load in both blood and lymph nodes, a lack of SIV-specific Ab development and no depletion of CD4+ T cells at mucosal sites. Interestingly, after control of viral replication was achieved, CD8+ T cells were depleted, and essentially no SIV DNA or RNA was present at necropsy, suggesting that complete clearance of virus may have been achieved (Hansen et al. 2011). However, this vaccine strategy was not effective at eradicating virus in chronically infected monkeys on ART (Picker 2017). A possible explanation for these results is that CD8+ T cells are capable of killing infected cells in primary infection as they transition into long-lived latently infected memory CD4+ cells. In contrast, the long-lived latently infected cells probably do not express viral antigen and therefore are not susceptible to killing by CD8+ T cells unless latency is reversed. These data suggest that persistent vectors such as CMV that maintain long term SIV or HIV-1 T cell responses may be promising candidates for preventative vaccine strategies. They may also be effective as therapeutic vaccines in conjunction with LRAs as part of shock and kill strategies.

## 4.4   CD8+ T Cell Based "Shock and Kill" Therapy

The goal of a CD8+ T cell based therapy would be to kill latently infected cells as they are reactivated. Recent studies have shown that intracellular HIV-1 mRNA is upregulated within 1 h after activation of latently infected cells with PMA and ionomycin and virus is produced within a 6 h time frame (Walker-Sperling et al. 2015). In spite of these rapid kinetics, CD8+ T cells from some viremic controllers are able to eliminate reactivated cells within 18 h. In contrast, CD8+ T cells from CPs do not have a significant effect on virus production after activation of CD4+ T cells (Walker-Sperling et al. 2015). Other studies have modeled shock and kill strategies by treating CD4+ T cells with LRAs. Jones et al., used HIV-1-specific CD8+ T cell clones as biosensors for antigen expression and demonstrated that treatment of latently infected cells with some LRAs resulted in the expression of antigen that was recognized by these clones (Jones et al. 2016). In one study, CD8+ T cells from CPs were unable to inhibit the production of virus after latently infected cells were treated with the combination of bryostatin-1 and romidepsin. However, in another study, CD8+ T cell lines from CPs were able to partially clear latently infected CD4+ T cells following exposure to the LRA, vorinostat. This

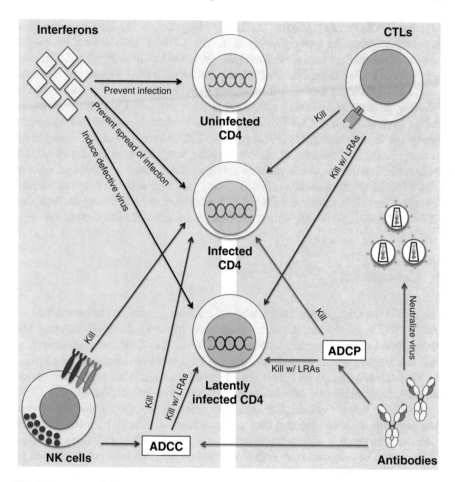

**Fig. 1** Summary of the antiviral immune response against HIV infection. *Left panel*, illustrates the innate immune response against HIV infection. Type I interferons are secreted in response to recognition of the virus by pattern recognition receptors. These cytokines then upregulate restriction factors that can prevent infection in uninfected CD4s, block the spread of infection in already infected CD4 T cells, and induce defective virus that will be integrated into the latent reservoir. NK cells can kill infected CD4s by recognizing the upregulation of stress signals and downregulation of which occurs with HIV infection. Additionally, NK cells can kill infected and reactivated latently infected cells via ADCC. *Right panel*, illustrates the adaptive immune response against HIV infection. HIV-specific antibodies are best known for their ability to neutralize virions and prevent infection. However, the effector functions of the antibody Fc domain are just as important and lead to the killing of HIV infected cells by ADCC and ADCP. Cytotoxic T-Lymphocytes (CTLs) kill HIV infected cells and reactivated latently infected cells though TCR engagement and release of catalytic proteins

discrepancy in results may be due to the fact that romidepsin (Jones et al. 2014; Walker-Sperling et al. 2016) and bryostatin (Walker-Sperling et al. 2016) have both been shown to inhibit CD8$^+$ T cell responses whereas vorinostat has less of an effect on CTLs (Jones et al. 2014; Walker-Sperling et al. 2016). However, it is clear that even without the adverse effect of LRAs, CD8$^+$ T cells from CPs are not capable of controlling viral replication (Migueles and Connors 2015) and will need to be primed before being able to effectively kill infected cells after latency reversal.

Overall, effective CD8$^+$ T cell responses may be necessary to eliminate HIV-1 infection, especially as virus is reactivated from the latent reservoir. However, studies have shown that CD8$^+$ T cells are excluded from lymphoid follicles where there is active viral replication (Folkvord et al. 2005; Connick et al. 2007) and in the monkey model of elite control, SIV replication was shown to be restricted to T follicular helper cells in B cell follicles in lymph nodes because of this exclusion of CD8$^+$ T cells (Fukazawa et al. 2015). Thus, eradication of HIV-1 may require strategies that both improve CD8$^+$ T cell responses, and facilitate the entry of HIV-1 specific CD8$^+$ T cells into anatomical sanctuaries of viral replication.

# 5  Conclusions

The HIV-1 latent reservoir represents a significant barrier to the eradication of HIV-1. While substantial advances have been made in the understanding and manipulation of the immune response to the virus, it is clear that a response that utilizes a single mechanism probably will not be capable of completely eliminating the reservoir (Fig. 1). Type I IFNs are potent inhibitors of viral replication but do not effectively eliminate infected cells and NK cells and HIV-1-specific CD8$^+$ T cells become dysfunctional with prolonged exposure to ongoing viral replication. Abs are limited by the fact that they cannot kill infected cells by themselves, the virus has developed many mechanisms to evade both the Ab and CTL response and CTLs tend to be excluded from critical reservoir sites. Therefore, our efforts should probably be focused on developing combinatorial efforts to eradicate the reservoir. If the best characteristics of each aspect of the immune system can be combined into a therapy perhaps, we can effectively target the reservoir. Fortunately, there is a substantial amount of work already devoted to accomplishing this goal. The development of a vaccine that activates both of the innate and adaptive systems using TLR7 agonist as an adjuvant coupled with a T cell vaccine has shown promising results in animal studies and is being tested in human clinical trials. The development of bispecific Abs has attempted to target both the humoral and cell-mediate arms of the adaptive system utilizing both the specificity and ability of an Ab to move freely through the immune system with the effector power of a T cell. The modification of the Fc domain of Abs to preferentially bind to certain Fc$\gamma$R and elicit a stronger innate effector cell response has tied together the innate and humoral systems. And finally, the development of Abs coupled to immuno-toxins has given Abs cytolytic function independent of effector cells. Though none of these developments have eradicated the reservoir thus far, they have had sub-stantial effects of reservoir size and are important for informing future studies that may lead to the eventual eradication of the reservoir and a cure for HIV-1.

# References

Ackerman ME, Mikhailova A, Brown EP, Dowell KG, Walker BD, Bailey-Kellogg C, Suscovich TJ, Alter G (2016) Polyfunctional HIV-1-specific antibody responses are associated with spontaneous HIV-1 control. PLoS Pathog 12(1):e1005315

Almeida JR, Price DA, Papagno L, Arkoub ZA, Sauce D, Bornstein E, Asher TE, Samri A, Schnuriger A, Theodorou I, Costagliola D, Rouzioux C, Agut H, Marcelin AG, Douek D, Autran B, Appay V (2007) Superior control of HIV-1 replication by CD8+ T cells is reflected by their avidity, polyfunctionality, and clonal turnover. J Exp Med 204(10):2473–2485

Altfeld M, Gale M Jr (2015) Innate immunity against HIV-1 infection. Nat Immunol 16(6):554–562

Altfeld M, Addo MM, Rosenberg ES, Hecht FM, Lee PK, Vogel M, Yu XG, Draenert R, Johnston MN, Strick D, Allen TM, Feeney ME, Kahn JO, Sekaly RP, Levy JA, Rockstroh JK, Goulder PJ, Walker BD (2003) Influence of HLA-B57 on clinical presentation and viral control during acute HIV-1 infection. AIDS (London, England) 17(18):2581–2591

Altman JD, Moss PA, Goulder PJ, Barouch DH, McHeyzer-Williams MG, Bell JI, McMichael AJ, Davis MM.2011 (1996) Phenotypic analysis of antigen-specific T lymphocytes. Science 274:94–96, J Immunol (Baltimore, Md.: 1950) 187(1):7–9

Alvarez RA, Hamlin RE, Monroe A, Moldt B, Hotta MT, Rodriguez Caprio G, Fierer DS, Simon V, Chen BK (2014) HIV-1 Vpu antagonism of tetherin inhibits antibody-dependent cellular cytotoxic responses by natural killer cells. J Virol 88(11):6031–6046

Andres C, Plana M, Guardo AC, Alvarez-Fernandez C, Climent N, Gallart T, Leon A, Clotet B, Autran B, Chomont N, Gatell JM, Sanchez-Palomino S, Garcia F (2015) HIV-1 reservoir dynamics after vaccination and antiretroviral therapy interruption are associated with dendritic cell vaccine-induced T cell responses. J Virol 89(18):9189–9199

Asmuth DM, Murphy RL, Rosenkranz SL, Lertora JJL, Kottilil S, Cramer Y, Chan ES, Schooley RT, Rinaldo CR, Thielman N, Li XD, Wahl SM, Shore J, Janik J, Lempicki RA, Simpson Y, Pollard RB & for the ACTG A5192 Team (2010) Safety, tolerability and mechanisms of antiretroviral activity of peginterferon alfa-2a in HIV-1-mono-infected subjects: a phase II clinical trial. J Infect Dis 201(11):1686–1696

Azzoni L, Foulkes AS, Papasavvas E, Mexas AM, Lynn KM, Mounzer K, Tebas P, Jacobson JM, Frank I, Busch MP, Deeks SG, Carrington M, O'Doherty U, Kostman J, Montaner LJ (2013) Pegylated interferon Alfa-2a monotherapy results in suppression of HIV-1 type 1 replication and decreased cell-associated HIV-1 DNA integration. J Infect Dis 207(2):213–222

Bailey JR, Williams TM, Siliciano RF, Blankson JN (2006) Maintenance of viral suppression in HIV-1-infected HLA-B*57 + elite suppressors despite CTL escape mutations. J Exp Med 203 (5):1357–1369

Bar KJ, Sneller MC, Harrison LJ, Justement JS, Overton ET, Petrone ME, Salantes DB, Seamon CA, Scheinfeld B, Kwan RW, Learn GH, Proschan MA, Kreider EF, Blazkova J, Bardsley M, Refsland EW, Messer M, Clarridge KE, Tustin NB, Madden PJ, Oden K, O'Dell SJ, Jarocki B, Shiakolas AR, Tressler RL, Doria-Rose NA, Bailer RT, Ledgerwood JE, Capparelli EV, Lynch RM, Graham BS, Moir S, Koup RA, Mascola JR, Hoxie JA, Fauci AS, Tebas P, Chun TW (2016) Effect of HIV-1 antibody VRC01 on viral rebound after treatment interruption. N Engl J Med 375(21):2037–2050

Barouch DH, Whitney JB, Moldt B, Klein F, Oliveira TY, Liu J, Stephenson KE, Chang HW, Shekhar K, Gupta S, Nkolola JP, Seaman MS, Smith KM, Borducchi EN, Cabral C, Smith JY, Blackmore S, Sanisetty S, Perry JR, Beck M, Lewis MG, Rinaldi W, Chakraborty AK, Poignard P, Nussenzweig MC, Burton DR (2013) Therapeutic efficacy of potent neutralizing HIV-1-specific monoclonal antibodies in SHIV-1-infected rhesus monkeys. Nature 503 (7475):224–228

Betts MR, Nason MC, West SM, De Rosa SC, Migueles SA, Abraham J, Lederman MM, Benito JM, Goepfert PA, Connors M, Roederer M, Koup RA (2006) HIV-1 nonprogressors preferentially maintain highly functional HIV-1-specific CD8+ T cells. Blood 107(12): 4781–4789

Blankson JN (2006) Viral reservoirs and HIV-1-specific immunity. Curr Opin HIV-1 AIDS 1 (2):147–151

Blankson JN, Bailey JR, Thayil S, Yang HC, Lassen K, Lai J, Gandhi SK, Siliciano JD, Williams TM, Siliciano RF (2007) Isolation and characterization of replication-competent human immunodeficiency virus type 1 from a subset of elite suppressors. J Virol 81(5): 2508–2518

Boasso A, Shearer GM (2008) Chronic innate immune activation as a cause of HIV-1 immunopathogenesis. Clin Immunol (Orlando, Fla.) 126(3):235–242

Bonilla FA, Oettgen HC (2010) Adaptive immunity. J Allergy Clin Immunol 125(2 Suppl 2): S33–S40

Borducchi EN, Cabral C, Stephenson KE, Liu J, Abbink P, Ng'ang'a D, Nkolola JP, Brinkman AL, Peter L, Lee BC, Jimenez J, Jetton D, Mondesir J, Mojta S, Chandrashekar A, Molloy K, Alter G, Gerold JM, Hill AL, Lewis MG, Pau MG, Schuitemaker H, Hesselgesser J, Geleziunas R, Kim JH, Robb ML, Michael NL, Barouch DH (2016) Ad26/MVA therapeutic vaccination with TLR7 stimulation in SIV-infected rhesus monkeys. Nature 540(7632): 284–287

Borrow P, Lewicki H, Hahn BH, Shaw GM, Oldstone MB (1994) Virus-specific CD8+ cytotoxic T-lymphocyte activity associated with control of viremia in primary human immunodeficiency virus type 1 infection. J Virol 68(9):6103–6110

Boue F, Reynes J, Rouzioux C, Emilie D, Souala F, Tubiana R, Goujard C, Lancar R, Costagliola D (2011) Alpha interferon administration during structured interruptions of combination antiretroviral therapy in patients with chronic HIV-1 infection: INTERVAC ANRS 105 trial. AIDS (London, England) 25(1):115–118

Boulet S, Kleyman M, Kim JY, Kamya P, Sharafi S, Simic N, Bruneau J, Routy JP, Tsoukas CM, Bernard NF (2008a) A combined genotype of KIR3DL1 high expressing alleles and HLA-B*57 is associated with a reduced risk of HIV-1 infection. AIDS (London, England) 22 (12):1487–1491

Boulet S, Sharafi S, Simic N, Bruneau J, Routy JP, Tsoukas CM, Bernard NF (2008b) Increased proportion of KIR3DS1 homozygotes in HIV-1-exposed uninfected individuals. AIDS (London, England) 22(5):595–599

Bruner KM, Murray AJ, Pollack RA, Soliman MG, Laskey SB, Capoferri AA, Lai J, Strain MC, Lada SM, Hoh R, Ho YC, Richman DD, Deeks SG, Siliciano JD, Siliciano RF (2016) Defective proviruses rapidly accumulate during acute HIV-1 infection. Nat Med 22(9): 1043–1049

Buckheit RW 3rd, Salgado M, Martins KO, Blankson JN (2013a) The implications of viral reservoirs on the elite control of HIV-1 infection. Cell Mol Life Sci: CMLS 70(6):1009–1019

Buckheit RW, Siliciano RF, Blankson JN (2013b) Primary CD8(+) T cells from elite suppressors effectively eliminate non-productively HIV-1 infected resting and activated CD4(+) T cells. Retrovirology 10(1):68

Cartwright EK, Spicer L, Smith SA, Lee D, Fast R, Paganini S, Lawson BO, Nega M, Easley K, Schmitz JE, Bosinger SE, Paiardini M, Chahroudi A, Vanderford TH, Estes JD, Lifson JD, Derdeyn CA, Silvestri G (2016) CD8(+) lymphocytes are required for maintaining viral suppression in SIV-infected macaques treated with short-term antiretroviral therapy. Immunity 20;45(3):656–668

Caskey M, Klein F, Nussenzweig MC (2016) Broadly neutralizing antibodies for HIV-1 prevention or immunotherapy. N Engl J Med 375(21):2019–2021

Catano G, Kulkarni H, He W, Marconi VC, Agan BK, Landrum M, Anderson S, Delmar J, Telles V, Song L, Castiblanco J, Clark RA, Dolan MJ, Ahuja SK (2008) HIV-1 disease-influencing effects associated with ZNRD1, HCP5 and HLA-C alleles are attributable mainly to either HLA-A10 or HLA-B*57 alleles. PLoS ONE 3(11):e3636

Chan AC, Carter PJ (2010) Therapeutic antibodies for autoimmunity and inflammation. Nat Rev Immunol 10(5):301–316

Chehimi J, Papasavvas E, Tomescu C, Gekonge B, Abdulhaqq S, Raymond A, Hancock A, Vinekar K, Carty C, Reynolds G, Pistilli M, Mounzer K, Kostman J, Montaner LJ (2010)

Inability of plasmacytoid dendritic cells to directly lyse HIV-1-Infected autologous CD4(+) T cells despite induction of tumor necrosis factor-related apoptosis-inducing ligand. J Virol 84 (6):2762–2773

Cheng L, Ma J, Li J, Li D, Li G, Li F, Zhang Q, Yu H, Yasui F, Ye C, Tsao LC, Hu Z, Su L, Zhang L (2017) Blocking type I interferon signaling enhances T cell recovery and reduces HIV-1 reservoirs. J Clin Investig 127(1):269–279

Cohen GB, Gandhi RT, Davis DM, Mandelboim O, Chen BK, Strominger JL, Baltimore D (1999) The selective downregulation of class I major histocompatibility complex proteins by HIV-1 protects HIV-1-infected cells from NK cells. Immunity 10(6):661–671

Collins KL, Chen BK, Kalams SA, Walker BD, Baltimore D (1998) HIV-1 Nef protein protects infected primary cells against killing by cytotoxic T lymphocytes. Nature 391(6665):397–401

Connick E, Mattila T, Folkvord JM, Schlichtemeier R, Meditz AL, Ray MG, McCarter MD, Mawhinney S, Hage A, White C, Skinner PJ (2007) CTL fail to accumulate at sites of HIV-1 replication in lymphoid tissue. J Immunol (Baltimore, Md.: 1950) 178(11):6975–6983

Dalmasso C, Carpentier W, Meyer L, Rouzioux C, Goujard C, Chaix ML, Lambotte O, Avettand-Fenoel V, Le Clerc S, de Senneville LD, Deveau C, Boufassa F, Debre P, Delfraissy JF, Broet P, Theodorou I, ANRS Genome Wide Association 01 (2008) Distinct genetic loci control plasma HIV-1-RNA and cellular HIV-1-DNA levels in HIV-1 infection: the ANRS genome wide association 01 study. PloS one 3(12):e3907

Deng K, Pertea M, Rongvaux A, Wang L, Durand CM, Ghiaur G, Lai J, McHugh HL, Hao H, Zhang H, Margolick JB, Gurer C, Murphy AJ, Valenzuela DM, Yancopoulos GD, Deeks SG, Strowig T, Kumar P, Siliciano JD, Salzberg SL, Flavell RA, Shan L, Siliciano RF (2015) Broad CTL response is required to clear latent HIV-1 due to dominance of escape mutations. Nature 517(7534):381–385

Denton PW, Long JM, Wietgrefe SW, Sykes C, Spagnuolo RA, Snyder OD, Perkey K, Archin NM, Choudhary SK, Yang K, Hudgens MG, Pastan I, Haase AT, Kashuba AD, Berger EA, Margolis DM, Garcia JV (2014) Targeted cytotoxic therapy kills persisting hiv-1 infected cells during ART. PLoS Pathog 10(1):e1003872. doi:10.1371/journal.ppat.1003872

Dianzani F, Rozera G, Abbate I, D'Offizi G, Abdeddaim A, Vlassi C, Antonucci G, Narciso P, Martini F, Capobianchi MR (2008) Interferon may prev HIV-1 viral rebound after HAART interruption in HIV-1 patients. J Interferon cytokine Res: Off J Int Interferon Cytokine Res 28 (1):1–3

Doyle T, Goujon C, Malim MH (2015) HIV-1 and interferons: who's interfering with whom? Nat Rev Microbiol 13(7):403–413

Emu B, Sinclair E, Hatano H, Ferre A, Shacklett B, Martin JN, McCune JM, Deeks SG (2008) HLA class I-restricted T-cell responses may contribute to the control of human immunodeficiency virus infection, but such responses are not always necessary for long-term virus control. J Virol 82(11):5398–5407

Euler Z, Alter G (2015) Exploring the potential of monoclonal antibody therapeutics for HIV-1 eradication. AIDS Res Hum Retroviruses 31(1):13–24

Fellay J, Shianna KV, Ge D, Colombo S, Ledergerber B, Weale M, Zhang K, Gumbs C, Castagna A, Cossarizza A, Cozzi-Lepri A, De Luca A, Easterbrook P, Francioli P, Mallal S, Martinez-Picado J, Miro JM, Obel N, Smith JP, Wyniger J, Descombes P, Antonarakis SE, Letvin NL, McMichael AJ, Haynes BF, Telenti A, Goldstein DB (2007) A whole-genome association study of major determinants for host control of HIV-1. Science (New York, N.Y.) 317(5840):944–947

Ferre AL, Hunt PW, Critchfield JW, Young DH, Morris MM, Garcia JC, Pollard RB, Yee HF Jr, Martin JN, Deeks SG, Shacklett BL (2009) Mucosal immune responses to HIV-1 in elite controllers: a potential correlate of immune control. Blood 113(17):3978–3989

Folkvord JM, Armon C, Connick E (2005) Lymphoid follicles are sites of heightened human immunodeficiency virus type 1 (HIV-1) replication and reduced antiretroviral effector mechanisms. AIDS Res Hum Retroviruses 21(5):363–370

Friedrich TC, Valentine LE, Yant LJ, Rakasz EG, Piaskowski SM, Furlott JR, Weisgrau KL, Burwitz B, May GE, Leon EJ, Soma T, Napoe G, Capuano SV 3rd, Wilson NA, Watkins DI

(2007) Subdominant CD8+ T-cell responses are involved in durable control of AIDS virus replication. J Virol 81(7):3465–3476

Fukazawa Y, Lum R, Okoye AA, Park H, Matsuda K, Bae JY, Hagen SI, Shoemaker R, Deleage C, Lucero C, Morcock D, Swanson T, Legasse AW, Axthelm MK, Hesselgesser J, Geleziunas R, Hirsch VM, Edlefsen PT, Piatak M Jr, Estes JD, Lifson JD, Picker LJ (2015) B cell follicle sanctuary permits persistent productive simian immunodeficiency virus infection in elite controllers. Nat Med 21(2):132–139

GarcÃ-a F, Climent N, Assoumou L, Gil C, GonzÃ¡lez N, AlcamÃ- J, LeÃ3n A, Romeu J, Dalmau J, MartÃ-nez-Picado J, Lifson J, Autran B, Costagliola D, Clotet B, Gatell JM, Plana M, Gallart T & for the DCV2/MANON07- AIDS Vaccine Research Objective Study Group (2011) A therapeutic dendritic cell-based vaccine for HIV-1 infection. J Infect Dis 203 (4):473–478

Goede V, Fischer K, Busch R, Engelke A, Eichhorst B, Wendtner CM, Chagorova T, de la Serna J, Dilhuydy MS, Illmer T, Opat S, Owen CJ, Samoylova O, Kreuzer KA, Stilgenbauer S, Dohner H, Langerak AW, Ritgen M, Kneba M, Asikanius E, Humphrey K, Wenger M, Hallek M (2014) Obinutuzumab plus chlorambucil in patients with CLL and coexisting conditions. N Engl J Med 370(12):1101–1110

Goff SP (2003) Death by deamination: a novel host restriction system for HIV-1. Cell 114(3): 281–283

Goonetilleke N, Liu MK, Salazar-Gonzalez JF, Ferrari G, Giorgi E, Ganusov VV, Keele BF, Learn GH, Turnbull EL, Salazar MG, Weinhold KJ, Moore S, CHAVI Clinical Core B, Letvin N, Haynes BF, Cohen MS, Hraber P, Bhattacharya T, Borrow P, Perelson AS, Hahn BH, Shaw GM, Korber BT, McMichael AJ (2009) The first T cell response to transmitted/founder virus contributes to the control of acute viremia in HIV-1 infection. J Exp Med 206(6):1253–1272

Gougeon ML (2003) Apoptosis as an HIV-1 strategy to escape immune attack. Nat Rev Immunol 3(5):392–404

Goujard C, Chaix ML, Lambotte O, Deveau C, Sinet M, Guergnon J, Courgnaud V, Rouzioux C, Delfraissy JF, Venet A, Meyer L & Agence Nationale de Recherche sur le Sida PRIMO Study Group (2009) Spontaneous control of viral replication during primary HIV-1 infection: when is "HIV-1 controller" status established? Clin Infect Dis: Off Publ Infect Dis Soc Am 49(6):982–986

Graf EH, Mexas AM, Yu JJ, Shaheen F, Liszewski MK, Di Mascio M, Migueles SA, Connors M, O'Doherty U (2011) Elite suppressors harbor low levels of integrated HIV-1 DNA and high levels of 2-LTR circular HIV-1 DNA compared to HIV-1+ patients on and off HAART. PLoS Pathog 7(2):e1001300

Graf EH, Pace MJ, Peterson BA, Lynch LJ, Chukwulebe SB, Mexas AM, Shaheen F, Martin JN, Deeks SG, Connors M, Migueles SA, O'Doherty U (2013) Gag-positive reservoir cells are susceptible to HIV-1-specific cytotoxic T lymphocyte mediated clearance in vitro and can be detected in vivo. PLoS ONE 8(8):e71879

Gulzar N, Copeland KF (2004) "CD8+ T-cells: function and response to HIV-1 infection". Curr HIV-1 Res 2(1):23–37

Halper-Stromberg A, Lu CL, Klein F, Horwitz JA, Bournazos S, Nogueira L, Eisenreich TR, Liu C, Gazumyan A, Schaefer U, Furze RC, Seaman MS, Prinjha R, Tarakhovsky A, Ravetch JV, Nussenzweig MC (2014) Broadly neutralizing antibodies and viral inducers decrease rebound from HIV-1 latent reservoirs in humanized mice. Cell 158(5):989–999

Han Y, Lai J, Barditch-Crovo P, Gallant JE, Williams TM, Siliciano, RF, Blankson JN (2008) The role of protective HCP5 and HLA-C associated polymorphisms in the control of HIV-1 replication in a subset of elite suppressors. AIDS (London, England) 22(4):541–544

Hansen SG, Ford JC, Lewis MS, Ventura AB, Hughes CM, Coyne-Johnson L, Whizin N, Oswald K, Shoemaker R, Swanson T, Legasse AW, Chiuchiolo MJ, Parks CL, Axthelm MK, Nelson JA, Jarvis MA, Piatak M Jr, Lifson JD, Picker LJ (2011) Profound early control of highly pathogenic SIV by an effector memory T-cell vaccine. Nature 473(7348):523–527

Hansen SG, Sacha JB, Hughes CM, Ford JC, Burwitz BJ, Scholz I, Gilbride RM, Lewis MS, Gilliam AN, Ventura AB, Malouli D, Xu G, Richards R, Whizin N, Reed JS, Hammond KB, Fischer M, Turner JM, Legasse AW, Axthelm MK, Edlefsen PT, Nelson JA, Lifson JD, Fruh K, Picker LJ (2013) Cytomegalovirus vectors violate CD8+ T cell epitope recognition paradigms. Science (New York, N.Y.) 340(6135):1237874

Hansen SG, Wu HL, Burwitz BJ, Hughes CM, Hammond KB, Ventura AB, Reed JS, Gilbride RM, Ainslie E, Morrow DW, Ford JC, Selseth AN, Pathak R, Malouli D, Legasse AW, Axthelm MK, Nelson JA, Gillespie GM, Walters LC, Brackenridge S, Sharpe HR, Lopez CA, Fruh K, Korber BT, McMichael AJ, Gnanakaran S, Sacha JB, Picker LJ (2016) Broadly targeted CD8(+) T cell responses restricted by major histocompatibility complex E. Science (New York, N.Y.) 351(6274):714–720

Hardy GAD, Sieg S, Rodriguez B, Anthony D, Asaad R, Jiang W, Mudd J, Schacker T, Funderburg N.T, Pilch-Cooper HA, Debernardo R, Rabin RL, Lederman MM & Harding CV (2013) Interferon-α the primary plasma type-I IFN in HIV-1 infection and correlates with immune activation and disease markers. PLoS ONE 8(2):e56527. doi:10.1371/journal.pone.0056527

Harris RS, Bishop KN, Sheehy AM, Craig HM, Petersen-Mahrt SK, Watt IN, Neuberger MS, Malim MH (2003) DNA deamination mediates innate immunity to retroviral infection. Cell 113(6):803–809

Haynes BF, Gilbert PB, McElrath MJ, Zolla-Pazner S, Tomaras GD, Alam SM, Evans DT, Montefiori DC, Karnasuta C, Sutthent R, Liao HX, DeVico AL, Lewis GK, Williams C, Pinter A, Fong Y, Janes H, DeCamp A, Huang Y, Rao M, Billings E, Karasavvas N, Robb ML, Ngauy V, de Souza MS, Paris R, Ferrari G, Bailer RT, Soderberg KA, Andrews C, Berman PW, Frahm N, De Rosa SC, Alpert MD, Yates NL, Shen X, Koup RA, Pitisuttithum P, Kaewkungwal J, Nitayaphan S, Rerks-Ngarm S, Michael NL, Kim JH (2012) Immune-correlates analysis of an HIV-1 vaccine efficacy trial. N Engl J Med 366(14):1275–1286

Hersperger AR, Pereyra F, Nason M, Demers K, Sheth P, Shin LY, Kovacs CM, Rodriguez B, Sieg SF, Teixeira-Johnson L, Gudonis D, Goepfert PA, Lederman MM, Frank I, Makedonas G, Kaul R, Walker BD, Betts MR (2010) Perforin expression directly ex vivo by HIV-1-specific CD8 T-cells is a correlate of HIV-1 elite control. PLoS Pathog 6(5):e1000917

Ho YC, Shan L, Hosmane NN, Wang J, Laskey SB, Rosenbloom DI, Lai J, Blankson JN, Siliciano JD, Siliciano RF (2013) Replication-competent noninduced proviruses in the latent reservoir increase barrier to HIV-1 cure. Cell 155(3):540–551

Horwitz JA, Halper-Stromberg A, Mouquet H, Gitlin AD, Tretiakova A, Eisenreich TR, Malbec M, Gravemann S, Billerbeck E, Dorner M, Buning H, Schwartz O, Knops E, Kaiser R, Seaman MS, Wilson JM, Rice CM, Ploss A, Bjorkman PJ, Klein F, Nussenzweig MC (2013) HIV-1 suppression and durable control by combining single broadly neutralizing antibodies and antiretroviral drugs in humanized mice. Proc Natl Acad Sci USA 110(41):16538–16543

International HIV-1 Controllers Study, Pereyra F, Jia X, McLaren PJ, Telenti A, de Bakker PI, Walker BD, Ripke S, Brumme CJ, Pulit SL, Carrington M, Kadie CM, Carlson JM, Heckerman D, Graham RR, Plenge RM, Deeks SG, Gianniny L, Crawford G, Sullivan J, Gonzalez E, Davies L, Camargo A, Moore JM, Beattie N, Gupta S, Crenshaw A, Burtt NP, Guiducci C, Gupta N, Gao X, Qi Y, Yuki Y, Piechocka-Trocha A, Cutrell E, Rosenberg R, Moss KL, Lemay P, O'Leary J, Schaefer T, Verma P, Toth I, Block B, Baker B, Rothchild A, Lian J, Proudfoot J, Alvino D.M, Vine S, Addo MM, Allen TM, Altfeld M, Henn MR, Le Gall S, Streeck H, Haas DW, Kuritzkes DR, Robbins GK, Shafer RW, Gulick RM, Shikuma CM, Haubrich R, Riddler S, Sax PE, Daar ES, Ribaudo HJ, Agan B, Agarwal S, Ahern RL, Allen BL, Altidor S, Altschuler EL, Ambardar S, Anastos K, Anderson B, Anderson V, Andrady U, Antoniskis D, Bangsberg D, Barbaro D, Barrie W, Bartczak J, Barton S, Basden P, Basgoz N, Bazner S, Bellos NC, Benson AM, Berger J, Bernard NF, Bernard AM, Birch C, Bodner SJ, Bolan RK, Boudreaux ET, Bradley M, Braun JF, Brndjar JE, Brown SJ, Brown K, Brown ST, Burack J, Bush LM, Cafaro V, Campbell O, Campbell J, Carlson RH, Carmichael JK, Casey KK, Cavacuiti C, Celestin G, Chambers ST, Chez N,

Chirch LM, Cimoch PJ, Cohen D, Cohn LE, Conway B, Cooper DA, Cornelson B, Cox DT, Cristofano MV, Cuchural G Jr, Czartoski JL, Dahman JM, Daly JS, Davis BT, Davis K, Davod SM, DeJesus E, Dietz CA, Dunham E, Dunn ME, Ellerin TB, Eron JJ, Fangman JJ, Farel CE, Ferlazzo H, Fidler S, Fleenor-Ford A, Frankel R, Freedberg KA, French NK, Fuchs JD, Fuller JD, Gaberman J, Gallant JE, Gandhi RT, Garcia E, Garmon D, Gathe JC Jr, Gaultier, CR, Gebre W, Gilman FD, Gilson I, Goepfert P, Gottlieb MS, Goulston C, Groger RK, Gurley TD, Haber S, Hardwicke R, Hardy WD, Harrigan PR, Hawkins, TN, Heath S, Hecht FM, Henry WK, Hladek M, Hoffman RP, Horton JM, Hsu RK, Huhn GD, Hunt P, Hupert MJ, Illeman ML, Jaeger H, Jellinger RM, John M, Johnson JA, Johnson KL, Johnson H, Johnson K, Joly J, Jordan WC, Kauffman CA, Khanlou H, Killian RK, Kim AY, Kim DD, Kinder CA, Kirchner JT, Kogelman L, Kojic EM, Korthuis PT, Kurisu W, Kwon DS, LaMar M, Lampiris H, Lanzafame M, Lederman MM, Lee DM, Lee JM, Lee MJ, Lee ET, Lemoine J, Levy JA, Llibre JM, Liguori MA, Little SJ, Liu AY, Lopez AJ, Loutfy MR, Loy D, Mohammed DY, Man A, Mansour MK, Marconi VC, Markowitz M, Marques R, Martin JN, Martin HL Jr, Mayer KH, McElrath MJ, McGhee TA, McGovern BH, McGowan K, McIntyre D, Mcleod GX, Menezes P, Mesa G, Metroka CE, Meyer-Olson D, Miller AO, Montgomery K, Mounzer KC, Nagami EH, Nagin I, Nahass RG, Nelson MO, Nielsen C, Norene DL, O'Connor DH, Ojikutu BO, Okulicz J, Oladehin OO, Oldfield EC 3rd, Olender SA, Ostrowski M, Owen WF Jr, Pae E, Parsonnet J, Pavlatos AM, Perlmutter AM, Pierce MN, Pincus JM, Pisani L, Price LJ, Proia L, Prokesch RC, Pujet HC, Ramgopal M, Rathod A, Rausch M, Ravishankar J, Rhame FS, Richards CS, Richman DD, Rodes B, Rodriguez M, Rose RC 3rd, Rosenberg ES, Rosenthal D, Ross PE, Rubin DS, Rumbaugh E, Saenz L, Salvaggio MR, Sanchez WC, Sanjana VM, Santiago S, Schmidt W, Schuitemaker H, Sestak PM, Shalit P, Shay W, Shirvani VN, Silebi VI, Sizemore JM Jr, Skolnik PR, Sokol-Anderson M, Sosman JM, Stabile P, Stapleton JT, Starrett S, Stein F, Stellbrink HJ, Sterman FL, Stone VE, Stone DR, Tambussi G, Taplitz RA, Tedaldi EM, Telenti A, Theisen W, Torres R, Tosiello L, Tremblay C, Tribble MA, Trinh PD, Tsao A, Ueda P, Vaccaro A, Valadas E, Vanig TJ, Vecino I, Vega VM, Veikley W, Wade BH, Walworth C, Wanidworanun C, Ward DJ, Warner DA, Weber RD, Webster D, Weis S, Wheeler DA, White DJ, Wilkins E, Winston A, Wlodaver CG, van't Wout A, Wright DP, Yang OO, Yurdin DL, Zabukovic BW, Zachary KC, Zeeman B, Zhao M (2010) The major genetic determinants of HIV-1 control affect HLA class I peptide presentation. Science (New York, N. Y.) 330(6010):1551–1557

Isitman G, Lisovsky I, Tremblay-McLean A, Kovacs C, Harris M, Routy JP, Bruneau J, Wainberg MA, Tremblay C, Bernard NF (2016) Antibody-dependent cellular cytotoxicity activity of effector cells from HIV-1-infected elite and viral controllers. AIDS Res Hum Retroviruses 32(10–11):1079–1088

Jin X, Bauer DE, Tuttleton SE, Lewin S, Gettie A, Blanchard J, Irwin CE, Safrit JT, Mittler J, Weinberger L, Kostrikis LG, Zhang L, Perelson AS, Ho DD (1999) Dramatic rise in plasma viremia after CD8(+) T cell depletion in simian immunodeficiency virus-infected macaques. J Exp Med 189(6):991–998

Jones RB, O'Connor R, Mueller S, Foley M, Szeto GL, Karel D, Lichterfeld M, Kovacs C, Ostrowski MA, Trocha A, Irvine DJ, Walker BD (2014) Histone deacetylase inhibitors impair the elimination of HIV-1-infected cells by cytotoxic T-lymphocytes. PLoS Pathog 10(8): e1004287

Jones RB, Mueller S, O'Connor R, Rimpel K, Sloan DD, Karel D, Wong HC, Jeng EK, Thomas AS, Whitney JB, Lim SY, Kovacs C, Benko E, Karandish S, Huang SH, Buzon MJ, Lichterfeld M, Irrinki A, Murry JP, Tsai A, Yu H, Geleziunas R, Trocha A, Ostrowski MA, Irvine DJ, Walker BD (2016) A subset of latency-reversing agents expose HIV-1-infected resting CD4+ T-cells to recognition by cytotoxic T-lymphocytes. PLoS Pathog 12(4): e1005545

Julg B, Pereyra F, Buzon MJ, Piechocka-Trocha A, Clark MJ, Baker BM, Lian J, Miura T, Martinez-Picado J, Addo MM, Walker BD (2010) Infrequent recovery of HIV-1 from but robust exogenous infection of activated CD4(+) T cells in HIV-1 elite controllers. Clin Infect Dis: Off Publ Infect Dis Soc Am 51(2):233–238

Keele BF, Giorgi EE, Salazar-Gonzalez JF, Decker JM, Pham KT, Salazar MG, Sun C, Grayson T, Wang S, Li H, Wei X, Jiang C, Kirchherr JL, Gao F, Anderson JA, Ping LH, Swanstrom R, Tomaras GD, Blattner WA, Goepfert PA, Kilby JM, Saag MS, Delwart EL, Busch MP, Cohen MS, Montefiori DC, Haynes BF, Gaschen B, Athreya GS, Lee HY, Wood N, Seoighe C, Perelson AS, Bhattacharya T, Korber BT, Hahn BH, Shaw GM (2008) Identification and characterization of transmitted and early founder virus envelopes in primary HIV-1 infection. Proc Natl Acad Sci USA 105(21):7552–7557

Khakoo SI, Thio CL, Martin MP, Brooks CR, Gao X, Astemborski J, Cheng J, Goedert JJ, Vlahov D, Hilgartner M, Cox S, Little AM, Alexander GJ, Cramp ME, O'Brien SJ, Rosenberg WM, Thomas DL, Carrington M (2004) HLA and NK cell inhibitory receptor genes in resolving hepatitis C virus infection. Science (New York, N.Y.) 305(5685):872–874

Kieffer TL, Kwon P, Nettles RE, Han Y, Ray SC, Siliciano RF (2005) G→ A hypermutation in protease and reverse transcriptase regions of human immunodeficiency virus type 1 residing in resting CD4+ T cells in vivo. J Virol 79(3):1975–1980

Klein F, Halper-Stromberg A, Horwitz JA, Gruell H, Scheid JF, Bournazos S, Mouquet H, Spatz LA, Diskin R, Abadir A, Zang T, Dorner M, Billerbeck E, Labitt RN, Gaebler C, Marcovecchio PM, Incesu RB, Eisenreich TR, Bieniasz PD, Seaman MS, Bjorkman PJ, Ravetch JV, Ploss A, Nussenzweig MC (2012) HIV-1 therapy by a combination of broadly neutralizing antibodies in humanized mice. Nature 492(7427):118–122

Koup RA, Safrit JT, Cao Y, Andrews CA, McLeod G, Borkowsky W, Farthing C, Ho DD (1994) Temporal association of cellular immune responses with the initial control of viremia in primary human immunodeficiency virus type 1 syndrome. J Virol 68(7):4650–4655

Lambotte O, Boufassa F, Madec Y, Nguyen A, Goujard C, Meyer L, Rouzioux C, Venet A, Delfraissy JF & SEROCO-HEMOCO Study Group (2005) HIV-1 controllers: a homogeneous group of HIV-1-infected patients with spontaneous control of viral replication. Clin infect dis: off publ Infect Dis Soc Am 41(7):1053–1056

Lambotte O, Ferrari G, Moog C, Yates NL, Liao HX, Parks RJ, Hicks CB, Owzar K, Tomaras GD, Montefiori DC, Haynes BF, Delfraissy JF (2009) Heterogeneous neutralizing antibody and antibody-dependent cell cytotoxicity responses in HIV-1 elite controllers. AIDS (London, England) 23(8):897–906

Lane HC, Kovacs JA, Feinberg J, Herpin B, Davey V, Walker R, Deyton L, Metcalf JA, Baseler M, Salzman N (1988) Anti-retroviral effects of interferon-alpha in AIDS-associated Kaposi's sarcoma. Lancet (London, England) 2(8622):1218–1222

Lavender KJ, Gibbert K, Peterson KE, Van Dis E, Francois S, Woods T, Messer RJ, Gawanbacht A, Müller JA, Müller J, Phillips K, Race B, Harper MS, Guo K, Lee EJ, Trilling M, Hengel H, Piehler J, Verheyen J, Wilson CC, Santiago ML, Hasenkrug KJ, Dittmer U (2016) Interferon alpha subtype-specific suppression of HIV-1 infection in vivo. J Virol 90(13):6001–6013

Lecossier D, Bouchonnet F, Clavel F, Hance AJ (2003) Hypermutation of HIV-1 DNA in the absence of the Vif protein. Science (New York, N.Y.) 300(5622):1112

Li SX, Barrett BS, Heilman KJ, Messer RJ, Liberatore RA, Bieniasz PD, Kassiotis G, Hasenkrug KJ, Santiago ML (2014) Tetherin promotes the innate and adaptive cell-mediated immune response against retrovirus infection in vivo. J Immunol (Baltimore, MD: 1950) 193 (1):306–316

Limou S, Le Clerc S, Coulonges C, Carpentier W, Dina C, Delaneau O, Labib T, Taing L, Sladek R, Deveau C, Ratsimandresy R, Montes M, Spadoni JL, Lelievre JD, Levy Y, Therwath A, Schachter F, Matsuda F, Gut I, Froguel P, Delfraissy JF, Hercberg S, Zagury JF & ANRS Genomic Group (2009) Genomewide association study of an AIDS-nonprogression cohort emphasizes the role played by HLA genes (ANRS Genomewide Association Study 02). J Infect Dis 199(3):419–426

Lynch RM, Boritz E, Coates EE, DeZure A, Madden P, Costner P, Enama ME, Plummer S, Holman L, Hendel CS, Gordon I, Casazza J, Conan-Cibotti M, Migueles SA, Tressler R,

Bailer RT, McDermott A, Narpala S, O'Dell S, Wolf G, Lifson JD, Freemire BA, Gorelick RJ, Pandey JP, Mohan S, Chomont N, Fromentin R, Chun TW, Fauci AS, Schwartz RM, Koup RA, Douek DC, Hu Z, Capparelli E, Graham BS, Mascola JR, Ledgerwood JE, VRC 601 Study Team (2015) Virologic effects of broadly neutralizing antibody VRC01 administration during chronic HIV-1 infection. Sci Transl Med 7(319):319ra206

Mangeat B, Turelli P, Caron G, Friedli M, Perrin L, Trono D (2003) Broad antiretroviral defence by human APOBEC3G through lethal editing of nascent reverse transcripts. Nature 424 (6944):99–103

Marras F, Nicco E, Bozzano F, Di Biagio A, Dentone C, Pontali E, Boni S, Setti M, Orofino G, Mantia E, Bartolacci V, Bisio F, Riva A, Biassoni R, Moretta L, De Maria A (2013) Natural killer cells in HIV-1 controller patients express an activated effector phenotype and do not up-regulate NKp44 on IL-2 stimulation. Proc Natl Acad Sci USA 110(29):11970–11975

Martin MP, Gao X, Lee JH, Nelson GW, Detels R, Goedert JJ, Buchbinder S, Hoots K, Vlahov D, Trowsdale J, Wilson M, O'Brien SJ, Carrington M (2002) Epistatic interaction between KIR3DS1 and HLA-B delays the progression to AIDS. Nat Genet 31(4):429–434

Martin MP, Qi Y, Gao X, Yamada E, Martin JN, Pereyra F, Colombo S, Brown EE, Shupert WL, Phair J, Goedert JJ, Buchbinder S, Kirk GD, Telenti A, Connors M, O'Brien SJ, Walker BD, Parham P, Deeks SG, McVicar DW, Carrington M (2007) Innate partnership of HLA-B and KIR3DL1 subtypes against HIV-1. Nat Genet 39(6):733–740

Migueles SA, Connors M (2015) Success and failure of the cellular immune response against HIV-1. Nat Immunol 16(6):563–570

Migueles SA, Sabbaghian MS, Shupert WL, Bettinotti MP, Marincola FM, Martino L, Hallahan CW, Selig SM, Schwartz D, Sullivan J, Connors M (2000) HLA B*5701 is highly associated with restriction of virus replication in a subgroup of HIV-1-infected long term nonprogressors. Proc Natl Acad Sci USA 97(6):2709–2714

Migueles SA, Laborico AC, Shupert WL, Sabbaghian MS, Rabin R, Hallahan CW, Van Baarle D, Kostense S, Miedema F, McLaughlin M, Ehler L, Metcalf J, Liu S, Connors M (2002) HIV-1-specific CD8+ T cell proliferation is coupled to perforin expression and is maintained in nonprogressors. Nat Immunol 3(11):1061–1068

Migueles SA, Osborne CM, Royce C, Compton AA, Joshi RP, Weeks KA, Rood JE, Berkley AM, Sacha JB, Cogliano-Shutta NA, Lloyd M, Roby G, Kwan R, McLaughlin M, Stallings S, Rehm C, O'Shea MA, Mican J, Packard BZ, Komoriya A, Palmer S, Wiegand AP, Maldarelli F, Coffin JM, Mellors JW, Hallahan CW, Follman DA, Connors M (2008) Lytic granule loading of CD8+ T cells is required for HIV-1-infected cell elimination associated with immune control. Immunity 29(6):1009–1021

Norman JM, Mashiba M, McNamara LA, Onafuwa-Nuga A, Chiari-Fort E, Shen W, Collins KL (2011) The anti-viral factor APOBEC3G enhances natural killer cell recognition of HIV-1-infected primary T cells. Nat Immunol 12(10):975–983

O'Connell KA, Han Y, Williams TM, Siliciano RF, Blankson JN (2009) Role of natural killer cells in a cohort of elite suppressors: low frequency of the protective KIR3DS1 allele and limited inhibition of human immunodeficiency virus type 1 replication in vitro. J Virol 83(10):5028–5034

Okulicz JF, Lambotte O (2011) Epidemiology and clinical characteristics of elite controllers. Curr Opin HIV-1 AIDS 6(3):163–168

Overbaugh J, Morris L (2012) The antibody response against HIV-1. Cold Spring Harb Perspect Med 2(1):a007039. doi:10.1101/cshperspect.a007039

Pandrea I, Gaufin T, Gautam R, Kristoff J, Mandell D, Montefiori D, Keele BF, Ribeiro RM, Veazey RS, Apetrei C (2011) Functional cure of SIVagm infection in rhesus macaques results in complete recovery of CD4+ T cells and is reverted by CD8+ cell depletion. PLoS Pathog 7 (8):e1002170

Pegu A, Asokan M, Wu L, Wang K, Hataye J, Casazza JP, Guo X, Shi W, Georgiev I, Zhou T, Chen X, O'Dell S, Todd JP, Kwong PD, Rao SS, Yang ZY, Koup RA, Mascola JR, Nabel GJ (2015) Activation and lysis of human CD4 cells latently infected with HIV-1. Nat commun 6:8447

Pereyra F, Addo MM, Kaufmann DE, Liu Y, Miura T, Rathod A, Baker B, Trocha A, Rosenberg R, Mackey E, Ueda P, Lu Z, Cohen D, Wrin T, Petropoulos CJ, Rosenberg ES, Walker BD (2008) Genetic and immunologic heterogeneity among persons who control HIV-1 infection in the absence of therapy. J Infect Dis 197(4):563–571

Pham TN, Lukhele S, Hajjar F, Routy JP, Cohen EA (2014) HIV-1 Nef and Vpu protect HIV-1-infected CD4+ T cells from antibody-mediated cell lysis through down-modulation of CD4 and BST2. Retrovirology 11:15-4690-11-15

Phillips RE, Rowland-Jones S, Nixon DF, Gotch FM, Edwards JP, Ogunlesi AO, Elvin JG, Rothbard JA, Bangham CR, Rizza CR (1991) Human immunodeficiency virus genetic variation that can escape cytotoxic T cell recognition. Nature 354(6353):453–459

Picker LJ (2017) Therapeutic vaccination for HIV-1/SIV: what will it take for cure? In: Visions of HIV-1 Cure, Conference on Retroviruses and Opportunistic Infections (CROI), Seattle, WA, February 13–16, 2017

Pollack RA, Jones RB, Pertea M, Bruner KM, Martin AR, Thomas AS, Capoferri AA, Beg SA, Huang SH, Karandish S, Hao H, Halper-Stromberg E, Yong PC, Kovacs C, Benko E, Siliciano RF, Ho YC (2017) Defective HIV-1 proviruses are expressed and can be recognized by cytotoxic T lymphocytes, which shape the proviral landscape. Cell host microbe 21(4): 494–506.e4

Richard J, Sindhu S, Pham TNQ, Belzile JP, Cohen Ã (2010) HIV-1 Vpr up-regulates expression of ligands for the activating NKG2D receptor and promotes NK cellâ€"mediated killing. Blood 115(7):1354–1363

Saez-Cirion A, Lacabaratz C, Lambotte O, Versmisse P, Urrutia A, Boufassa F, Barre-Sinoussi F, Delfraissy JF, Sinet M, Pancino G, Venet A & Agence Nationale de Recherches sur le Sida EP36 HIV-1 Controllers Study Group (2007) HIV-1 controllers exhibit potent CD8 T cell capacity to suppress HIV-1 infection ex vivo and peculiar cytotoxic T lymphocyte activation phenotype. Proc Natl Acad Sci USA 104(16):6776–6781

Saez-Cirion A, Sinet M, Shin SY, Urrutia A, Versmisse P, Lacabaratz C, Boufassa F, Avettand-Fenoel V, Rouzioux C, Delfraissy JF, Barre-Sinoussi F, Lambotte O, Venet A, Pancino G, ANRS EP36 HIV-1 Controllers Study Group (2009) Heterogeneity in HIV-1 suppression by CD8 T cells from HIV-1 controllers: association with Gag-specific CD8 T cell responses. J Immunol (Baltimore, MD: 1950) 182(12):7828–7837

Sandler NG, Bosinger SE, Estes JD, Zhu RT, Tharp GK, Boritz E, Levin D, Wijeyesinghe S, Makamdop KN, del Prete GQ, Hill BJ, Timmer JK, Reiss E, Yarden G, Darko S, Contijoch E, Todd JP, Silvestri G, Nason M, Norgren RB Jr, Keele BF, Rao S, Langer JA, Lifson JD, Schreiber G, Douek DC (2014) Type I interferon responses in rhesus macaques prevent SIV infection and slow disease progression. Nature 511(7511):601–605

Scheid JF, Horwitz JA, Bar-On Y, Kreider EF, Lu CL, Lorenzi JCC, Feldmann A, Braunschweig M, Nogueira L, Oliveira T, Shimeliovich I, Patel R, Burke L, Cohen YZ, Hadrigan S, Settler A, Witmer-Pack M, West AP, Juelg B, Keler T, Hawthorne T, Zingman B, Gulick RM, Pfeifer N, Learn GH, Seaman MS, Bjorkman PJ, Klein F, Schlesinger SJ, Walker BD, Hahn BH, Nussenzweig MC, Caskey M (2016) HIV-1 antibody 3BNC117 suppresses viral rebound in humans during treatment interruption. Nature 535(7613):556–560

Schmitz JE, Kuroda MJ, Santra S, Sasseville VG, Simon MA, Lifton MA, Racz P, Tenner-Racz K, Dalesandro M, Scallon BJ, Ghrayeb J, Forman MA, Montefiori DC, Rieber EP, Letvin NL, Reimann KA (1999) Control of viremia in simian immunodeficiency virus infection by CD8 + lymphocytes. Science (New York, N.Y.) 283(5403):857–860

Scully E, Alter G (2016) NK Cells in HIV-1 Disease. Curr HIV-1/AIDS Rep 13:85–94

Shah AH, Sowrirajan B, Davis ZB, Ward JP, Campbell EM, Planelles V, Barker E (2010) Degranulation of natural killer cells following interaction with HIV-1-infected cells is hindered by downmodulation of NTB-A by Vpu. Cell Host Microbe 8(5):397–409

Shan L, Deng K, Shroff NS, Durand CM, Rabi SA, Yang HC, Zhang H, Margolick JB, Blankson JN, Siliciano RF (2012) Stimulation of HIV-1-specific cytolytic T lymphocytes facilitates elimination of latent viral reservoir after virus reactivation. Immunity 36(3):491–501

Sheehy AM, Gaddis NC, Choi JD, Malim MH (2002) Isolation of a human gene that inhibits HIV-1 infection and is suppressed by the viral Vif protein. Nature 418(6898):646–650

Shibata-Koyama M, Iida S, Misaka H, Mori K, Yano K, Shitara K, Satoh M (2009) Nonfucosylated rituximab potentiates human neutrophil phagocytosis through its high binding for FcgammaRIIIb and MHC class II expression on the phagocytotic neutrophils. Exp Hematol 37(3):309–321

Shingai M, Nishimura Y, Klein F, Mouquet H, Donau OK, Plishka R, Buckler-White A, Seaman M, Piatak M Jr, Lifson JD, Dimitrov DS, Nussenzweig MC, Martin MA (2013) Antibody-mediated immunotherapy of macaques chronically infected with SHIV-1 suppresses viraemia. Nature 503(7475):277–280

Siliciano JD, Kajdas J, Finzi D, Quinn TC, Chadwick K, Margolick JB, Kovacs C, Gange SJ, Siliciano RF (2003) Long-term follow-up studies confirm the stability of the latent reservoir for HIV-1 in resting CD4+ T cells. Nat Med 9(6):727–728

Sloan DD, Lam CY, Irrinki A, Liu L, Tsai A, Pace CS, Kaur J, Murry JP, Balakrishnan M, Moore PA, Johnson S, Nordstrom JL, Cihlar T, Koenig S (2015) Targeting HIV-1 reservoir in infected CD4 T cells by dual-affinity re-targeting molecules (DARTs) that bind HIV-1 envelope and recruit cytotoxic T cells. PLoS Pathog 11(11):e1005233

Smalls-Mantey A, Doria-Rose N, Klein R, Patamawenu A, Migueles SA, Ko SY, Hallahan CW, Wong H, Liu B, You L, Scheid J, Kappes JC, Ochsenbauer C, Nabel GJ, Mascola JR, Connors M (2012) Antibody-dependent cellular cytotoxicity against primary HIV-1-infected CD4+ T cells is directly associated with the magnitude of surface IgG binding. J Virol 86 (16):8672–8680

Specht A, DeGottardi MQ, Schindler M, Hahn B, Evans DT, Kirchhoff F (2008) Selective downmodulation of HLA-A and -B by Nef alleles from different groups of primate lentiviruses. Virology 373(1):229–237

Sung JA, Pickeral J, Liu L, Stanfield-Oakley SA, Lam CY, Garrido C, Pollara J, LaBranche C, Bonsignori M, Moody MA, Yang Y, Parks R, Archin N, Allard B, Kirchherr J, Kuruc JD, Gay CL, Cohen MS, Ochsenbauer C, Soderberg K, Liao HX, Montefiori D, Moore P, Johnson S, Koenig S, Haynes BF, Nordstrom JL, Margolis DM, Ferrari G (2015) Dual-affinity re-targeting proteins direct T cell-mediated cytolysis of latently HIV-1-infected cells. J Clin Investig 125(11):4077–4090

Tomaras GD, Yates NL, Liu P, Qin L, Fouda GG, Chavez LL, Decamp AC, Parks RJ, Ashley VC, Lucas JT, Cohen M, Eron J, Hicks CB, Liao HX, Self SG, Landucci G, Forthal DN, Weinhold KJ, Keele BF, Hahn BH, Greenberg ML, Morris L, Karim SS, Blattner WA, Montefiori DC, Shaw GM, Perelson AS, Haynes BF (2008) Initial B-cell responses to transmitted human immunodeficiency virus type 1: virion-binding immunoglobulin M (IgM) and IgG antibodies followed by plasma anti-gp41 antibodies with ineffective control of initial viremia. J Virol 82(24):12449–12463

Tomescu C, Chehimi J, Maino VC, Montaner LJ (2007) NK cell lysis of HIV-1-infected autologous CD4 primary T cells: requirement for IFN-mediated NK activation by plasmacytoid dendritic cells. Journal of immunology (Baltimore, Md.: 1950) 179(4):2097–2104

van Manen D, Kootstra NA, Boeser-Nunnink B, Handulle MA, van't Wout AB, Schuitemaker H (2009) Association of HLA-C and HCP5 gene regions with the clinical course of HIV-1 infection. AIDS (London, England) 23(1):19–28

Walker-Sperling VE, Cohen VJ, Tarwater PM, Blankson JN (2015) Reactivation kinetics of HIV-1 and susceptibility of reactivated latently infected CD4+ T Cells to HIV-1-specific CD8+ T Cells. J Virol 89(18):9631–9638

Walker-Sperling VE, Pohlmeyer CW, Tarwater PM, Blankson JN (2016) The effect of latency reversal agents on primary CD8+ T cells: implications for shock and kill strategies for human immunodeficiency virus eradication. EBioMedicine 8:217–229

Walker-Sperling VE, Pohlmeyer CW, Veenhuis RT, May M, Luna KA, Kirkpatrick AR, Laeyendecker O, Cox AL, Carrington M, Bailey JR, Arduino RC, Blankson JN (2017) Factors associated with the control of viral replication and virologic breakthrough in a recently infected hiv-1 controller. EBioMedicine 16:141–149

Wren LH, Chung AW, Isitman G, Kelleher AD, Parsons MS, Amin J, Cooper DA, ADCC study collaboration investigators, Stratov I, Navis M, Kent SJ (2013) Specific antibody-dependent cellular cytotoxicity responses associated with slow progression of HIV-1 infection. Immunology 138(2):116–123

Yu Q, Konig R, Pillai S, Chiles K, Kearney M, Palmer S, Richman D, Coffin JM, Landau NR (2004) Single-strand specificity of APOBEC3G accounts for minus-strand deamination of the HIV-1 genome. Nat Struct Mol Biol 11(5):435–442

Zhang H, Yang B, Pomerantz RJ, Zhang C, Arunachalam SC, Gao L (2003) The cytidine deaminase CEM15 induces hypermutation in newly synthesized HIV-1 DNA. Nature 424 (6944):94–98

# Nonhuman Primate Models for Studies of AIDS Virus Persistence During Suppressive Combination Antiretroviral Therapy

**Gregory Q. Del Prete and Jeffrey D. Lifson**

**Abstract** Nonhuman primate (NHP) models of AIDS represent a potentially powerful component of the effort to understand in vivo sources of AIDS virus that persist in the setting of suppressive combination antiretroviral therapy (cART) and to develop and evaluate novel strategies for more definitive treatment of HIV infection (i.e., viral eradication "cure", or sustained off-cART remission). Multiple different NHP models are available, each characterized by a particular NHP species, infecting virus, and cART regimen, and each with a distinct capacity to recapitulate different aspects of HIV infection. Given these different biological characteristics, and their associated strengths and limitations, different models may be preferred to address different questions pertaining to virus persistence and cure research, or to evaluate different candidate intervention approaches. Recent developments in improved cART regimens for use in NHPs, new viruses, a wider array of sensitive virologic assay approaches, and a better understanding of pathogenesis should allow even greater contributions from NHP models to this important area of HIV research in the future.

## Contents

G.Q. Del Prete (✉) · J.D. Lifson (✉)
AIDS and Cancer Virus Program, Frederick National Laboratory for Cancer Research,
Leidos Biomedical Research, Inc., Frederick, MD, USA
e-mail: delpretegq@mail.nih.gov

J.D. Lifson
e-mail: lifsonj@mail.nih.gov

Current Topics in Microbiology and Immunology (2018) 417:69–110
DOI 10.1007/82_2017_73
© Springer International Publishing AG 2017
Published Online: 13 October 2017

# 1   Introduction

Animal models can be valuable tools for studying human disease. Although they have specific limitations and typically cannot recapitulate every aspect of the human diseases they are used to study, if applied thoughtfully and with an appreciation of these limitations, animal models can provide important insights into complex in vivo biology. Nevertheless, caution must always be exercised when seeking to draw direct conclusions about human disease based solely on data generated in an animal model system, particularly when using animal species that are separated from humans by a great phylogenetic distance. Due to their genetic relatedness to humans and the attendant similarities in host anatomy, physiology, and immunology, nonhuman primates (NHPs) represent highly attractive and relevant research animals for the study of human disease.

NHP models of HIV infection and AIDS are powerful experimental systems that have made critical contributions to multiple areas of AIDS research and have greatly improved our understanding of HIV infection, disease, immune responses, and both prevention and treatment approaches. These models, which may make use of multiple different available viruses and several different primate species, can recapitulate the key features of pathogenic HIV infection, provided that the right combination of virus and host primate species is used. Important discoveries about viral transmission and dissemination within an infected host (Miller et al. 2005; Barouch et al. 2016; Liu et al. 2016), early virus-mediated damage to the gut (Veazey et al. 1998; Brenchley et al. 2006; Estes et al. 2010), and the role that chronic immune activation plays in HIV pathogenesis (Silvestri et al. 2003, 2005; Sumpter et al. 2007; Estes et al. 2008) were greatly facilitated by the use of NHP/AIDS models, and some of the first evaluations HIV treatment (Tsai et al.

1995, 1998) and prevention approaches (Desrosiers et al. 1989; Daniel et al. 1992; Arthur et al. 1995) have been performed within these systems. In this chapter, we will discuss NHP models of AIDS, and the strengths, limitations, and applications of such models for studies of HIV persistence and latency in the setting of suppressive combination antiretroviral therapy (cART), and for evaluations of proposed strategies for HIV functional cure, control, and eradication.

There are number of generalizable animal model features, and features of NHP/AIDS models, more specifically, that are relevant for HIV cure research that we will consider in this chapter (Fig. 1). These features include: (1) degree of model characterization, extent of historical data, and the associated species-specific and virus-specific reagents available, (2) consistency, predictability, and clinical relevance of the natural history of infection in the absence of therapy, (3) practical considerations such as costs, study feasibility, and drug and animal availability, (4) the ability to durably suppress viral replication to clinically relevant levels using antiretroviral drugs, and (5) sensitive and precise assays to measure virologic parameters.

We note that humanized mouse models, which involve the use of mice that have been engineered so that specific components of the mouse immune system are experimentally replaced by the corresponding human immune system components, are alternative animal models of HIV infection that are also being actively used and developed for studies of HIV latency and viral eradication. However, these models, which have their own sets of inherent advantages and limitations, have been recently reviewed elsewhere (Garcia 2016) and will not be discussed in this chapter.

## 2  NHP/AIDS Models

The availability and use of several different NHP species and subspecies coupled with numerous available viruses has resulted in the development of multiple different NHP models of AIDS, each of which potentially could be harnessed for HIV cure research (Fig. 1). Each combination of primate species and virus has a specific set of scientific strengths and weaknesses, including varying degrees of prior utilization and characterization, which must be carefully considered when designing an NHP study.

## 2.1  HIV-1 Infection of Nonhuman Primates

Initial efforts to model HIV-1 infection in animals included the use of experimental infection of chimpanzees with HIV-1 itself, which can establish a persistent infection in chimps (Fultz et al. 1989). However, the current protected status of chimpanzees, ethical, practical, and logistical challenges for their use, a lack of pathogenesis in the vast majority of experimentally HIV-1-infected animals, and a very long window of

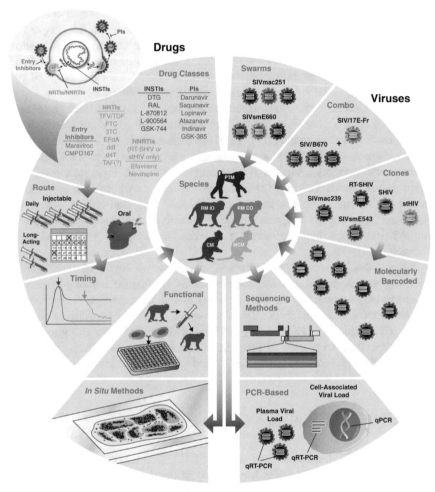

**Fig. 1** Schematic representation of key study parameters to consider for nonhuman primate (NHP) models of AIDS virus persistence during suppressive combination antiretroviral therapy (cART). Relevant parameters include: **a** choice of virus, including the use of viral swarms, infectious molecular clones, or molecularly barcoded infectious molecular clones, or the use of chimeric and modified viruses such as SHIV, RT-SHIV, or stHIV; **b** combination antiretroviral therapy regimen, including selection of specific regimen drugs, route of drug administration, timing of therapy initiation, and duration of therapy; **c** host primate species; and **d** virologic assays, including PCR-based methods to quantify viral nucleic acid content, sequencing methods for use with molecularly barcoded viruses, functional virus outgrowth assays, and in situ approaches. SHIV, simian-human immunodeficiency virus; RT-SHIV, reverse-transcriptase simian-human immunodeficiency virus; stHIV, simian-tropic HIV; qPCR, quantitative polymerase chain reaction; qRT-PCR, quantitative reverse-transcriptase polymerase chain reaction; DTG, dolutegravir; RAL, raltegravir; TFV/TDF, tenofovir/tenofovir disoproxil fumarate; FTC, emtricitabine; 3TC, lamivudine; EFdA, 4'-ethynyl-2-fluoro-2'-deoxyadenosine; ddI, didanosine; d4t, stavudine; TAF, tenofovir alafenamide; PTM, pigtail macaque; RM IO, Indian-origin rhesus macaque; RM CO, Chinese-origin rhesus macaque; CM, cynomolgus macaque; MCM, Mauritian cynomolgus macaque

time (up to a decade) between initial infection and disease progression in those rare animals that did progress (Novembre et al. 1997; O'Neil et al. 2000) led to the discontinuation of experimental HIV-1 infection of chimpanzees. It was later determined that chimpanzees infected with simian immunodeficiency virus (SIV)cpz in the wild were in fact the source of HIV-1 infection following zoonotic transmission of SIVcpz into humans (Keele et al. 2006), perhaps explaining why some chimpanzees could be productively infected with HIV-1. Notably, chimpanzees naturally infected with SIVcpz in the wild have a higher mortality rate than uninfected animals, with evidence of AIDS-like immunopathology (Keele et al. 2009), though disease progression likely occurs on timescales that are impractical for experimental modeling. As experimental HIV-1 infection of chimpanzees did not represent a feasible model for AIDS-related studies, AIDS researchers thus turned to the use of experimental infection of several Asian macaque species, rhesus macaques (*Macaca mulatta*), pigtail macaques (*Macaca nemestrina*), and cynomolgus macaques (*Macaca fascicularis*), for the vast majority of NHP/AIDS research, which in turn necessitated the use of simian immunodeficiency viruses (SIVs) to model HIV infection in NHPs.

Although HIV-1 can establish persistent, productive infections in humans and chimpanzees, it cannot productively infect macaques, due in large measure to the expression of species-specific host restriction factors (Hatziioannou and Bieniasz 2011). These multiple factors, which are expressed by potential viral target cells, and which are often stimulated by type I interferons and thus upregulated in response to viral infection, can potently interdict retroviral infection at varying points in the viral replication cycle. While humans, like monkeys, also express such host restriction factors, HIV-1 is capable of evading and counteracting the human versions of these factors but cannot do so for many of the macaque restriction factors. For example, the rhesus TRIM5α protein interacts with incoming HIV-1 virion capsid proteins soon after viral entry into a new target cell, interfering with successful progression to reverse transcription and the subsequent steps of viral replication (Stremlau et al. 2004, 2006). By contrast, the human TRIM5α does not recognize HIV-1 capsid proteins due to specific amino acid residues within the HIV-1 capsid that preclude this interaction (Owens et al. 2003; Hatziioannou et al. 2004; Ikeda et al. 2004). Similarly, the HIV-1 Vpu accessory gene product can specifically counteract the human tetherin protein restriction factor, but not rhesus macaque tetherin, which can therefore successfully tether HIV-1 virions to the surface of producer cells, inhibiting virion release and the ability of the virus to spread to new target cells in macaques (Neil et al. 2008; McNatt et al. 2009).

The fact that such restrictions preclude sustained, pathogenic HIV-1 infection in macaques has prevented the direct use of HIV-1 in macaque-based models, necessitating for most studies over the past three decades of the use of parallel models, such as simian immunodeficiency virus (SIV) infection of macaques (discussed below). However, the potential advantages of a model using a virus that contains the authentic HIV-1 targets for pharmacological or immunological interventions, rather than their SIV homologs, along with an emerging understanding of the underlying mechanistic basis for at least some of the restriction of HIV-1 replication in macaque cells and macaques, has led more recently to efforts to generate engineered HIV-1 viruses containing minimal specific changes that allow HIV-1 to avoid and counteract macaque restriction factors (Hatziioannou et al. 2009; Thippeshappa et al. 2011; Hatziioannou et al. 2014). These efforts have resulted in the generation of minimally modified HIV-1 adapted to pigtail macaques that are capable of sustaining high viral loads and inducing AIDS-defining clinical endpoints in animals that are experimentally transiently depleted of CD8$^+$ T cells at the time of infection (Hatziioannou et al. 2014). The ability to use a minimally altered HIV-1 for cure research conducted in NHPs has intriguing potential, but additional work will be required to determine the suitability of this model for such studies.

## 2.2   Discovery of Simian Immunodeficiency Virus (SIV) and Simian AIDS

During the 1960s, efforts to identify the etiologic agent of kuru, a rare transmissible spongiform encephalopathy, included the passage of human brain material into African sooty mangabey monkeys and then serially from sooty mangabeys into Indian-origin rhesus macaques (Gardner 2016). Many years later, a number of the rhesus macaques involved in these experiments developed evidence of clinically significant immunosuppression, including lymphomas and opportunistic infections, ultimately leading to the discovery of simian immunodeficiency virus (SIV) (Letvin et al. 1985; Kanki et al. 1985; Daniel et al. 1985), a primate lentivirus closely related to but distinct from HIV (Chakrabarti et al. 1987). Later studies showed that sooty mangabeys, like most species of African nonhuman primates, are naturally infected with a type of SIV (Fultz et al. 1986; Hirsch et al. 1989; Marx et al. 1991) but that like other Asian macaques, rhesus macaques are not (Lowenstine et al. 1986; Ohta et al. 1988), and that incidental transmission of the virus from sooty mangabeys into rhesus macaque nonnatural hosts ultimately led to a pathogenic infection course and the development of simian AIDS within the rhesus animals (Murphey-Corb et al. 1986; Baskin et al. 1988; Hirsch and Johnson 1994; Sharp et al. 1995). We now appreciate that this scenario was analogous to the emergence of HIV-1, where a similar cross-species transmission of a primate lentivirus (from chimpanzees into humans in the case of pandemic HIV-1) resulted in a pathogenic infection in the new, nonnatural host (Keele et al. 2006). In the same way that spread of HIV-1 infection into the human population represents transmission of virus into a nonnatural host, where it leads to clinical disease, infection of Asian macaques with SIV similarly represents a cross-species transmission of virus into a nonnatural host that then develops disease because of the infection.

## 2.3 SIV Infection in Different NHP Species

Because rhesus macaques of Indian origin were the first recognized hosts of pathogenic SIV infection, the initial development of SIV involved serially passaging virus from animal to animal within Indian-origin rhesus macaques, with isolation of virus swarms and molecular cloning of infectious viral genomes associated with disease development (Daniel et al. 1985; Chakrabarti et al. 1987; Naidu et al. 1988). Thus, the first isolated, best characterized, and most widely used pathogenic SIVs, of the SIVmac lineage, so called due to their derivation in rhesus macaques (*Macaca mulatta*), are exquisitely well adapted to Indian-origin rhesus monkeys. When infected with the uncloned viral isolate SIVmac251 or the infectious molecular clone SIVmac239 (Fig. 1), the vast majority of Indian-origin rhesus macaques develop high peak viral loads, typically in excess of $10^7$ viral RNA copies per ml of plasma, and sustained, high chronic viral loads (Hirsch and Lifson 2000). Nearly all of the hallmarks of HIV-1 pathogenesis in humans are recapitulated in this model, including viral targeting of $CD4^+CCR5^+$ T cells, very early depletion of $CD4^+$ T cells at mucosal sites and associated damage to the gut epithelium, translocation of microbial products from the gut lumen into the host, chronically elevated levels of immune activation, progressive depletion of $CD4^+$ T cells in the periphery, and the eventual development of opportunistic infections and malignancies (Whetter et al. 1999; Brenchley et al. 2006; Estes et al. 2010). While these key features of HIV-1 infection are captured in SIVmac-infected rhesus macaques, the timescale of pathogenesis is substantially accelerated compared to human HIV infection, consistent with the higher viral replication levels observed in these animal models, with untreated animals manifesting AIDS-defining clinical endpoints within 1–2 years. The consistency and magnitude of viral replication, coupled with a relatively abbreviated disease course make SIVmac/Indian-origin rhesus macaque models robust, feasible, and attractive models for AIDS research, however, as will be further discussed below, these very features may have ironically presented obstacles for adapting these models for HIV cure research.

As discussed earlier, the SIVmac viruses were derived from SIV circulating in African sooty mangabeys, known as the SIVsm lineage, following cross-species transmission of virus into rhesus macaques. There are several other lentiviruses from the SIVsm lineage that are also pathogenic in Indian-origin rhesus macaques and that also recapitulate the key features of HIV-1 infection in humans. Most notable and extensively utilized among these are the uncloned isolate SIVsmE660 and the genetically related infectious molecular clone SIVsmE543-3 (Fig. 1). Like SIVmac251 and SIVmac239, SIVsmE660 and SIVsmE543-3 can replicate to high peak viral loads and can maintain high chronic phase viremia, with early depletion of $CD4^+$ T cells at mucosal sites, the establishment of chronic immune activation, the eventual depletion of peripheral $CD4^+$ T cells, and the development of AIDS-defining clinical endpoints (Hirsch and Johnson 1994; Hirsch et al. 1997). While the best-characterized SIVmac viruses are exceptionally resistant to neutralization by antibodies (Burns et al. 1993; Means et al. 1997; Mason et al. 2016), the SIVsmE660 virus swarm contains individual virus clones with a range of neutralization sensitivities, a majority of which are more sensitive to antibody-mediated neutralization than the SIVmac viruses (Lopker et al. 2013). This feature of SIVsmE660, combined with its genetic divergence from the SIVmac lineage made SIVsmE660 a common choice as a heterologous challenge virus for vaccine studies conducted in NHPs (Shedlock et al. 2009; Manrique et al. 2013; Lee et al. 2015). Although greater overall neutralization sensitivity might also make SIVsmE600 better suited than SIVmac for evaluations of functional cure strategies involving the elicitation or passive administration of neutralizing antibodies, the E660/rhesus macaque model has not thus far been implemented for this purpose or more generally for studies involving cART-mediated virologic suppression.

An additional issue with the use of SIVsmE660/543-3 that has hindered their use for cure research is the relatively more recent appreciation that, unlike the SIVmac lineage viruses, E660 and E543-3 are highly susceptible to restriction by rhesus TRIM5$\alpha$ in animals with certain TRIM5$\alpha$ alleles (Kirmaier et al. 2010; de Groot et al. 2011; Reynolds et al. 2011; Yeh et al. 2011; Letvin et al. 2011; Sheppard et al. 2014). The impact of restrictive TRIM5$\alpha$ alleles on cART-mediated suppression, latent reservoir establishment, and viral recrudescence upon cART cessation are currently unclear but would likely need to be accounted for in any studies utilizing SIVsmE660/E543-3. Thus, further research using TRIM5-genotyped rhesus would be needed to adapt this model for use in viral latency research involving cART-mediated suppression.

Like rhesus macaques, pigtail macaques are not naturally infected with SIV in the wild. When experimentally infected with SIVmac239/251 or SIVsmE660/E543-3, pigtail macaques experience a disease course similar to rhesus macaques (Klatt et al. 2012) and they have thus also been utilized as a host species for AIDS research (Fig. 1), albeit to a lesser extent than rhesus macaques. In general, much of the work in pigtail macaques has focused on transmission and prophylactic studies in female macaques, as the vagina in pigtail macaques appears to better recapitulate human anatomy and biology in this area than rhesus macaques (Blakley et al. 1981; Kersh et al. 2014, 2015a, b). When compared with rhesus macaques, in the absence of infection pigtail macaques have been shown to have higher levels of baseline immune activation, in association with greater intestinal epithelial permeability (Klatt et al. 2010). The nature and stability of viral latency and persistent viral sources may thus prove to be different in this host species, with the possibility of residual immune activation levels that are higher than in SIV-infected rhesus or HIV-infected humans in the setting of cART. To date, the use of pigtail macaques specifically for HIV cure research has been primarily within the context of experimental systems designed to evaluate the potential for persistent residual virus sources to reside within the central nervous system (CNS) (discussed further below) (Clements et al. 2011), and thus the number of individual antiretroviral drugs and cART regimens that have been evaluated within this species has been relatively limited.

The use of pigtail macaques for cure research may increase in the future, however, because this species of macaque expresses a variant form of TRIM5, which contains a transposed cyclophilin A domain, that renders it unable to restrict HIV-1 infection (Liao et al. 2007; Brennan et al. 2008; Virgen et al. 2008; Newman et al. 2008). This absence of a key restriction factor has led scientists developing the minimally modified HIV-1 for use in NHPs discussed earlier to focus on pigtail macaques as the initial host species of choice (Hatziioannou et al. 2009; Thippeshappa et al. 2011; Hatziioannou et al. 2014) and any successful future adaptation of these models for studies of persistent residual virus in the setting of suppressive cART will likely utilize pigtail macaques.

Scientists, primarily those in European laboratories, have also made use of cynomolgus macaques for AIDS research (Fig. 1), though like pigtail macaques their use has been less extensive than rhesus macaques. When infected with the best-characterized SIVs available, cynomolgus macaques have an infection course with less consistent pathogenesis, characterized by more variable viral loads that may be substantially lower than those measured in similarly infected Indian rhesus (Reimann et al. 2005), likely because the viruses were specifically adapted for infection of rhesus macaques (Kaizu et al. 2003). This potential limitation for the use of cynomolgus macaques might therefore be surmountable by using viruses specifically adapted for infection of this primate species (Shinohara et al. 1999; Kaizu et al. 2003; Borsetti et al. 2008). Cynomolgus macaques from the island of Mauritius, located off the coast of Madagascar, have recently become an NHP subspecies of particular interest for AIDS researchers. Mauritius contains a population of cynomolgus macaques that descended from a small number of founder animals that since their establishment have been geographically isolated from other cynomolgus macaques. As a result, this population of animals has limited MHC class I and II diversity (Krebs et al. 2005; O'Connor et al. 2007; Mee et al. 2009) and has become an intriguing tool for studies seeking to more feasibly control for MHC haplotypes than might be possible for other, more outbred NHP populations (Cain et al. 2013; Mohns et al. 2015). Though the use of Mauritian cynomolgus macaques for HIV cure research and studies of latent virus has thus far been limited, this macaque subspecies has intriguing potential for future studies in which a potential role for specific MHC-restricted immune responses are of interest.

# 3   Advantages of NHP Models for HIV/AIDS Cure Research

In general, NHP models provide a number of key scientific advantages that make them attractive tools for HIV/AIDS research. First, there is a much higher degree of experimental control, and thus fewer unforeseen and/or unknown confounding variables than is typically achievable in human clinical studies. Given the complexities of virus–host interactions and immune responses, which cannot be accurately modeled in cell culture or other more reductive systems, the ability to conduct experiments in an in vivo system with a high degree of experimental control is highly useful. All aspects of the initial viral infection, including the precise identity and diversity of the infecting virus, and the dose, route, and timing of the initial viral inoculation, are all knowable and controllable by the researcher within the context of NHP/AIDS models. The timing of all subsequent study events, including sample collections and treatments can also be carefully controlled, including the ability to initiate cART very early post-infection, which may be an important component of future successful off-cART remission strategies (Steingrover et al. 2008; von Wyl et al. 2011; Hamlyn et al. 2012; Saez-Cirion et al. 2013; Ananworanich et al. 2016; Crowell et al. 2016; Frange et al. 2016).

The use of research animals also allows for the collection of a much broader range of specimen types, including fluids and, most importantly, tissues, many of which are difficult or impossible to sample in HIV-infected humans but which represent key sites of viral replication and pathogenesis. With appropriate animal care and use committee (ACUC) oversight and with appropriate veterinary support and expertise, fluids such as blood, cerebrospinal fluid, bronchoalveolar lavages, and bone marrow aspirates, and tissues such as lymph nodes and gastrointestinal tissues can be routinely and longitudinally collected from NHPs used for AIDS research and may be collected with greater frequency than is feasible in human trials. More involved and comprehensive sample collections are also possible through the use of timed necropsies, when appropriate. Specific tissue sites may represent anatomic and pharmacologic sanctuaries that may support the persistence of virus even in the setting apparently effective antiretroviral therapy (as measured by peripheral viral loads) (Fletcher et al. 2014) or effective immune responses (Fukazawa et al. 2015), and the ability to sample these sites is highly important to further our understanding of viral persistence and recrudescence upon cART cessation and to identify potential targets for viral eradication strategies.

Current antiretroviral therapies are highly effective and well tolerated in the majority of treated HIV-1-infected people. Although there are issues with global access to drugs, costs, and treatment compliance (Cleary and McIntyre 2010; Al-Dakkak et al. 2013) that must be addressed, along with an elevated incidence of non-AIDS morbidities and mortality in long-term treated HIV-infected people (Hunt 2012; Vinikoor et al. 2013; Rajasuriar et al. 2015), the fact remains that well-suppressed treated individuals can now enjoy a lifespan approaching that of HIV-uninfected people (Antiretroviral Therapy Cohort 2017). Given this well established, safe, effective and well-tolerated standard-of-care for daily cART, the bars for safety and for efficacy are both very high for clinical evaluation of any proposed viral functional cure or eradication strategy. However, many of the HIV functional cure/eradication strategies being proposed involve novel approaches for which in vivo proof-of-concept feasibility has not been demonstrated and which involve an uncertain degree of potential patient risk, presenting a hurdle for initial evaluations in humans.

NHP studies thus provide a key means for evaluation of both safety and proof-of-concept activity for promising interventions. While various ex vivo laboratory measurements may be informative, the gold standard evaluation criteria for successful functional cure treatments will ultimately require cART cessation, with delays in the emergence of recrudescent viremia and/or control of viral replication levels after cART discontinuation serving as essential experimental readouts. It is unclear, however, if there are potential long-term ramifications for cART interruptions in HIV-1-infected humans, particularly for prolonged off-cART observation, which may be critical for assessing off-cART immunologic control of viral replication absent any substantial changes in the size of the viral reservoir and the associated time to measurable viral recrudescence. However, in NHP models, cART can be halted readily with appropriate scientific justification. By conducting preclinical proof-of-concept evaluations in NHPs, scientists can therefore evaluate the in vivo functional effectiveness of a given treatment as well as assess its safety and tolerability. NHP models are thus well positioned to provide critical preclinical information as part of the broader efforts to identify definitive treatment for HIV infection.

# 4 Challenges for Using NHP/AIDS Models for HIV Cure Research

## 4.1 Logistical Challenges

Despite the demonstrated advantages of NHP models of AIDS for studying HIV latency and persistence, there are several challenges and limitations for the application of these models for cure research. First, there are practical and feasibility challenges that must be considered. NHP study animals are costly to acquire and their proper care and use necessitate considerable space, labor, and dedicated expert care. These requirements, combined with the relatively long durations of studies involving suppression of viral replication to clinically relevant levels using cART, results in high overall study costs which often limit the number of animals that can be accommodated on a given study and which, in turn, can limit the statistical power of the studies. Second, although NHPs are relatively large when compared with other animal species typically used for biological research, they are still considerably smaller than humans. At 5–8 kg, the average sexually mature rhesus macaque is approximately the size of the average 3–6-month-old human infant. For studies of residual and/or latent virus in the setting of suppressive cART, where the biological events of interest are rare (Chun et al. 1997; Dinoso et al. 2009), limitations on the volume of blood that can be collected during routine survival bleeds limits the effective sensitivity of virologic assays and also limits of the number of experiments and/or evaluations that can be performed with a given blood sample. Although this limitation may be overcome through the use of leukapheresis procedures, which allow for the collection of PBMC from large volumes of blood while the plasma fraction is returned to the animal, the relatively small size of macaques can also pose challenges for these labor-intensive, technically demanding procedures (Pathiraja et al. 2013; Donahue et al. 2014).

## 4.2 Challenges in cART Administration to AIDS Virus Infected NHP

The vast majority of proposed HIV-1 functional cure and eradication strategies are predicated on first achieving effective suppression of ongoing viral replication through the administration of antiretroviral drugs, with an adjunctive treatment then used to hopefully activate, eliminate, and/or induce control of the limited number of remaining long-lived viral sources that remain in the face of suppressive therapy (Durand et al. 2012). Thus, essential to the use of NHP/models of AIDS for studies of residual virus is the ability to achieve durable, clinically relevant levels of virologic suppression (i.e., <50 viral RNA [vRNA] copies per ml of plasma) using antiretroviral drugs. However, identifying feasible cART regimens capable of effecting clinically relevant levels of suppression in SIV-infected macaques has proven challenging and these challenges initially hindered the development of such models.

Licensed antiretroviral drugs were specifically developed to inhibit HIV-1 replication in humans. Although HIV and SIV are genetically related, substantial differences in drug potency against SIV compared with HIV have been noted, with many drugs having a lower demonstrable potency against SIV. As a result, to achieve effective suppression of SIV replication in NHPs, antiretroviral drugs must often maintain in vivo drug concentrations that exceed those required to inhibit HIV replication in infected people. Reduced activity against SIV for antiretroviral drugs is perhaps most strikingly illustrated by the non-nucleoside reverse-transcriptase inhibitor (NNRTI) class of drugs, which have essentially no activity against SIV, thereby precluding the use of this entire class of drugs in SIV/NHP studies (Uberla et al. 1995). Although it is currently unclear if the specific antiretroviral drug components of suppressive cART regimens used influence the establishment and maintenance of residual virus sources, or if this will impact the effectiveness of future functional cure/eradication approaches, there are some data to suggest that the specific component drugs used may be important. In a clinical study assessing whether treatment intensification, i.e., the addition of a fourth antiretroviral drug to a standard-of-care three-drug regimen, in the setting of apparently effective suppression with the original regimen, can lead to further declines in measures of apparent viral replication, the addition of an integrase strand transfer inhibitor (INSTI) led to a significant transient increase in 2-LTR circles. Because transient accumulation of 2-LTR circles occurs in the setting of AIDS virus infection when viral integration is blocked, in this case, by the addition of the INSTI to the ART regimen, this finding was consistent with residual viral replication on the standard-of-care regimen (Buzon et al. 2010). However, this result was only seen in patients receiving protease inhibitor-containing cART regimens, but not those receiving NNRTI-containing regimens. Thus, while identifying cART regimens that can suppress plasma viral loads to clinically relevant levels, regardless of the number of drugs required or specific drug classes used, has been the primary focus for developing usable NHP/AIDS models for studies of viral latency and persistence, continuing efforts to identify additional effective regimens for use in these models, and attention to the clinical relevance of different regimens will likely be important.

In addition to challenges posed by lower antiretroviral drug potency against the viruses used in NHP/AIDS research, differences in drug pharmacology between macaques and humans, the species for which the available drugs were specifically developed, can also limit in vivo antiretroviral drug effectiveness. Caution must therefore be exercised when it comes to presuming pharmacological similarities between humans and macaques. For example, in humans, the protease inhibitor ritonavir (RTV) is known to inhibit the cytochrome P450-3A4 metabolic pathway (Zeldin and Petruschke 2004), which is responsible for the clearance of a number of other protease inhibitors and coreceptor antagonists in humans. RTV is therefore often administered to humans at sub-therapeutic doses in combination with other protease inhibitors and coreceptor antagonists, not solely for its antiviral activity but as a boosting agent to increase the concentration of these other drugs by inhibiting their metabolism. Based on these practices in humans and the presumption of comparable pharmacology in macaques, RTV has been empirically administered as a putative boosting agent to rhesus macaques receiving protease inhibitors or coreceptor antagonists, but there have been no published demonstrations of a boosting effect for RTV in macaques. When RTV was withdrawn from a cART regimen containing the protease inhibitor darunavir (DRV), no clear changes in DRV plasma concentrations were observed prior to and after RTV withdrawal in a small number of animals (Del Prete et al. 2015). Differences in drug pharmacology between macaque species are also possible. The NNRTI nevirapine, for example, was ineffective in pigtail macaques against a chimeric virus that was sensitive to nevirapine both in vitro and in cynomolgus macaques in vivo, with different drug pharmacology and bioavailability in pigtail macaques as the likely explanation (Zuber et al. 2001; Ambrose et al. 2004, 2007).

When identifying component antiretroviral drugs to use in a cART regimen in an NHP model, it is therefore important to first assess the sensitivity of the specific virus to be used against the individual component drugs, including defining a plasma-adjusted dose-response curve and/or assessing the drug's instantaneous inhibitory potential (Shen et al. 2008; Deng et al. 2012). By comparing drug concentrations required for inhibition of viral replication in vitro with in vivo pharmacokinetic measurements of drug concentrations in plasma of the specific NHP species to be used, investigators can identify individual drugs, and dosages and dose frequencies of those drugs that may be effective cART components. Monotherapy evaluations of individual component drugs in infected animals can also be a highly useful approach to evaluate the suppressive capacity of individual component drugs, with both declines in plasma viral loads and the selection of drug resistance mutations used as readouts indicating effective drug activity in vivo.

Reduced drug potency coupled with altered drug pharmacology for the viruses and host species used in NHP models of AIDS have often necessitated the need to administer more than three antiretroviral drugs and to administer the drugs at elevated doses relative to those used in HIV-infected humans to effect sufficient levels of virologic suppression. This requirement for relatively elevated drug dosages in macaques increases concerns over potential drug toxicity, which has the capacity to seriously compromise long-term animal studies. Routine monitoring of animal serum chemistries, along with urinalysis and complete blood count (CBC), are therefore critical to identify early evidence of renal, hepatic, or other drug toxicities. Both short and long-term overdosing with the nucleos(t)ide RT inhibitor (NRTI) tenofovir (TFV, PMPA), a drug that has been a keystone component of most cART regimens used in NHPs, has been associated with renal toxicity (Sanders-Beer et al. 2011), though well tolerated, prolonged TFV administration can be achieved with appropriate dosing (Van Rompay et al. 2008). Prompt dose reduction and or discontinuation at the earliest signs of toxicity can limit progressive adverse effects, but some damage may not be reversible. Drug toxicity, manifested as the development of diabetes, was also observed in a subset of animals receiving long-term (>6 months) treatment with the NRTIs didanosine (ddI) and stavudine (d4T), likely due to ddI toxicity (Dunham et al. 2012; Vaccari et al. 2012). Drug–drug interactions can also lead to toxicity issues, as well as alterations of drug levels or effectiveness, which may be of particular concern for cure studies that involve the addition of one or more treatments on top of a cART regimen. An example of severe drug–drug interactions when adjunctive treatments were combined with cART in macaques was observed when animals receiving a cART regimen that included ddI also received the indoleamine 2,3-dioxygenase inhibitor 1-methyl-D-tryptophan and the cytotoxic-T-lymphocyte-associated antigen 4 (CTLA-4) blocking antibody MDX-010, a combination of treatments intended to reduce immune activation and boost responses to a therapeutic vaccine (Hryniewicz et al. 2006; Cecchinato et al. 2008). All animals that received this combination of treatments succumbed to an unanticipated fulminant diabetes and fatal acute pancreatitis, likely an exacerbation of ddI-mediated toxicity induced by the other treatments used (Vaccari et al. 2012).

## 5  Administration of Antiretroviral Drugs to NHPs

Routes of drug dosing can also pose challenges for administration in NHP study animals and are an important consideration when selecting component drugs for a combination regimen. Although orally bioavailable drugs that can be administered in pill form are highly desirable for human therapies, oral administration in NHPs can be difficult for several reasons. First, for drugs that are not available as bulk active pharmaceutical ingredients (APIs), pills must be ground/pulverized for hiding in food or treats and these efforts can be extremely labor-intensive. Second, for any oral drug administration, the drug must be hidden in food or treats that mask any smells or flavors deemed unappealing by the study animals, with frequent rotation of the food/treat to maintain reliable consumption of full intended doses. Third, oral drug administration requires observation of individual animals for dose consumption and may result in only partial dose administration due to the vagaries of ad libitum treat consumption. For these reasons, injectable drug administrations, particularly subcutaneous injections that can be easily performed cageside, are often preferred.

## 5.1   Antiretroviral Drugs That Have Been Utilized in NHPs

Drugs from each of the antiretroviral drug classes used in the clinic to treat HIV-infected humans have been tested in NHP models of AIDS (Fig. 1). NRTI drugs were among those first evaluated in SIV-infected macaques and several of these drugs remain cornerstone components of the most effective, benchmark combination regimens currently in use. The NRTI TFV, initially shown to prevent SIV infection of NHPs in pre-exposure and post-exposure prophylaxis studies (Tsai et al. 1995, 1998) has been used in numerous cART regimens and across multiple different NHP models (North et al. 2005; Dinoso et al. 2009; Radzio et al. 2012; Shytaj et al. 2012; Del Prete et al. 2014b; Ling et al. 2014; Whitney et al. 2014; Del Prete et al. 2015; Fennessey et al. 2017). More recently, a prodrug form of TFV, tenofovir disoproxil fumarate (TDF) has been advanced as an attractive replacement for TFV in cART regimens for NHPs (Del Prete et al. 2016b), as TDF can achieve therapeutic potency with reduced dosages, mitigating concerns over TFV-associated renal toxicities in study animals. The more recently developed tenofovir alafenamide (TAF), another prodrug form of TFV which in humans has greater antiviral potency and a superior renal safety profile than TDF (Markowitz et al. 2014; Sax et al. 2014; Post et al. 2017), may represent an even more attractive option for NHP/AIDS models, though TAF has not yet been evaluated empirically in macaques.

Additional members of the NRTI drug class, such as lamivudine (3TC) and emtricitabine (FTC) have also demonstrated effectiveness to suppress SIV replication in NHPs, as evidenced by reductions in plasma viral loads and the emergence of drug resistance mutations in monotherapy experiments (Van Rompay et al. 2002). Importantly, when TFV was added to FTC monotherapy in animals infected with an FTC resistant SIV, the FTC resistance mutation reverted, indicating that TFV selects against FTC resistance (Murry et al. 2003). This finding, coupled with the relative ease of co-formulating TFV and FTC bulk APIs for administration by a single, daily subcutaneous injection led to the early establishment of TFV and FTC as foundational components of many different cART regimens utilized in NHP/AIDS research. In recent early studies, the NRTI 4′-ethynyl-2-fluoro-2′-deoxyadenosine (EFdA) has also demonstrated both promising potent antiviral activity and favorable pharmacological properties (Murphey-Corb et al. 2012; Stoddart et al. 2015) and shows great potential for future use in NHP cART regimens.

Several different protease inhibitors, including lopinavir (Brandin et al. 2006), indinavir (Bourry et al. 2010; Sellier et al. 2010; Moreau et al. 2012; Fennessey et al. 2017), atazanavir (Dinoso et al. 2009; Graham et al. 2011; Gama et al. 2017) and saquinavir (Dinoso et al. 2009; Graham et al. 2011) have been administered to SIV-infected macaques as components of cART regimens, often as orally administered agents. However, few protease inhibitors have been tested as monotherapy agents, so distinguishing their specific contributions to the overall level of suppression observed for cART regimens in which they are included is problematic. When the protease inhibitor DRV was evaluated in a small monotherapy experiment, transient declines in plasma viral loads were measured with subsequent selection for variants containing a known protease inhibitor resistance mutation (Del Prete et al. 2016b), denoting clear in vivo anti-SIV activity for this protease inhibitor. Several different INSTIs have also demonstrated activity against SIV replication, including the clinically approved drug raltegravir (Lewis et al. 2010) and the unlicensed investigational compound L-870812 (Hazuda et al. 2004). These drugs, as well as the unlicensed investigational compound L-900564 (Del Prete et al. 2014b) and the approved INSTI dolutegravir (DTG) (Whitney et al. 2014; Del Prete et al. 2015, 2016b) have all been effectively administered to SIV-infected rhesus macaques. Recently, a highly effective well tolerated, three-drug cART regimen was developed in which the component drugs were co-formulated for administration of all three agents via a single once daily subcutaneous injection (Del Prete et al. 2016b). When initiated during the acute phase of infection, this regimen, which contains the NRTIs TDF and FTC plus the INSTI DTG, effectively and durably suppressed SIVmac239 replication in Indian-origin rhesus macaques to <15 vRNA copies/ml plasma. In a recent monotherapy evaluation, DTG was effective against SIV replication in rhesus macaques (K. Van Rompay, personal communication), and it is likely that DTG contributes substantially to the effectiveness of TDF/FTC/DTG regimen, as two NRTIs alone are usually insufficient to suppress SIVmac replication to below plasma viral load assay detection limits and drug resistance to DTG results in a substantial replicative fitness cost for SIV (Hassounah et al. 2014).

## 5.2 Future cART Regimen Development for NHPs

The development of a feasible, well-tolerated cART regimen that can effect clinically relevant levels of virologic suppression in macaques infected with pathogenic SIVs has been a critically important step toward accelerating the establishment and utilization of NHP models for studies of in vivo viral latency and persistence and to evaluate proposed viral remission strategies. However, further efforts to develop additional cART regimens, utilizing alternative drugs and drug classes with formulations well suited for use in NHP studies will likely continue to be important; while several drugs from the entry inhibitor class, such as the CCR5 antagonists maraviroc and CMPD167, have also been utilized in NHP models (Fig. 1), these compounds have been used most often in the setting of infection prevention studies, with drug delivery through topical microbicides and intravaginal rings (Veazey et al. 2005a, b, 2010; Tsibris et al. 2011; Malcolm et al. 2012; Dobard et al. 2015).

Systemic administration of entry inhibitors in animals already virally infected has been more limited, and it is not clear if entry inhibitors can meaningfully contribute to virologic suppression with feasible dosage levels and frequencies, particularly if administered orally.

Long-acting formulations of inhibitors, comprising nanosuspensions of antiretroviral drugs that may be able to maintain therapeutic drug concentrations with much more infrequent dosing than traditional drug formulations, may also be of practical value for NHP research involving long-term administration of cART (Fig. 1). In addition to allowing for less frequent administrations of one or more components of a cART regimen, long-acting drug formulations may allow for the use of "maintenance" drug regimens with considerable practical advantages. Conceptually, such a regimen would involve long-acting drugs plus additional component drugs that require daily or even twice daily administrations during initial treatment ("induction phase"). The conventional drugs might then be discontinued and only the long-acting drugs administered once animals have been stably suppressed to desirable levels ("maintenance phase"). In such a scenario, animals could theoretically remain well suppressed with drug administered as infrequently as once every several weeks, though the feasibility of such an approach has not yet been formally demonstrated. However, recent successful evaluations of systemically administered long-acting drug formulations for pre-exposure prophylaxis have suggested promise for the pharmacokinetic properties of long-acting drug formulations in NHPs (Andrews et al. 2014, 2015; Radzio et al. 2015; Andrews et al. 2017). Prolonged and variable decay of drug levels following discontinuation of administration of long-acting formulations may complicate interpretation of results in studies involving evaluation of viral rebound parameters after cART discontinuation as a key readout.

# 6  Alternative Models

## 6.1  Chinese-Origin Rhesus Macaques

Virologic suppression on cART is the net result of all forces acting against viral replication, including both pharmacologic and immunologic pressures (Cartwright et al. 2016). Although the consistently high levels of viral replication and associated consistent, robust pathogenesis in an experimentally tractable timeframe in the best-characterized NHP/AIDS models have been viewed as some of the more attractive features of these models for researchers, they may have hindered the identification and development of effective cART regimens for use in these models, as pretreatment viral loads have been inversely associated with the ability to suppress viral replication (Phillips et al. 2001; Matthews et al. 2002; Maldarelli et al. 2007; European Collaborative et al. 2007; Read et al. 2012; Tanner et al. 2016; Snippenburg et al. 2017). Plasma viral loads in SIVmac239/251-infected

Indian-origin rhesus macaques are generally 10–100-fold higher than the typical untreated HIV-1-infected human and thus suppression of viral replication to clinically relevant levels represents a significant challenge. Efforts to circumvent this problem have included selecting already-infected animals with atypically low pretreatment plasma viral loads or animal models where untreated animals tend to have lower viral loads. Following infection with the most commonly used SIVs, rhesus macaques of Chinese origin (Fig. 1) have acute and chronic phase plasma viral loads that are often several orders of magnitude lower than similarly infected Indian-origin rhesus macaques (Joag et al. 1994; Marthas et al. 2001; Ling et al. 2002; Trichel et al. 2002; Jasny et al. 2012), likely because the available, best-characterized SIVs have been specifically adapted for infection of Indian-origin rhesus (Burdo et al. 2005). Although the biological mechanisms underlying the reduced SIV replication levels observed in Chinese-origin rhesus have not been clearly delineated, suppression of viral replication in these animals is routinely achievable using three or even just two antiretroviral drugs (Vagenas et al. 2010; Jasny et al. 2012; Ling et al. 2014). The greater ease of achieving clinically relevant suppression within this rhesus macaque subspecies represents a benefit for study feasibility, however this is balanced by the far lower availability of Chinese rhesus animals, particularly in US primate facilities, the much lower historical use of Chinese rhesus, and the associated far more limited availability of historical data, immunologic reagents and genetic characterization.

## 6.2   RT-SHIV Infections

Additional efforts to improve the practical feasibility of cART-mediated suppression of viral replication in NHPs have included the development of chimeric viruses comprising an SIV genome with the RT gene region replaced by the corresponding gene from HIV. The incorporation of an HIV RT gene renders these chimeric RT simian-human immunodeficiency viruses (RT-SHIVs) susceptible to suppression by NNRTI drugs (Uberla et al. 1995; Balzarini et al. 1995, 1997; Ambrose et al. 2004), thereby enabling the use of this drug class in NHP/AIDS studies in which the animals are RT-SHIV-infected (Fig. 1). Monotherapy evaluations of the NNRTI efavirenz demonstrated clear virologic activity, evidenced by viral load declines and the emergence of drug resistance mutations, in rhesus and pigtail macaques infected with species-specific RT-SHIVs (Hofman et al. 2004; Ambrose et al. 2007). In the setting of combination therapy, efavirenz plus two NRTI drugs (TFV and either FTC or 3TC) can effectively suppress RT-SHIV replication to clinically relevant levels (North et al. 2005; Ambrose et al. 2007; Deere et al. 2010; North et al. 2010). However, plasma viral loads in untreated RT-SHIV-infected macaques tend to be lower and more inconsistent than SIVmac-infected animals, especially in the chronic phase of infection, and while some relevant data have been generated (Deere et al. 2014) further work is required to establish the nature and size of the residual virus pool in RT-SHIV-infected animals on suppressive therapy as well as assess RT-SHIV recrudescence when cART is halted. Minimally chimeric HIV viruses, which consist of a genome predominantly composed of HIV sequence (Hatziioannou et al. 2009; Thippeshappa et al. 2011; Hatziioannou et al. 2014), may also allow for the use of all antiretroviral drug classes, including NNRTIs (unpublished data), that are effective against HIV in NHP/AIDS models, though further model development and characterization is required.

## 6.3 SIVagm.sab Infections

Some researchers are also pursuing the use of NHP models where viral infection is spontaneously controlled without any interventions or antiretroviral drugs, thereby obviating the need for antiretroviral therapy altogether. When Indian-origin rhesus macaques are infected with SIVagm.sab, an SIV derived from African green monkeys, the virus replicates robustly during the acute phase of infection, with peak viral loads $>10^8$ vRNA copies/ml plasma, and with marked depletion of $CD4^+$ T cells in gut-associated lymphoid tissue, as is seen is classically pathogenic HIV and SIV infections (Pandrea et al. 2011). However, over the ensuing 2–3 months, plasma viral loads decline in all animals to below assay quantification limits, with progressive recovery of $CD4^+$ T cells in the gut, in the absence of any treatment. Despite this profound degree of virologic control, plasma viremia transiently returns following experimental depletion of $CD8^+$ T cells (Pandrea et al. 2011), and transfer of this recrudescent plasma virus to naïve animals results in a productive infection (Ma et al. 2015), indicating that the SIVagm.sab-infected macaques control but do not eradicate their infection. Although the SIVagm.sab/rhesus macaque system may better model elite control of HIV-1 infection than cART-mediated control of viral infection, it may prove useful for evaluations of approaches designed to reactivate latent sources of virus and to examine the impact of viral reactivation in the setting of an effective antiviral immune response on the size of residual virus pool (Policicchio et al. 2016).

## 6.4 CNS Infection Models

There have also been efforts to adapt NHP AIDS models originally developed for studying narrower, specific areas of HIV biology for use in studies involving cART-mediated suppression and persistent viral reservoirs. The development of CNS disease and established CNS infection is inconsistent or understudied in most models consisting of rhesus macaques infected with standard SIVmac viruses. The possibility that virally infected cells within the CNS may represent a significant source of latent and persistent virus that may be particularly difficult to purge or eliminate, coupled with the challenges of sampling CNS tissue from HIV-infected people, has provided the impetus to adapt an NHP model of lentivirus-mediated

CNS disease for cure research. One model system, which results in extremely rapid disease progression in the absence of cART, characterized by consistent brain infection and encephalitis in the majority of infected animals, involves the co-infection of pigtail macaques with a mixed inoculum containing the aggressively immunosuppressive viral swarm SIV/deltaB670 and the macrophage-tropic infectious clone SIV/17E-Fr (Zink et al. 1999), (Fig. 1). Although the accelerated pathogenesis of deltaB670/17E-Fr infection in pigtail macaques necessitates the use of early cART initiation (i.e., 12 days post-infection) (Dinoso et al. 2009; Zink et al. 2010; Graham et al. 2011), the consistent brain infection achieved with this model make it well suited for feasibly assessing potential long-lived CNS reservoirs of infectious virus during suppressive therapy (Clements et al. 2005; Queen et al. 2011; Gama et al. 2017), more so than other current NHP models of HIV infection where CNS involvement is less consistent, predictable, or frequent.

## 6.5   Envelope SHIVs

Still other NHP models of AIDS are being developed and adapted for use in HIV cure research because they enable researchers to interrogate types of interventions that are specific for HIV and would be less effective or ineffective against SIV. Envelope SHIVs (Fig. 1), which in an oversimplification can be considered as containing an HIV-1 env gene in an SIV backbone and are typically referred to simply as "SHIVs", were originally developed to provide in vivo NHP-based experimental platforms for the direct testing of vaccination and treatment approaches targeting the HIV-1-envelope, in particular antibodies (Del Prete et al. 2016a). Initially, these models were used almost exclusively in vaccination/prevention studies where acquisition of infection was the primary endpoint. While many of the developed and available SHIVs are mucosally transmissible in macaques, they often establish inconsistent chronic phase levels of viral replication and viremia with inconsistent pathogenicity. However, for studies evaluating the impact of strategies intended to target relevant replication-competent viral reservoirs consistently capable of leading to recrudescent viral infection when cART is withdrawn, SHIVs that can consistently maintain elevated viral loads in the absence of therapy, with authentic pathogenesis and target cell infection, are critical. The limited number of early generation CCR5-tropic SHIVs typically contained HIV-1 env genes obtained from chronically infected humans and initially replicated poorly in macaques, requiring multiple animal-to-animal serial passages to improve in vivo viral replication (Tan et al. 1999; Song et al. 2006; Nishimura et al. 2010). Although several of these first-generation animal-passaged SHIVs have been very useful for in vivo infection prevention studies, they represent limited HIV-1 Env diversity and, despite multiple serial passages in macaques, many of these adapted SHIVs nevertheless demonstrate inconsistent chronic phase replication levels. As such, the use of SHIVs for NHP-based HIV cure research has been relatively limited, despite the field's ever-growing and ever-improving collection of potent, broadly neutralizing anti-HIV-1 Env antibodies that could play a role in a successful off-cART virus remission or eradication strategy.

More recently, several groups have generated new SHIVs containing transmitted-founder Envs (i.e., those that initiated authentic human infections), with and without targeted modifications, from multiple different HIV-1 subtypes and demonstrated their in vivo infectivity in the absence of animal-to-animal serial passage, allowing for the preservation of their original neutralization sensitivity, antigenicity, and other properties (Del Prete et al. 2014a; Asmal et al. 2015; Chang et al. 2015; Li et al. 2016; Tartaglia et al. 2016). Several of these SHIVs appear promising for their ability to maintain chronic phase viremia (Del Prete et al. 2014a; Li et al. 2016) with authentic pathogenesis, and may be useful for future HIV eradication/functional cure research evaluating interventions that target the HIV-1 Env.

# 7 Virologic Assays

## 7.1 PCR and RT-PCR-Based Assays

As is the case for HIV cure research conducted in humans, sensitive and accurate virologic assays are a critical element for studies of residual virus and for the assessment of viral eradication and functional cure strategies in NHPs (Fig. 1). There are multiple different assays and approaches that are used to quantify the residual virus pool in HIV-infected people on cART, and a similar suite of methods, plus some additional approaches not feasible in the context of human studies, have been developed to quantify the residual virus pool in SIV-infected macaques. The most commonly used assays for assessments of SIV viral loads are real-time quantitative PCR (qPCR) and RT-PCR (qRT-PCR) methods, which are used to quantify viral DNA (vDNA) and vRNA, respectively, in plasma, cells, and tissues (Hansen et al. 2011; Li et al. 2016; Del Prete et al. 2014b, 2015). Although standard clinical assays used for quantification of HIV RNA in plasma have sensitivity cutoffs of 50 vRNA copies/ml, more sensitive assays, including single copy assays, have been developed for experimental purposes (Palmer et al. 2003; Maldarelli et al. 2007; Somsouk et al. 2014; Caskey et al. 2015; Elliott et al. 2015; Lynch et al. 2015). These more sensitive assays have played an important role in furthering our understanding of residual virus sources and have revealed that even in well-suppressed patients, low-level viremia is typically present (Dornadula et al. 1999; Fischer et al. 2000; Maldarelli et al. 2007; Palmer et al. 2008), likely the result of occasional reactivations of latent virus sources or persistent production of virus from pharmacologic or immunologic sanctuary sites. The development of similarly sensitive assays, which achieve greater clinical sensitivity by assaying larger volumes of source plasma and by using digital assay approaches (Hansen et al. 2013), has thus been critically important for the development of usable NHP models for studies of residual virus. Standard qRT-PCR assays can now quantify plasma vRNA in macaque plasma down to 15 vRNA copies/ml using less than 1 ml of input plasma volume (Li et al. 2016), and ultrasensitive assays have been developed that can quantify vRNA below 1 vRNA copy/ml (Hansen et al. 2013), provided enough plasma volume is available from a given study time point.

Sensitive quantitative PCR and RT-PCR assays, including nested amplification assays, for measuring cell- and tissue-associated viral loads in samples from SIV-infected NHPs have also been developed and applied effectively (Hansen et al. 2011, 2013; Del Prete et al. 2014b; Liu et al. 2016). These assays, which can determine vRNA and vDNA copy numbers within the same sample, with both values normalized to the number of assayed cells, based on quantification of a cellular gene in the same extracted nucleic acid sample, allow for several important assessments. First, by determining cell-associated vRNA and vDNA, scientists can evaluate the overall virus burden within infected animals, in PBMCs and across the multiple tissues that can be readily sampled in the context of NHP studies. In addition, by assessing changes within these parameters, the impact of proposed latency reversing agents and other reservoir reactivating agents can be evaluated, with changes in the ratio of vRNA and vDNA signals serving as an approach to identify signal changes that are likely due to changes in viral transcription (Del Prete et al. 2014b). It is important to note that the sensitive, specific detection of viral nucleic acid sequences in cells or tissues does not necessarily imply the presence of a replication-competent proviral genome, due to the frequent occurrence and accumulation of deletions and/or missense mutations via APOBEC mediated hypermutation during viral replication (Bruner et al. 2016), and results of such testing must be interpreted in this context. Newer multiplexed assays testing multiple regions of the viral genome are in development and may help to address this challenge in part.

## 7.2 In Situ Approaches

Although PCR-based approaches are robust and powerful, they require homogenization of tissues and thus a loss of tissue architecture and anatomical structures housing virus signals, which may be important and informative. In situ hybridization (ISH) approaches provide this complementary information and have been developed to probe for SIV nucleic acids in fixed tissues, where the specific anatomic compartmentalization of virus-infected cells and virions may be examined (Brenchley et al. 2012; Fukazawa et al. 2015; Deleage et al. 2016). When coupled in multiplex formats with immunohistochemical or immunofluorescence based approaches to identify the phenotype of cells harboring vRNA and vDNA, these methods can be used to identify both active and latent viral reservoirs and their specific anatomical locations. Such approaches are particularly valuable for identifying immunologic and/or pharmacologic sanctuary sites for virus in the setting of suppressive cART and allow for the evaluation of approaches designed to specifically target certain sites of virus persistence, including T follicular helper cells, the follicular dendritic cell network, CNS reservoirs, and others. Recently, highly sensitive next generation ISH approaches that provide exquisite sensitivity and specificity, with dramatically improved assay throughput have been described for the quantitative detection of both vRNA (RNAScope) and vDNA (DNAScope) (Wang et al. 2012; Fukazawa et al. 2015; Deleage et al. 2016). This advance in tissue analysis is particularly valuable for NHP studies, where tissue samples can be abundantly available, greatly enhancing one of the primary strengths of animal studies.

## 7.3　Other Virologic Assessments

By applying sequencing approaches combined with fragmentation of genomic DNA, recent studies have identified multiple instances of identical proviral integration sites in samples from HIV-infected people on cART (Maldarelli et al. 2014; Wagner et al. 2014; Cohn et al. 2015; Kearney et al. 2015; Simonetti et al. 2016). This finding represents convincing evidence for clonal expansion of virally infected cells because independent instances of integration into the identical site following multiple de novo infections is exceedingly unlikely. Similar integration site analysis based evaluations of clonal expansion of infected cells are also feasible in NHPs on cART, with such efforts greatly facilitated by recent improvements in the annotation of the published rhesus macaque genome (Rosenbloom et al. 2015). Initial efforts to examine clonal expansion of infected cells in NHPs have determined that this biological phenomenon is recapitulated in SIV-infected rhesus macaques on cART, with evidence that expanded clones are established early post-infection (as early as 4 weeks post-infection) and that identical clones can be found in blood and in tissues (Hughes SL et al., manuscript in preparation). These findings and the associated methods set the stage for examinations of the mechanisms underlying clonal expansion of infected cells, the role that expanded clones play in viral rebound following cART cessation, and the evaluation of interventions designed to target the establishment and outgrowth of this newly identified and challenging viral reservoir.

These various ex vivo evaluations of viral nucleic acids are important for comprehensive assessments of the residual virus pool in virologically suppressed NHPs and play important roles for understanding the mechanisms underlying interventions intended to activate, purge, or control viral reservoirs. However, none of these methods specifically examine the replication-competent residual virus pool that can lead to recrudescent infection when suppressive therapy is discontinued. To this end, quantitative, limiting dilution ex vivo virus outgrowth assays have been developed that allow for an estimation of the frequency of cells harboring replication-competent viral genomes (Shen et al. 2003, 2007; Hansen et al. 2013). A costlier, but perhaps more sensitive approach that has been effectively implemented in NHP models involves the adoptive transfer of large numbers of cells from suppressed/putatively cured animals into naïve hosts to reveal the presence of replication-competent virus capable of initiating de novo infection (Hansen et al. 2013). Although these assays have merits, they are costly and time-consuming and still may not necessarily capture virus sources that are likely to lead to actual viral recrudescence nor do they allow for evaluations of the frequency of cells harboring functional viral genomes that may be impacted by proposed virus reactivation methods. Limiting dilution culture assays that measure only extracellular vRNA following stimulation or treatment with proposed virus reactivation compounds may represent a more feasible option that is intermediate to functional virus outgrowth and strict nucleic acid based methods (Beliakova-Bethell et al. 2017; Del Prete et al., unpublished data). The recent application of a high sensitivity digital bead based immunoassay approach has allowed highly sensitive detection of HIV-1 p24 protein in the supernatants of viral outgrowth cultures (Cabrera et al. 2015; Passaes et al. 2017; Bosque et al. 2017). Because the adaptive immune system typically recognizes protein, not nucleic acid, even though viral RNA in high speed centrifugation pellets from culture supernatants may reasonably be expected to be virion associated, measurements of viral protein may provide a complementary and arguably more relevant measure of virus induction for agents intended to induce viral expression leading to immunological clearance.

Ultimately, cessation of cART with longitudinal monitoring for virologic rebound remains the most germane experimental readout available to evaluate functional residual virus and perturbations of the replication-competent residual virus pool. While this approach can reveal definitive evidence of a functional cure or viral remission—a lack of viral recrudescence in the absence of cART—it can also reveal reductions in the size of the residual virus pool indicated by significant delays in the time to viral recrudescence. However, mathematical modeling studies have suggested that substantial reductions, on the order of several logs, in the functional virus reservoir may be needed to measurably alter and reliably interpret delays in the time to measurable rebound viremia (Hill et al. 2014; Pinkevych et al. 2015; Hill et al. 2016). Therefore, interventions that are partially effective may not be evident, even if they effect sizable reductions in the residual virus pool. Given the likely stochastic nature of viral reactivation events, very large numbers of study animals are thus required to demonstrate a statistically significant change in the time to rebound and the associated functional size of the residual virus pool.

The recent development of SIVmac239M, a molecularly barcoded synthetic viral swarm based on SIVmac239 may address this issue (Fennessey et al. 2017). This virus allows for the convenient and feasible use of next generation deep sequencing methods to quantify the size of the residual virus reservoir and highlights the potential experimental utility of NHP models of AIDS for HIV cure research. By infecting animals with hundreds to thousands of individual virus clones that are isogenic apart from a small inserted sequence of random nucleotides (i.e., a "barcode"), researchers can track chains of infection events involving individual virus variants within each infected animal through the use of deep sequencing technology on this small region of the viral genome. Importantly, because there are no apparent functional differences between the various barcoded clones within the swarm, researchers can use the number and relative proportion of each virus clone within the recrudescent plasma viremia as a surrogate marker for the relative timing of reactivation for that clone, with the highest proportion clone likely the first virus to have reactivated. This ability to distinguish multiple reactivation events within each animal allows for greater statistical power to be derived from a smaller number of animals. Notably, the approach of inserting a barcode into the SIV genome could potentially be applied to other viruses of interest, including SHIVs.

## 8  Concluding Remarks

Until very recently, the use of NHP models for studies of AIDS virus persistence in the face of suppressive therapy has been relatively limited, due in large part to the challenges in using these models discussed earlier. However, concerted efforts by multiple laboratories to develop and establish appropriate models and reagents, feasible and effective drug regimens, and a range of sensitive virologic assays have greatly improved the potential utility of NHP models for HIV cure research. Initial studies have been performed to evaluate candidate latency reversing agents in the setting of cART suppression, with broadly promising initial agreement between human and NHP studies evaluating the same agents (Del Prete et al. 2014b, 2015). With these new tools and methods, NHP models are well positioned to play a key role in improving our understanding of the nature, quantity, and distribution of residual sources of virus in vivo and in evaluating the safety and activity of proposed strategies designed to reduce, eliminate, or control these sources in the absence of antiretroviral therapy.

**Financial Support** The authors were supported with federal funds from the National Cancer Institute, National Institutes of Health under contract HHSN261200800001E. The content of this publication does not necessarily reflect the views or policies of the Department of Health and Human Services, nor does mention of trade names, commercial products, or organizations imply endorsement by the US Government.

## References

Al-Dakkak I, Patel S, McCann E, Gadkari A, Prajapati G, Maiese EM (2013) The impact of specific HIV treatment-related adverse events on adherence to antiretroviral therapy: a systematic review and meta-analysis. AIDS care 25(4):400–414. doi:10.1080/09540121.2012. 712667

Ambrose Z, Boltz V, Palmer S, Coffin JM, Hughes SH, Kewalramani VN (2004) In vitro characterization of a simian immunodeficiency virus-human immunodeficiency virus (HIV) chimera expressing HIV type 1 reverse transcriptase to study antiviral resistance in pigtail macaques. J Virol 78(24):13553–13561

Ambrose Z, Palmer S, Boltz VF, Kearney M, Larsen K, Polacino P, Flanary L, Oswald K, Piatak M Jr, Smedley J, Shao W, Bischofberger N, Maldarelli F, Kimata JT, Mellors JW, Hu SL, Coffin JM, Lifson JD, KewalRamani VN (2007) Suppression of viremia and evolution of human immunodeficiency virus type 1 drug resistance in a macaque model for antiretroviral therapy. J Virol 81(22):12145–12155

Ananworanich J, Chomont N, Eller LA, Kroon E, Tovanabutra S, Bose M, Nau M, Fletcher JL, Tipsuk S, Vandergeeten C, O'Connell RJ, Pinyakorn S, Michael N, Phanuphak N, Robb ML, Rv, groups RSs (2016) HIV DNA set point is rapidly established in acute HIV infection and dramatically reduced by early ART. EBioMedicine 11:68–72. doi:10.1016/j.ebiom.2016.07. 024

Andrews CD, Spreen WR, Mohri H, Moss L, Ford S, Gettie A, Russell-Lodrigue K, Bohm RP, Cheng-Mayer C, Hong Z, Markowitz M, Ho DD (2014) Long-acting integrase inhibitor protects macaques from intrarectal simian/human immunodeficiency virus. Science 343 (6175):1151–1154. doi:10.1126/science.1248707

Andrews CD, Yueh YL, Spreen WR, St Bernard L, Boente-Carrera M, Rodriguez K, Gettie A, Russell-Lodrigue K, Blanchard J, Ford S, Mohri H, Cheng-Mayer C, Hong Z, Ho DD, Markowitz M (2015) A long-acting integrase inhibitor protects female macaques from repeated high-dose intravaginal SHIV challenge. Sci Transl Med 7(270):270ra274. doi:10.1126/ scitranslmed.3010298

Andrews CD, Bernard LS, Poon AY, Mohri H, Gettie N, Spreen WR, Gettie A, Russell-Lodrigue K, Blanchard J, Hong Z, Ho DD, Markowitz M (2017) Cabotegravir long acting injection protects macaques against intravenous challenge with SIVmac251. AIDS 31(4):461–467. doi:10.1097/QAD.0000000000001343

Antiretroviral Therapy Cohort Collaboration (2017) Survival of HIV-positive patients starting antiretroviral therapy between 1996 and 2013: a collaborative analysis of cohort studies. Lancet HIV. doi:10.1016/S2352-3018(17)30066-8

Arthur LO, Bess JW Jr, Urban RG, Strominger JL, Morton WR, Mann DL, Henderson LE, Benveniste RE (1995) Macaques immunized with HLA-DR are protected from challenge with simian immunodeficiency virus. J Virol 69(5):3117–3124

Asmal M, Luedemann C, Lavine CL, Mach LV, Balachandran H, Brinkley C, Denny TN, Lewis MG, Anderson H, Pal R, Sok D, Le K, Pauthner M, Hahn BH, Shaw GM, Seaman MS, Letvin NL, Burton DR, Sodroski JG, Haynes BF, Santra S (2015) Infection of monkeys by simian-human immunodeficiency viruses with transmitted/founder clade C HIV-1 envelopes. Virology 475:37–45. doi:10.1016/j.virol.2014.10.032

Balzarini J, Weeger M, Camarasa MJ, De Clercq E, Uberla K (1995) Sensitivity/resistance profile of a simian immunodeficiency virus containing the reverse transcriptase gene of human immunodeficiency virus type 1 (HIV-1) toward the HIV-1-specific non-nucleoside reverse transcriptase inhibitors. Biochem Biophys Res Commun 211(3):850–856

Balzarini J, De Clercq E, Uberla K (1997) SIV/HIV-1 hybrid virus expressing the reverse transcriptase gene of HIV-1 remains sensitive to HIV-1-specific reverse transcriptase inhibitors after passage in rhesus macaques. J Acquir Immune Defic Syndr Hum Retrovirol 15(1):1–4

Barouch DH, Ghneim K, Bosche WJ, Li Y, Berkemeier B, Hull M, Bhattacharyya S, Cameron M, Liu J, Smith K, Borducchi E, Cabral C, Peter L, Brinkman A, Shetty M, Li H, Gittens C, Baker C, Wagner W, Lewis MG, Colantonio A, Kang HJ, Li W, Lifson JD, Piatak M Jr, Sekaly RP (2016) Rapid inflammasome activation following mucosal siv infection of rhesus monkeys. Cell 165(3):656–667. doi:10.1016/j.cell.2016.03.021

Baskin GB, Murphey-Corb M, Watson EA, Martin LN (1988) Necropsy findings in rhesus monkeys experimentally infected with cultured simian immunodeficiency virus (SIV)/delta. Vet Pathol 25(6):456–467. doi:10.1177/030098588802500609

Beliakova-Bethell N, Hezareh M, Wong JK, Strain MC, Lewinski MK, Richman DD, Spina CA (2017) Relative efficacy of T cell stimuli as inducers of productive HIV-1 replication in latently infected CD4 lymphocytes from patients on suppressive cART. Virology 508:127–133. doi:10. 1016/j.virol.2017.05.008

Blakley GB, Beamer TW, Dukelow WR (1981) Characteristics of the menstrual cycle in nonhuman primates. IV. Timed mating in Macaca nemestrina. Lab Anim 15(4):351–353

Borsetti A, Baroncelli S, Maggiorella MT, Bellino S, Moretti S, Sernicola L, Belli R, Ridolfi B, Farcomeni S, Negri DR, Cafaro A, Ensoli B, Titti F (2008) Viral outcome of simian-human immunodeficiency virus SHIV-89.6P adapted to cynomolgus monkeys. Adv Virol 153(3):463–472. doi:10.1007/s00705-007-0009-2

Bosque A, Nilson KA, Macedo AB, Spivak AM, Archin NM, Van Wagoner RM, Martins LJ, Novis CL, Szaniawski MA, Ireland CM, Margolis DM, Price DH, Planelles V (2017) Benzotriazoles reactivate latent HIV-1 through inactivation of STAT5 SUMOylation. Cell Rep 18(5):1324–1334. doi:10.1016/j.celrep.2017.01.022

Bourry O, Mannioui A, Sellier P, Roucairol C, Durand-Gasselin L, Dereuddre-Bosquet N, Benech H, Roques P, Le Grand R (2010) Effect of a short-term HAART on SIV load in macaque tissues is dependent on time of initiation and antiviral diffusion. Retrovirology 7:78. doi:10.1186/1742-4690-7-78

Brandin E, Thorstensson R, Bonhoeffer S, Albert J (2006) Rapid viral decay in simian immunodeficiency virus-infected macaques receiving quadruple antiretroviral therapy. J Virol 80(19):9861–9864

Brenchley JM, Price DA, Schacker TW, Asher TE, Silvestri G, Rao S, Kazzaz Z, Bornstein E, Lambotte O, Altmann D, Blazar BR, Rodriguez B, Teixeira-Johnson L, Landay A, Martin JN, Hecht FM, Picker LJ, Lederman MM, Deeks SG, Douek DC (2006) Microbial translocation is a cause of systemic immune activation in chronic HIV infection. Nat Med 12(12):1365–1371. doi:10.1038/nm1511

Brenchley JM, Vinton C, Tabb B, Hao XP, Connick E, Paiardini M, Lifson JD, Silvestri G, Estes JD (2012) Differential infection patterns of CD4+ T cells and lymphoid tissue viral burden distinguish progressive and nonprogressive lentiviral infections. Blood 120(20):4172–4181

Brennan G, Kozyrev Y, Hu SL (2008) TRIMCyp expression in old world primates macaca nemestrina and macaca fascicularis. Proc Natl Acad Sci USA 105(9):3569–3574. doi:10.1073/pnas.0709511105

Bruner KM, Murray AJ, Pollack RA, Soliman MG, Laskey SB, Capoferri AA, Lai J, Strain MC, Lada SM, Hoh R, Ho YC, Richman DD, Deeks SG, Siliciano JD, Siliciano RF (2016) Defective proviruses rapidly accumulate during acute HIV-1 infection. Nat Med 22(9):1043–1049. doi:10.1038/nm.4156

Burdo TH, Marcondes MC, Lanigan CM, Penedo MC, Fox HS (2005) Susceptibility of Chinese rhesus monkeys to SIV infection. Aids 19(15):1704–1706

Burns DP, Collignon C, Desrosiers RC (1993) Simian immunodeficiency virus mutants resistant to serum neutralization arise during persistent infection of rhesus monkeys. J Virol 67(7):4104–4113

Buzon MJ, Massanella M, Llibre JM, Esteve A, Dahl V, Puertas MC, Gatell JM, Domingo P, Paredes R, Sharkey M, Palmer S, Stevenson M, Clotet B, Blanco J, Martinez-Picado J (2010) HIV-1 replication and immune dynamics are affected by raltegravir intensification of HAART-suppressed subjects. Nat Med 16(4):460–465. doi:10.1038/nm.2111

Cabrera C, Chang L, Stone M, Busch M, Wilson DH (2015) Rapid, fully automated digital immunoassay for p24 protein with the sensitivity of nucleic acid amplification for detecting acute HIV infection. Clin Chem 61(11):1372–1380. doi:10.1373/clinchem.2015.243287

Cain BT, Pham NH, Budde ML, Greene JM, Weinfurter JT, Scarlotta M, Harris M, Chin E, O'Connor SL, Friedrich TC, O'Connor DH (2013) T cell response specificity and magnitude against SIVmac239 are not concordant in major histocompatibility complex-matched animals. Retrovirology 10:116. doi:10.1186/1742-4690-10-116

Cartwright EK, Spicer L, Smith SA, Lee D, Fast R, Paganini S, Lawson BO, Nega M, Easley K, Schmitz JE, Bosinger SE, Paiardini M, Chahroudi A, Vanderford TH, Estes JD, Lifson JD, Derdeyn CA, Silvestri G (2016) CD8(+) lymphocytes are required for maintaining viral suppression in siv-infected macaques treated with short-term antiretroviral therapy. Immunity 45(3):656–668. doi:10.1016/j.immuni.2016.08.018

Caskey M, Klein F, Lorenzi JC, Seaman MS, West AP Jr, Buckley N, Kremer G, Nogueira L, Braunschweig M, Scheid JF, Horwitz JA, Shimeliovich I, Ben-Avraham S, Witmer-Pack M, Platten M, Lehmann C, Burke LA, Hawthorne T, Gorelick RJ, Walker BD, Keler T, Gulick RM, Fatkenheuer G, Schlesinger SJ, Nussenzweig MC (2015) Viraemia suppressed in HIV-1-infected humans by broadly neutralizing antibody 3BNC117. Nature 522(7557):487–491. doi:10.1038/nature14411

Cecchinato V, Tryniszewska E, Ma ZM, Vaccari M, Boasso A, Tsai WP, Petrovas C, Fuchs D, Heraud JM, Venzon D, Shearer GM, Koup RA, Lowy I, Miller CJ, Franchini G (2008) Immune activation driven by CTLA-4 blockade augments viral replication at mucosal sites in simian immunodeficiency virus infection. J Immunol 180(8):5439–5447

Chakrabarti L, Guyader M, Alizon M, Daniel MD, Desrosiers RC, Tiollais P, Sonigo P (1987) Sequence of simian immunodeficiency virus from macaque and its relationship to other human and simian retroviruses. Nature 328(6130):543–547. doi:10.1038/328543a0

Chang HW, Tartaglia LJ, Whitney JB, Lim SY, Sanisetty S, Lavine CL, Seaman MS, Rademeyer C, Williamson C, Ellingson-Strouss K, Stamatatos L, Kublin J, Barouch DH (2015) Generation and evaluation of clade C simian-human immunodeficiency virus challenge stocks. J Virol 89(4):1965–1974. doi:10.1128/JVI.03279-14

Chun TW, Carruth L, Finzi D, Shen X, DiGiuseppe JA, Taylor H, Hermankova M, Chadwick K, Margolick J, Quinn TC, Kuo YH, Brookmeyer R, Zeiger MA, Barditch-Crovo P, Siliciano RF (1997) Quantification of latent tissue reservoirs and total body viral load in HIV-1 infection. Nature 387(6629):183–188

Cleary S, McIntyre D (2010) Financing equitable access to antiretroviral treatment in South Africa. BMC Health Serv Res 10(Suppl 1):S2. doi:10.1186/1472-6963-10-s1-s2

Clements JE, Li M, Gama L, Bullock B, Carruth LM, Mankowski JL, Zink MC (2005) The central nervous system is a viral reservoir in simian immunodeficiency virus–infected macaques on combined antiretroviral therapy: a model for human immunodeficiency virus patients on highly active antiretroviral therapy. J Neurovirol 11(2):180–189

Clements JE, Gama L, Graham DR, Mankowski JL, Zink MC (2011) A simian immunodeficiency virus macaque model of highly active antiretroviral treatment: viral latency in the periphery and the central nervous system. Curr Opin HIV AIDS 6(1):37–42

Cohn LB, Silva IT, Oliveira TY, Rosales RA, Parrish EH, Learn GH, Hahn BH, Czartoski JL, McElrath MJ, Lehmann C, Klein F, Caskey M, Walker BD, Siliciano JD, Siliciano RF, Jankovic M, Nussenzweig MC (2015) HIV-1 integration landscape during latent and active infection. Cell 160(3):420–432. doi:10.1016/j.cell.2015.01.020

Crowell TA, Fletcher JL, Sereti I, Pinyakorn S, Dewar R, Krebs SJ, Chomchey N, Rerknimitr R, Schuetz A, Michael NL, Phanuphak N, Chomont N, Ananworanich J, Group RSS (2016) Initiation of antiretroviral therapy before detection of colonic infiltration by HIV reduces viral reservoirs, inflammation and immune activation. J Int AIDS Soc 19(1):21163. doi:10.7448/IAS.19.1.21163

Daniel MD, Letvin NL, King NW, Kannagi M, Sehgal PK, Hunt RD, Kanki PJ, Essex M, Desrosiers RC (1985) Isolation of T-cell tropic HTLV-III-like retrovirus from macaques. Science 228(4704):1201–1204

Daniel MD, Kirchhoff F, Czajak SC, Sehgal PK, Desrosiers RC (1992) Protective effects of a live attenuated SIV vaccine with a deletion in the nef gene. Science 258(5090):1938–1941

de Groot NG, Heijmans CM, Koopman G, Verschoor EJ, Bogers WM, Bontrop RE (2011) TRIM5 allelic polymorphism in macaque species/populations of different geographic origins: its impact on SIV vaccine studies. Tissue Antigens 78(4):256–262. doi:10.1111/j.1399-0039.2011.01768.x

Deere JD, Higgins J, Cannavo E, Villalobos A, Adamson L, Fromentin E, Schinazi RF, Luciw PA, North TW (2010) Viral decay kinetics in the highly active antiretroviral therapy-treated rhesus macaque model of AIDS. PLoS ONE 5(7):e11640

Deere JD, Kauffman RC, Cannavo E, Higgins J, Villalobos A, Adamson L, Schinazi RF, Luciw PA, North TW (2014) Analysis of multiply spliced transcripts in lymphoid tissue reservoirs of rhesus macaques infected with RT-SHIV during HAART. PLoS ONE 9(2): e87914. doi:10.1371/journal.pone.0087914

Del Prete GQ, Ailers B, Moldt B, Keele BF, Estes JD, Rodriguez A, Sampias M, Oswald K, Fast R, Trubey CM, Chertova E, Smedley J, LaBranche CC, Montefiori DC, Burton DR, Shaw GM, Markowitz M, Piatak M Jr, KewalRamani VN, Bieniasz PD, Lifson JD, Hatziioannou T (2014a) Selection of unadapted, pathogenic SHIVs encoding newly transmitted HIV-1 envelope proteins. Cell Host Microbe 16(3):412–418. doi:10.1016/j.chom.2014.08.003

Del Prete GQ, Shoemaker R, Oswald K, Lara A, Trubey CM, Fast R, Schneider DK, Kiser R, Coalter V, Wiles A, Wiles R, Freemire B, Keele BF, Estes JD, Quinones OA, Smedley J, Macallister R, Sanchez RI, Wai JS, Tan CM, Alvord WG, Hazuda DJ, Piatak M Jr, Lifson JD (2014b) Effect of suberoylanilide hydroxamic acid (SAHA) administration on the residual virus pool in a model of combination antiretroviral therapy-mediated suppression in sivmac239-infected indian rhesus macaques. Antimicrob Agents Chemother 58(11):6790–6806. doi:10.1128/aac.03746-14

Del Prete GQ, Oswald K, Lara A, Shoemaker R, Smedley J, Macallister R, Coalter V, Wiles A, Wiles R, Li Y, Fast R, Kiser R, Lu B, Zheng J, Alvord WG, Trubey CM, Piatak M Jr, Deleage C, Keele BF, Estes JD, Hesselgesser J, Geleziunas R, Lifson JD (2015) Elevated plasma viral loads in romidepsin-treated simian immunodeficiency virus-infected rhesus macaques on suppressive combination antiretroviral therapy. Antimicrob Agents Chemother 60 (3):1560–1572. doi:10.1128/AAC.02625-15

Del Prete GQ, Lifson JD, Keele BF (2016a) Nonhuman primate models for the evaluation of HIV-1 preventive vaccine strategies: model parameter considerations and consequences. Curr Opin HIV AIDS 11(6):546–554. doi:10.1097/COH.0000000000000311

Del Prete GQ, Smedley J, Macallister R, Jones GS, Li B, Hattersley J, Zheng J, Piatak M Jr, Keele BF, Hesselgesser J, Geleziunas R, Lifson JD (2016b) Short communication: comparative evaluation of coformulated injectable combination antiretroviral therapy regimens in simian immunodeficiency virus-infected rhesus macaques. AIDS Res Hum Retroviruses 32(2):163–168. doi:10.1089/AID.2015.0130

Deleage C, Wietgrefe SW, Del Prete G, Morcock DR, Hao XP, Piatak M Jr, Bess J, Anderson JL, Perkey KE, Reilly C, McCune JM, Haase AT, Lifson JD, Schacker TW, Estes JD (2016) Defining HIV and SIV reservoirs in lymphoid tissues. Pathog Immun 1(1):68–106

Deng K, Zink MC, Clements JE, Siliciano RF (2012) A quantitative measurement of antiviral activity of anti-human immunodeficiency virus type 1 drugs against simian immunodeficiency virus infection: dose-response curve slope strongly influences class-specific inhibitory potential. J Virol 86(20):11368–11372

Desrosiers RC, Wyand MS, Kodama T, Ringler DJ, Arthur LO, Sehgal PK, Letvin NL, King NW, Daniel MD (1989) Vaccine protection against simian immunodeficiency virus infection. Proc Natl Acad Sci USA 86(16):6353–6357

Dinoso JB, Rabi SA, Blankson JN, Gama L, Mankowski JL, Siliciano RF, Zink MC, Clements JE (2009) A simian immunodeficiency virus-infected macaque model to study viral reservoirs that persist during highly active antiretroviral therapy. J Virol 83(18):9247–9257

Dobard CW, Taylor A, Sharma S, Anderson PL, Bushman LR, Chuong D, Pau CP, Hanson D, Wang L, Garcia-Lerma JG, McGowan I, Rohan L, Heneine W (2015) Protection against rectal chimeric simian/human immunodeficiency virus transmission in macaques by rectal-specific gel formulations of maraviroc and tenofovir. J Infect Dis 212(12):1988–1995. doi:10.1093/infdis/jiv334

Donahue RE, Metzger ME, Bonifacino AC, Parker DM, Leitman SF, Cullis H, Lienesch M (2014) Optimization of the spectra optia for leukapheresis of very small subjects. Blood 124:3857–3857

Dornadula G, Zhang H, VanUitert B, Stern J, Livornese L Jr, Ingerman MJ, Witek J, Kedanis RJ, Natkin J, DeSimone J, Pomerantz RJ (1999) Residual HIV-1 RNA in blood plasma of patients taking suppressive highly active antiretroviral therapy. JAMA 282(17):1627–1632

Dunham RM, Gordon SN, Vaccari M, Piatak M, Huang Y, Deeks SG, Lifson J, Franchini G, McCune JM (2012) Preclinical evaluation of HIV eradication strategies in the simian immunodeficiency virus-infected rhesus macaque: a pilot study testing inhibition of indoleamine 2,3-Dioxygenase. AIDS Res Hum Retroviruses

Durand CM, Blankson JN, Siliciano RF (2012) Developing strategies for HIV-1 eradication. Trends Immunol 33(11):554–562

Elliott JH, McMahon JH, Chang CC, Lee SA, Hartogensis W, Bumpus N, Savic R, Roney J, Hoh R, Solomon A, Piatak M, Gorelick RJ, Lifson J, Bacchetti P, Deeks SG, Lewin SR (2015) Short-term administration of disulfiram for reversal of latent HIV infection: a phase 2 dose-escalation study. Lancet HIV 2(12):e520–e529. doi:10.1016/S2352-3018(15)00226-X

Estes JD, Gordon SN, Zeng M, Chahroudi AM, Dunham RM, Staprans SI, Reilly CS, Silvestri G, Haase AT (2008) Early resolution of acute immune activation and induction of PD-1 in SIV-infected sooty mangabeys distinguishes nonpathogenic from pathogenic infection in rhesus macaques. J Immunol 180(10):6798–6807

Estes JD, Harris LD, Klatt NR, Tabb B, Pittaluga S, Paiardini M, Barclay GR, Smedley J, Pung R, Oliveira KM, Hirsch VM, Silvestri G, Douek DC, Miller CJ, Haase AT, Lifson J, Brenchley JM (2010) Damaged intestinal epithelial integrity linked to microbial translocation in pathogenic simian immunodeficiency virus infections. PLoS Pathog 6(8):e1001052. doi:10.1371/journal.ppat.1001052

European Collaborative S, Patel D, Cortina-Borja M, Thorne C, Newell ML (2007) Time to undetectable viral load after highly active antiretroviral therapy initiation among HIV-infected pregnant women. Clin Infect Dis: Off Publ Infect Dis Soc Am 44(12):1647–1656. doi:10.1086/518284

Fennessey CM, Pinkevych M, Immonen TT, Reynaldi A, Venturi V, Nadella P, Reid C, Newman L, Lipkey L, Oswald K, Bosche WJ, Trivett MT, Ohlen C, Ott DE, Estes JD, Del Prete GQ, Lifson JD, Davenport MP, Keele BF (2017) Genetically-barcoded SIV facilitates enumeration of rebound variants and estimation of reactivation rates in nonhuman primates following interruption of suppressive antiretroviral therapy. PLoS Pathog 13(5):e1006359. doi:10.1371/journal.ppat.1006359

Fischer M, Gunthard HF, Opravil M, Joos B, Huber W, Bisset LR, Ott P, Boni J, Weber R, Cone RW (2000) Residual HIV-RNA levels persist for up to 2.5 years in peripheral blood mononuclear cells of patients on potent antiretroviral therapy. AIDS Res Hum Retroviruses 16(12):1135–1140

Fletcher CV, Staskus K, Wietgrefe SW, Rothenberger M, Reilly C, Chipman JG, Beilman GJ, Khoruts A, Thorkelson A, Schmidt TE, Anderson J, Perkey K, Stevenson M, Perelson AS, Douek DC, Haase AT, Schacker TW (2014) Persistent HIV-1 replication is associated with lower antiretroviral drug concentrations in lymphatic tissues. Proc Natl Acad Sci USA 111(6):2307–2312. doi:10.1073/pnas.1318249111

Frange P, Faye A, Avettand-Fenoel V, Bellaton E, Descamps D, Angin M, David A, Caillat-Zucman S, Peytavin G, Dollfus C, Le Chenadec J, Warszawski J, Rouzioux C, Saez-Cirion A, Cohort AE-CP, the AEPVsg (2016) HIV-1 virological remission lasting more than 12 years after interruption of early antiretroviral therapy in a perinatally infected teenager enrolled in the French ANRS EPF-CO10 paediatric cohort: a case report. Lancet HIV 3(1):e49–e54. doi:10.1016/S2352-3018(15)00232-5

Fukazawa Y, Lum R, Okoye AA, Park H, Matsuda K, Bae JY, Hagen SI, Shoemaker R, Deleage C, Lucero C, Morcock D, Swanson T, Legasse AW, Axthelm MK, Hesselgesser J, Geleziunas R, Hirsch VM, Edlefsen PT, Piatak M Jr, Estes JD, Lifson JD, Picker LJ (2015) B cell follicle sanctuary permits persistent productive simian immunodeficiency virus infection in elite controllers. Nat Med 21(2):132–139. doi:10.1038/nm.3781

Fultz PN, McClure HM, Anderson DC, Swenson RB, Anand R, Srinivasan A (1986) Isolation of a T-lymphotropic retrovirus from naturally infected sooty mangabey monkeys (Cercocebus atys). Proc Natl Acad Sci USA 83(14):5286–5290

Fultz PN, McClure HM, Swenson RB, Anderson DC (1989) HIV infection of chimpanzees as a model for testing chemotherapeutics. Intervirology 30(suppl 1):51–58

Gama L, Abreu CM, Shirk EN, Price SL, Li M, Laird GM, Pate KA, Wietgrefe SW, O'Connor SL, Pianowski L, Haase AT, Van Lint C, Siliciano RF, Clements JE, Group L-SS (2017) Reactivation of simian immunodeficiency virus reservoirs in the brain of virally suppressed macaques. AIDS 31(1):5–14. doi:10.1097/QAD.0000000000001267

Garcia JV (2016) In vivo platforms for analysis of HIV persistence and eradication. J Clin Investig 126(2):424–431. doi:10.1172/JCI80562

Gardner M (2016) A historical perspective: simian AIDS-an accidental windfall. J Med Primatol 45(5):212–214. doi:10.1111/jmp.12234

Graham DR, Gama L, Queen SE, Li M, Brice AK, Kelly KM, Mankowski JL, Clements JE, Zink MC (2011) Initiation of HAART during acute simian immunodeficiency virus infection rapidly controls virus replication in the CNS by enhancing immune activity and preserving protective immune responses. J Neurovirol 17(1):120–130

Hamlyn E, Ewings FM, Porter K, Cooper DA, Tambussi G, Schechter M, Pedersen C, Okulicz JF, McClure M, Babiker A, Weber J, Fidler S (2012) Plasma HIV viral rebound following protocol-indicated cessation of ART commenced in primary and chronic HIV infection. PLoS ONE 7(8):e43754

Hansen SG, Ford JC, Lewis MS, Ventura AB, Hughes CM, Coyne-Johnson L, Whizin N, Oswald K, Shoemaker R, Swanson T, Legasse AW, Chiuchiolo MJ, Parks CL, Axthelm MK, Nelson JA, Jarvis MA, Piatak M Jr, Lifson JD, Picker LJ (2011) Profound early control of highly pathogenic SIV by an effector memory T-cell vaccine. Nature 473(7348):523–527

Hansen SG, Piatak M Jr, Ventura AB, Hughes CM, Gilbride RM, Ford JC, Oswald K, Shoemaker R, Li Y, Lewis MS, Gilliam AN, Xu G, Whizin N, Burwitz BJ, Planer SL, Turner JM, Legasse AW, Axthelm MK, Nelson JA, Fruh K, Sacha JB, Estes JD, Keele BF, Edlefsen PT, Lifson JD, Picker LJ (2013) Immune clearance of highly pathogenic SIV infection. Nature 502(7469):100–104. doi:10.1038/nature12519

Hassounah SA, Mesplede T, Quashie PK, Oliveira M, Sandstrom PA, Wainberg MA (2014) Effect of HIV-1 integrase resistance mutations when introduced into SIVmac239 on susceptibility to integrase strand transfer inhibitors. J Virol 88(17):9683–9692. doi:10.1128/jvi.00947-14

Hatziioannou T, Bieniasz PD (2011) Antiretroviral restriction factors. Curr Opin Virol 1(6):526–532. doi:10.1016/j.coviro.2011.10.007

Hatziioannou T, Cowan S, Von Schwedler UK, Sundquist WI, Bieniasz PD (2004) Species-specific tropism determinants in the human immunodeficiency virus type 1 capsid. J Virol 78(11):6005–6012. doi:10.1128/JVI.78.11.6005-6012.2004

Hatziioannou T, Ambrose Z, Chung NP, Piatak M Jr, Yuan F, Trubey CM, Coalter V, Kiser R, Schneider D, Smedley J, Pung R, Gathuka M, Estes JD, Veazey RS, KewalRamani VN, Lifson JD, Bieniasz PD (2009) A macaque model of HIV-1 infection. Proc Natl Acad Sci USA 106(11):4425–4429

Hatziioannou T, Del Prete GQ, Keele BF, Estes JD, McNatt MW, Bitzegeio J, Raymond A, Rodriguez A, Schmidt F, Mac Trubey C, Smedley J, Piatak M Jr, KewalRamani VN, Lifson JD, Bieniasz PD (2014) HIV-1-induced AIDS in monkeys. Science 344(6190):1401–1405. doi:10.1126/science.1250761

Hazuda DJ, Young SD, Guare JP, Anthony NJ, Gomez RP, Wai JS, Vacca JP, Handt L, Motzel SL, Klein HJ, Dornadula G, Danovich RM, Witmer MV, Wilson KA, Tussey L, Schleif WA, Gabryelski LS, Jin L, Miller MD, Casimiro DR, Emini EA, Shiver JW (2004) Integrase inhibitors and cellular immunity suppress retroviral replication in rhesus macaques. Science 305(5683):528–532

Hill AL, Rosenbloom DI, Fu F, Nowak MA, Siliciano RF (2014) Predicting the outcomes of treatment to eradicate the latent reservoir for HIV-1. Proc Natl Acad Sci USA 111(37):13475–13480. doi:10.1073/pnas.1406663111

Hill AL, Rosenbloom DI, Goldstein E, Hanhauser E, Kuritzkes DR, Siliciano RF, Henrich TJ (2016) Real-time predictions of reservoir size and rebound time during antiretroviral therapy interruption trials for HIV. PLoS Pathog 12(4):e1005535. doi:10.1371/journal.ppat.1005535

Hirsch VM, Johnson PR (1994) Pathogenic diversity of simian immunodeficiency viruses. Virus Res 32(2):183–203

Hirsch VM, Lifson JD (2000) Simian immunodeficiency virus infection of monkeys as a model system for the study of AIDS pathogenesis, treatment, and prevention. Adv Pharmacol 49:437–477

Hirsch VM, Olmsted RA, Murphey-Corb M, Purcell RH, Johnson PR (1989) An African primate lentivirus (SIVsm) closely related to HIV-2. Nature 339(6223):389–392. doi:10.1038/339389a0

Hirsch V, Adger-Johnson D, Campbell B, Goldstein S, Brown C, Elkins WR, Montefiori DC (1997) A molecularly cloned, pathogenic, neutralization-resistant simian immunodeficiency virus, SIVsmE543-3. J Virol 71(2):1608–1620

Hofman MJ, Higgins J, Matthews TB, Pedersen NC, Tan C, Schinazi RF, North TW (2004) Efavirenz therapy in rhesus macaques infected with a chimera of simian immunodeficiency virus containing reverse transcriptase from human immunodeficiency virus type 1. Antimicrob Agents Chemother 48(9):3483–3490

Hryniewicz A, Boasso A, Edghill-Smith Y, Vaccari M, Fuchs D, Venzon D, Nacsa J, Betts MR, Tsai WP, Heraud JM, Beer B, Blanset D, Chougnet C, Lowy I, Shearer GM, Franchini G (2006) CTLA-4 blockade decreases TGF-beta, IDO, and viral RNA expression in tissues of SIVmac251-infected macaques. Blood 108(12):3834–3842

Hunt PW (2012) HIV and inflammation: mechanisms and consequences. Curr HIV/AIDS Rep 9 (2):139–147. doi:10.1007/s11904-012-0118-8

Ikeda Y, Ylinen LM, Kahar-Bador M, Towers GJ (2004) Influence of gag on human immunodeficiency virus type 1 species-specific tropism. J Virol 78(21):11816–11822. doi:10.1128/JVI.78.21.11816-11822.2004

Jasny E, Geer S, Frank I, Vagenas P, Aravantinou M, Salazar AM, Lifson JD, Piatak M Jr, Gettie A, Blanchard JL, Robbiani M (2012) Characterization of peripheral and mucosal immune responses in rhesus macaques on long-term tenofovir and emtricitabine combination antiretroviral therapy. J Acquir Immune Defic Syndr 61(4):425–435

Joag SV, Stephens EB, Adams RJ, Foresman L, Narayan O (1994) Pathogenesis of SIVmac infection in Chinese and Indian rhesus macaques: effects of splenectomy on virus burden. Virology 200(2):436–446. doi:10.1006/viro.1994.1207

Kaizu M, Ami Y, Nakasone T, Sasaki Y, Izumi Y, Sato H, Takahashi E, Sakai K, Shinohara K, Nakanishi K, Honda M (2003) Higher levels of IL-18 circulate during primary infection of monkeys with a pathogenic SHIV than with a nonpathogenic SHIV. Virology 313(1):8–12

Kanki PJ, McLane MF, King NW Jr, Letvin NL, Hunt RD, Sehgal P, Daniel MD, Desrosiers RC, Essex M (1985) Serologic identification and characterization of a macaque T-lymphotropic retrovirus closely related to HTLV-III. Science 228(4704):1199–1201

Kearney MF, Wiegand A, Shao W, Coffin JM, Mellors JW, Lederman M, Gandhi RT, Keele BF, Li JZ (2015) Origin of rebound plasma HIV includes cells with identical proviruses that are transcriptionally active before stopping of antiretroviral therapy. J Virol 90(3):1369–1376. doi:10.1128/JVI.02139-15

Keele BF, Van Heuverswyn F, Li Y, Bailes E, Takehisa J, Santiago ML, Bibollet-Ruche F, Chen Y, Wain LV, Liegeois F, Loul S, Ngole EM, Bienvenue Y, Delaporte E, Brookfield JF, Sharp PM, Shaw GM, Peeters M, Hahn BH (2006) Chimpanzee reservoirs of pandemic and nonpandemic HIV-1. Science 313(5786):523–526. doi:10.1126/science.1126531

Keele BF, Jones JH, Terio KA, Estes JD, Rudicell RS, Wilson ML, Li Y, Learn GH, Beasley TM, Schumacher-Stankey J, Wroblewski E, Mosser A, Raphael J, Kamenya S, Lonsdorf EV, Travis DA, Mlengeya T, Kinsel MJ, Else JG, Silvestri G, Goodall J, Sharp PM, Shaw GM, Pusey AE, Hahn BH (2009) Increased mortality and AIDS-like immunopathology in wild chimpanzees infected with SIVcpz. Nature 460(7254):515–519. doi:10.1038/nature08200

Kersh EN, Henning T, Vishwanathan SA, Morris M, Butler K, Adams DR, Guenthner P, Srinivasan P, Smith J, Radzio J, Garcia-Lerma JG, Dobard C, Heneine W, McNicholl J (2014) SHIV susceptibility changes during the menstrual cycle of pigtail macaques. J Med Primatol 43 (5):310–316. doi:10.1111/jmp.12124

Kersh EN, Henning TR, Dobard C, Heneine W, McNicholl JM (2015a) Short communication: practical experience with analysis and design of repeat low-dose SHIVSF162P3 exposure studies in female pigtail macaques with varying susceptibility during menstrual cycling. AIDS Res Hum Retroviruses 31(11):1166–1169. doi:10.1089/AID.2014.0373

Kersh EN, Ritter J, Butler K, Ostergaard SD, Hanson D, Ellis S, Zaki S, McNicholl JM (2015b) Relationship of estimated SHIV acquisition time points during the menstrual cycle and thinning of vaginal epithelial layers in pigtail macaques. Sex Transm Dis 42(12):694–701. doi:10.1097/OLQ.0000000000000367

Kirmaier A, Wu F, Newman RM, Hall LR, Morgan JS, O'Connor S, Marx PA, Meythaler M, Goldstein S, Buckler-White A, Kaur A, Hirsch VM, Johnson WE (2010) TRIM5 suppresses cross-species transmission of a primate immunodeficiency virus and selects for emergence of resistant variants in the new species. PLoS Biol 8(8). doi:10.1371/journal.pbio.1000462

Klatt NR, Harris LD, Vinton CL, Sung H, Briant JA, Tabb B, Morcock D, McGinty JW, Lifson JD, Lafont BA, Martin MA, Levine AD, Estes JD, Brenchley JM (2010) Compromised gastrointestinal integrity in pigtail macaques is associated with increased microbial translocation, immune activation, and IL-17 production in the absence of SIV infection. Mucosal Immunol 3(4):387–398. doi:10.1038/mi.2010.14

Klatt NR, Canary LA, Vanderford TH, Vinton CL, Engram JC, Dunham RM, Cronise HE, Swerczek JM, Lafont BA, Picker LJ, Silvestri G, Brenchley JM (2012) Dynamics of simian immunodeficiency virus SIVmac239 infection in pigtail macaques. J Virol 86(2):1203–1213. doi:10.1128/JVI.06033-11

Krebs KC, Jin Z, Rudersdorf R, Hughes AL, O'Connor DH (2005) Unusually high frequency MHC class I alleles in Mauritian origin cynomolgus macaques. J Immunol 175(8):5230–5239

Lee FH, Mason R, Welles H, Learn GH, Keele BF, Roederer M, Bar KJ (2015) Breakthrough virus neutralization resistance as a correlate of protection in a nonhuman primate heterologous simian immunodeficiency virus vaccine challenge study. J Virol 89(24):12388–12400. doi:10.1128/JVI.01531-15

Letvin NL, Daniel MD, Sehgal PK, Desrosiers RC, Hunt RD, Waldron LM, MacKey JJ, Schmidt DK, Chalifoux LV, King NW (1985) Induction of AIDS-like disease in macaque monkeys with T-cell tropic retrovirus STLV-III. Science 230(4721):71–73

Letvin NL, Rao SS, Montefiori DC, Seaman MS, Sun Y, Lim SY, Yeh WW, Asmal M, Gelman RS, Shen L, Whitney JB, Seoighe C, Lacerda M, Keating S, Norris PJ, Hudgens MG, Gilbert PB, Buzby AP, Mach LV, Zhang J, Balachandran H, Shaw GM, Schmidt SD, Todd JP, Dodson A, Mascola JR, Nabel GJ (2011) Immune and genetic correlates of vaccine protection against mucosal infection by SIV in monkeys. Sci Transl Med 3(81):81ra36. doi:10.1126/scitranslmed.3002351

Lewis MG, Norelli S, Collins M, Barreca ML, Iraci N, Chirullo B, Yalley-Ogunro J, Greenhouse J, Titti F, Garaci E, Savarino A (2010) Response of a simian immunodeficiency virus (SIVmac251) to raltegravir: a basis for a new treatment for simian AIDS and an animal model for studying lentiviral persistence during antiretroviral therapy. Retrovirology 7:21

Li H, Wang S, Kong R, Ding W, Lee FH, Parker Z, Kim E, Learn GH, Hahn P, Policicchio B, Brocca-Cofano E, Deleage C, Hao X, Chuang GY, Gorman J, Gardner M, Lewis MG, Hatziioannou T, Santra S, Apetrei C, Pandrea I, Alam SM, Liao HX, Shen X, Tomaras GD, Farzan M, Chertova E, Keele BF, Estes JD, Lifson JD, Doms RW, Montefiori DC, Haynes BF, Sodroski JG, Kwong PD, Hahn BH, Shaw GM (2016) Envelope residue 375 substitutions in simian-human immunodeficiency viruses enhance CD4 binding and replication in rhesus macaques. Proc Natl Acad Sci USA 113(24):E3413–E3422. doi:10.1073/pnas.1606636113

Liao CH, Kuang YQ, Liu HL, Zheng YT, Su B (2007) A novel fusion gene, TRIM5-Cyclophilin A in the pig-tailed macaque determines its susceptibility to HIV-1 infection. AIDS 21(Suppl 8): S19–S26. doi:10.1097/01.aids.0000304692.09143.1b

Ling B, Veazey RS, Luckay A, Penedo C, Xu K, Lifson JD, Marx PA (2002) SIV(mac) pathogenesis in rhesus macaques of Chinese and Indian origin compared with primary HIV infections in humans. Aids 16(11):1489–1496

Ling B, Piatak M Jr, Rogers L, Johnson AM, Russell-Lodrigue K, Hazuda DJ, Lifson JD, Veazey RS (2014) Effects of treatment with suppressive combination antiretroviral drug therapy and the histone deacetylase inhibitor suberoylanilide hydroxamic acid; (SAHA) on

SIV-infected Chinese rhesus macaques. PLoS ONE 9(7):e102795. doi:10.1371/journal.pone. 0102795

Liu J, Ghneim K, Sok D, Bosche WJ, Li Y, Chipriano E, Berkemeier B, Oswald K, Borducchi E, Cabral C, Peter L, Brinkman A, Shetty M, Jimenez J, Mondesir J, Lee B, Giglio P, Chandrashekar A, Abbink P, Colantonio A, Gittens C, Baker C, Wagner W, Lewis MG, Li W, Sekaly RP, Lifson JD, Burton DR, Barouch DH (2016) Antibody-mediated protection against SHIV challenge includes systemic clearance of distal virus. Science 353(6303):1045–1049. doi:10.1126/science.aag0491

Lopker M, Easlick J, Sterrett S, Decker JM, Barbian H, Learn G, Keele BF, Robinson JE, Li H, Hahn BH, Shaw GM, Bar KJ (2013) Heterogeneity in neutralization sensitivities of viruses comprising the simian immunodeficiency virus SIVsmE660 isolate and vaccine challenge stock. J Virol 87(10):5477–5492. doi:10.1128/JVI.03419-12

Lowenstine LJ, Pedersen NC, Higgins J, Pallis KC, Uyeda A, Marx P, Lerche NW, Munn RJ, Gardner MB (1986) Seroepidemiologic survey of captive Old-World primates for antibodies to human and simian retroviruses, and isolation of a lentivirus from sooty mangabeys (Cercocebus atys). Int J Cancer 38(4):563–574

Lynch RM, Boritz E, Coates EE, DeZure A, Madden P, Costner P, Enama ME, Plummer S, Holman L, Hendel CS, Gordon I, Casazza J, Conan-Cibotti M, Migueles SA, Tressler R, Bailer RT, McDermott A, Narpala S, O'Dell S, Wolf G, Lifson JD, Freemire BA, Gorelick RJ, Pandey JP, Mohan S, Chomont N, Fromentin R, Chun TW, Fauci AS, Schwartz RM, Koup RA, Douek DC, Hu Z, Capparelli E, Graham BS, Mascola JR, Ledgerwood JE, Team VRCS (2015) Virologic effects of broadly neutralizing antibody VRC01 administration during chronic HIV-1 infection. Science Transl Med 7(319):319ra206. doi:10.1126/scitranslmed.aad5752

Ma D, Xu C, Cillo AR, Policicchio B, Kristoff J, Haret-Richter G, Mellors JW, Pandrea I, Apetrei C (2015) Simian immunodeficiency virus SIVsab infection of rhesus macaques as a model of complete immunological suppression with persistent reservoirs of replication-competent virus: implications for cure research. J Virol 89(11):6155–6160. doi:10.1128/JVI.00256-15

Malcolm RK, Veazey RS, Geer L, Lowry D, Fetherston SM, Murphy DJ, Boyd P, Major I, Shattock RJ, Klasse PJ, Doyle LA, Rasmussen KK, Goldman L, Ketas TJ, Moore JP (2012) Sustained release of the CCR5 inhibitors CMPD167 and maraviroc from vaginal rings in rhesus macaques. Antimicrob Agents Chemother 56(5):2251–2258. doi:10.1128/AAC.05810-11

Maldarelli F, Palmer S, King MS, Wiegand A, Polis MA, Mican J, Kovacs JA, Davey RT, Rock-Kress D, Dewar R, Liu S, Metcalf JA, Rehm C, Brun SC, Hanna GJ, Kempf DJ, Coffin JM, Mellors JW (2007) ART suppresses plasma HIV-1 RNA to a stable set point predicted by pretherapy viremia. PLoS Pathog 3(4):e46

Maldarelli F, Wu X, Su L, Simonetti FR, Shao W, Hill S, Spindler J, Ferris AL, Mellors JW, Kearney MF, Coffin JM, Hughes SH (2014) HIV latency. Specific HIV integration sites are linked to clonal expansion and persistence of infected cells. Science 345(6193):179–183. doi:10.1126/science.1254194

Manrique J, Piatak M, Lauer W, Johnson W, Mansfield K, Lifson J, Desrosiers R (2013) Influence of mismatch of Env sequences on vaccine protection by live attenuated simian immunodeficiency virus. J Virol 87(13):7246–7254. doi:10.1128/JVI.00798-13

Markowitz M, Zolopa A, Squires K, Ruane P, Coakley D, Kearney B, Zhong L, Wulfsohn M, Miller MD, Lee WA (2014) Phase I/II study of the pharmacokinetics, safety and antiretroviral activity of tenofovir alafenamide, a new prodrug of the HIV reverse transcriptase inhibitor tenofovir, in HIV-infected adults. J Antimicrob Chemother 69(5):1362–1369. doi:10.1093/jac/dkt532

Marthas ML, Lu D, Penedo MC, Hendrickx AG, Miller CJ (2001) Titration of an SIVmac251 stock by vaginal inoculation of Indian and Chinese origin rhesus macaques: transmission efficiency, viral loads, and antibody responses. AIDS Res Hum Retroviruses 17(15):1455–1466. doi:10.1089/088922201753197123

Marx PA, Li Y, Lerche NW, Sutjipto S, Gettie A, Yee JA, Brotman BH, Prince AM, Hanson A, Webster RG et al (1991) Isolation of a simian immunodeficiency virus related to human immunodeficiency virus type 2 from a west African pet sooty mangabey. J Virol 65(8):4480–4485

Mason RD, Welles HC, Adams C, Chakrabarti BK, Gorman J, Zhou T, Nguyen R, O'Dell S, Lusvarghi S, Bewley CA, Li H, Shaw GM, Sheng Z, Shapiro L, Wyatt R, Kwong PD, Mascola JR, Roederer M (2016) Targeted isolation of antibodies directed against major sites of SIV env vulnerability. PLoS Pathog 12(4):e1005537. doi:10.1371/journal.ppat.1005537

Matthews GV, Sabin CA, Mandalia S, Lampe F, Phillips AN, Nelson MR, Bower M, Johnson MA, Gazzard BG (2002) Virological suppression at 6 months is related to choice of initial regimen in antiretroviral-naive patients: a cohort study. AIDS 16(1):53–61

McNatt MW, Zang T, Hatziioannou T, Bartlett M, Fofana IB, Johnson WE, Neil SJ, Bieniasz PD (2009) Species-specific activity of HIV-1 Vpu and positive selection of tetherin transmembrane domain variants. PLoS Pathog 5(2):e1000300. doi:10.1371/journal.ppat.1000300

Means RE, Greenough T, Desrosiers RC (1997) Neutralization sensitivity of cell culture-passaged simian immunodeficiency virus. J Virol 71(10):7895–7902

Mee ET, Badhan A, Karl JA, Wiseman RW, Cutler K, Knapp LA, Almond N, O'Connor DH, Rose NJ (2009) MHC haplotype frequencies in a UK breeding colony of Mauritian cynomolgus macaques mirror those found in a distinct population from the same geographic origin. J Med Primatol 38(1):1–14. doi:10.1111/j.1600-0684.2008.00299.x

Miller CJ, Li Q, Abel K, Kim EY, Ma ZM, Wietgrefe S, La Franco-Scheuch L, Compton L, Duan L, Shore MD, Zupancic M, Busch M, Carlis J, Wolinsky S, Haase AT (2005) Propagation and dissemination of infection after vaginal transmission of simian immunodeficiency virus. J Virol 79(14):9217–9227. doi:10.1128/JVI.79.14.9217-9227.2005

Mohns MS, Greene JM, Cain BT, Pham NH, Gostick E, Price DA, O'Connor DH (2015) Expansion of simian immunodeficiency virus (SIV)-specific CD8 T cell lines from SIV-naive Mauritian cynomolgus macaques for adoptive transfer. J Virol 89(19):9748–9757. doi:10.1128/JVI.00993-15

Moreau M, Le Tortorec A, Deleage C, Brown C, Denis H, Satie AP, Bourry O, Deureuddre-Bosquet N, Roques P, Le Grand R, Dejucq-Rainsford N (2012) Impact of short-term HAART initiated during the chronic stage or shortly post-exposure on SIV infection of male genital organs. PLoS ONE 7(5):e37348

Murphey-Corb M, Martin LN, Rangan SR, Baskin GB, Gormus BJ, Wolf RH, Andes WA, West M, Montelaro RC (1986) Isolation of an HTLV-III-related retrovirus from macaques with simian AIDS and its possible origin in asymptomatic mangabeys. Nature 321(6068):435–437. doi:10.1038/321435a0

Murphey-Corb M, Rajakumar P, Michael H, Nyaundi J, Didier PJ, Reeve AB, Mitsuya H, Sarafianos SG, Parniak MA (2012) Response of simian immunodeficiency virus to the novel nucleoside reverse transcriptase inhibitor 4′-ethynyl-2-fluoro-2′-deoxyadenosine in vitro and in vivo. Antimicrob Agents Chemother 56(9):4707–4712. doi:10.1128/AAC.00723-12

Murry JP, Higgins J, Matthews TB, Huang VY, Van Rompay KK, Pedersen NC, North TW (2003) Reversion of the M184 V mutation in simian immunodeficiency virus reverse transcriptase is selected by tenofovir, even in the presence of lamivudine. J Virol 77(2):1120–1130

Naidu YM, Kestler HW 3rd, Li Y, Butler CV, Silva DP, Schmidt DK, Troup CD, Sehgal PK, Sonigo P, Daniel MD, Desrosiers RC (1988) Characterization of infectious molecular clones of simian immunodeficiency virus (SIVmac) and human immunodeficiency virus type 2: persistent infection of rhesus monkeys with molecularly cloned SIVmac. J Virol 62(12):4691–4696

Neil SJ, Zang T, Bieniasz PD (2008) Tetherin inhibits retrovirus release and is antagonized by HIV-1 Vpu. Nature 451(7177):425–430. doi:10.1038/nature06553

Newman RM, Hall L, Kirmaier A, Pozzi LA, Pery E, Farzan M, O'Neil SP, Johnson W (2008) Evolution of a TRIM5-CypA splice isoform in old world monkeys. PLoS Pathog 4(2):e1000003. doi:10.1371/journal.ppat.1000003

Nishimura Y, Shingai M, Willey R, Sadjadpour R, Lee WR, Brown CR, Brenchley JM, Buckler-White A, Petros R, Eckhaus M, Hoffman V, Igarashi T, Martin MA (2010) Generation of the pathogenic R5-tropic simian/human immunodeficiency virus SHIVAD8 by serial passaging in rhesus macaques. J Virol 84(9):4769–4781. doi:10.1128/JVI.02279-09

North TW, Van Rompay KK, Higgins J, Matthews TB, Wadford DA, Pedersen NC, Schinazi RF (2005) Suppression of virus load by highly active antiretroviral therapy in rhesus macaques infected with a recombinant simian immunodeficiency virus containing reverse transcriptase from human immunodeficiency virus type 1. J Virol 79(12):7349–7354

North TW, Higgins J, Deere JD, Hayes TL, Villalobos A, Adamson L, Shacklett BL, Schinazi RF, Luciw PA (2010) Viral sanctuaries during highly active antiretroviral therapy in a nonhuman primate model for AIDS. J Virol 84(6):2913–2922

Novembre FJ, Saucier M, Anderson DC, Klumpp SA, O'Neil SP, Brown CR 2nd, Hart CE, Guenthner PC, Swenson RB, McClure HM (1997) Development of AIDS in a chimpanzee infected with human immunodeficiency virus type 1. J Virol 71(5):4086–4091

O'Connor SL, Blasky AJ, Pendley CJ, Becker EA, Wiseman RW, Karl JA, Hughes AL, O'Connor DH (2007) Comprehensive characterization of MHC class II haplotypes in Mauritian cynomolgus macaques. Immunogenetics 59(6):449–462. doi:10.1007/s00251-007-0209-7

Ohta Y, Masuda T, Tsujimoto H, Ishikawa K, Kodama T, Morikawa S, Nakai M, Honjo S, Hayami M (1988) Isolation of simian immunodeficiency virus from African green monkeys and seroepidemiologic survey of the virus in various non-human primates. Int J Cancer 41 (1):115–122

O'Neil SP, Novembre FJ, Hill AB, Suwyn C, Hart CE, Evans-Strickfaden T, Anderson DC, deRosayro J, Herndon JG, Saucier M, McClure HM (2000) Progressive infection in a subset of HIV-1-positive chimpanzees. J Infect Dis 182(4):1051–1062. doi:10.1086/315823

Owens CM, Yang PC, Gottlinger H, Sodroski J (2003) Human and simian immunodeficiency virus capsid proteins are major viral determinants of early, postentry replication blocks in simian cells. J Virol 77(1):726–731

Palmer S, Wiegand AP, Maldarelli F, Bazmi H, Mican JM, Polis M, Dewar RL, Planta A, Liu S, Metcalf JA, Mellors JW, Coffin JM (2003) New real-time reverse transcriptase-initiated PCR assay with single-copy sensitivity for human immunodeficiency virus type 1 RNA in plasma. J Clin Microbiol 41(10):4531–4536

Palmer S, Maldarelli F, Wiegand A, Bernstein B, Hanna GJ, Brun SC, Kempf DJ, Mellors JW, Coffin JM, King MS (2008) Low-level viremia persists for at least 7 years in patients on suppressive antiretroviral therapy. Proc Natl Acad Sci USA 105(10):3879–3884

Pandrea I, Gaufin T, Gautam R, Kristoff J, Mandell D, Montefiori D, Keele BF, Ribeiro RM, Veazey RS, Apetrei C (2011) Functional cure of SIVagm infection in rhesus macaques results in complete recovery of CD4$^+$ T cells and is reverted by CD8$^+$ cell depletion. PLoS Pathog 7 (8):e1002170

Passaes CP, Bruel T, Decalf J, David A, Angin M, Monceaux V, Muller-Trutwin M, Noel N, Bourdic K, Lambotte O, Albert ML, Duffy D, Schwartz O, Saez-Cirion A, Consortium AR (2017) Ultrasensitive HIV-1 p 24 assay detects single infected cells and differences in reservoir induction by latency reversal agents. J Virol 91(6). doi:10.1128/JVI.02296-16

Pathiraja V, Matar AJ, Gusha A, Huang CA, Duran-Struuck R (2013) Leukapheresis protocol for nonhuman primates weighing less than 10 kg. J Am Assoc Lab Anim Sci 52(1):70–77

Phillips AN, Staszewski S, Weber R, Kirk O, Francioli P, Miller V, Vernazza P, Lundgren JD, Ledergerber B, Swiss HIVCS, Frakfurt HIVCC, Euro SSG (2001) HIV viral load response to antiretroviral therapy according to the baseline CD4 cell count and viral load. JAMA 286 (20):2560–2567

Pinkevych M, Cromer D, Tolstrup M, Grimm AJ, Cooper DA, Lewin SR, Sogaard OS, Rasmussen TA, Kent SJ, Kelleher AD, Davenport MP (2015) HIV reactivation from latency after treatment interruption occurs on average every 5–8 days-implications for HIV remission. PLoS Pathog 11(7):e1005000. doi:10.1371/journal.ppat.1005000

Policicchio BB, Xu C, Brocca-Cofano E, Raehtz KD, He T, Ma D, Li H, Sivanandham R, Haret-Richter GS, Dunsmore T, Trichel A, Mellors JW, Hahn BH, Shaw GM, Ribeiro RM, Pandrea I, Apetrei C (2016) Multi-dose romidepsin reactivates replication competent SIV in post-antiretroviral rhesus macaque controllers. PLoS Pathog 12(9):e1005879. doi:10.1371/journal.ppat.1005879

Post FA, Yazdanpanah Y, Schembri G, Lazzarin A, Reynes J, Maggiolo F, Yan M, Abram ME, Tran-Muchowski C, Cheng A, Rhee MS (2017) Efficacy and safety of emtricitabine/tenofovir alafenamide (FTC/TAF) vs. emtricitabine/tenofovir disoproxil fumarate (FTC/TDF) as a backbone for treatment of HIV-1 infection in virologically suppressed adults: subgroup analysis by third agent of a randomized, double-blind, active-controlled phase 3 trial. HIV Clin Trials:1–6. doi:10.1080/15284336.2017.1291867

Queen SE, Mears BM, Kelly KM, Dorsey JL, Liao Z, Dinoso JB, Gama L, Adams RJ, Zink MC, Clements JE, Kent SJ, Mankowski JL (2011) Replication-competent simian immunodeficiency virus (SIV) Gag escape mutations archived in latent reservoirs during antiretroviral treatment of SIV-infected macaques. J Virol 85(17):9167–9175

Radzio J, Aung W, Holder A, Martin A, Sweeney E, Mitchell J, Bachman S, Pau CP, Heneine W, Garcia-Lerma JG (2012) Prevention of vaginal SHIV transmission in macaques by a coitally-dependent Truvada regimen. PLoS ONE 7(12):e50632

Radzio J, Spreen W, Yueh YL, Mitchell J, Jenkins L, Garcia-Lerma JG, Heneine W (2015) The long-acting integrase inhibitor GSK744 protects macaques from repeated intravaginal SHIV challenge. Sci Transl Med 7(270):270ra275. doi:10.1126/scitranslmed.3010297

Rajasuriar R, Wright E, Lewin SR (2015) Impact of antiretroviral therapy (ART) timing on chronic immune activation/inflammation and end-organ damage. Curr Opin HIV AIDS 10(1):35–42. doi:10.1097/coh.0000000000000118

Read PJ, Mandalia S, Khan P, Harrisson U, Naftalin C, Gilleece Y, Anderson J, Hawkins DA, Taylor GP, de Ruiter A, London HIVPRG (2012) When should HAART be initiated in pregnancy to achieve an undetectable HIV viral load by delivery? AIDS 26(9):1095–1103. doi:10.1097/QAD.0b013e3283536a6c

Reimann KA, Parker RA, Seaman MS, Beaudry K, Beddall M, Peterson L, Williams KC, Veazey RS, Montefiori DC, Mascola JR, Nabel GJ, Letvin NL (2005) Pathogenicity of simian-human immunodeficiency virus SHIV-89.6P and SIVmac are attenuated in cynomolgus macaques and associated with early T-lymphocyte responses. J Virol 79(14):8878–8885. doi:10.1128/JVI.79.14.8878-8885.2005

Reynolds MR, Sacha JB, Weiler AM, Borchardt GJ, Glidden CE, Sheppard NC, Norante FA, Castrovinci PA, Harris JJ, Robertson HT, Friedrich TC, McDermott AB, Wilson NA, Allison DB, Koff WC, Johnson WE, Watkins DI (2011) The TRIM5α genotype of rhesus macaques affects acquisition of simian immunodeficiency virus SIVsmE660 infection after repeated limiting-dose intrarectal challenge. J Virol 85(18):9637–9640. doi:10.1128/JVI.05074-11

Rosenbloom KR, Armstrong J, Barber GP, Casper J, Clawson H, Diekhans M, Dreszer TR, Fujita PA, Guruvadoo L, Haeussler M, Harte RA, Heitner S, Hickey G, Hinrichs AS, Hubley R, Karolchik D, Learned K, Lee BT, Li CH, Miga KH, Nguyen N, Paten B, Raney BJ, Smit AF, Speir ML, Zweig AS, Haussler D, Kuhn RM, Kent WJ (2015) The UCSC genome browser database: 2015 update. Nucleic Acids Res 43(Database issue):D670–681. doi:10.1093/nar/gku1177

Saez-Cirion A, Bacchus C, Hocqueloux L, Avettand-Fenoel V, Girault I, Lecuroux C, Potard V, Versmisse P, Melard A, Prazuck T, Descours B, Guergnon J, Viard JP, Boufassa F, Lambotte O, Goujard C, Meyer L, Costagliola D, Venet A, Pancino G, Autran B, Rouzioux C, Group AVS (2013) Post-treatment HIV-1 controllers with a long-term virological remission after the interruption of early initiated antiretroviral therapy ANRS VISCONTI study. PLoS Pathog 9(3):e1003211. doi:10.1371/journal.ppat.1003211

Sanders-Beer BE, Spano YY, Golighty D, Lara A, Hebblewaite D, Nieves-Duran L, Rhodes L, Mansfield KG (2011) Clinical monitoring and correlates of nephropathy in SIV-infected macaques during high-dose antiretroviral therapy. AIDS Res Ther 8(1):3

Sax PE, Zolopa A, Brar I, Elion R, Ortiz R, Post F, Wang H, Callebaut C, Martin H, Fordyce MW, McCallister S (2014) Tenofovir alafenamide vs. tenofovir disoproxil fumarate in single tablet regimens for initial HIV-1 therapy: a randomized phase 2 study. J Acquir Immune Defic Syndr 67(1):52–58. doi:10.1097/QAI.0000000000000225

Sellier P, Mannioui A, Bourry O, Dereuddre-Bosquet N, Delache B, Brochard P, Calvo J, Prevot S, Roques P (2010) Antiretroviral treatment start-time during primary SIV(mac) infection in macaques exerts a different impact on early viral replication and dissemination. PLoS ONE 5(5):e10570

Sharp PM, Robertson DL, Hahn BH (1995) Cross-species transmission and recombination of 'AIDS' viruses. Philos Trans R Soc Lond B Biol Sci 349(1327):41–47. doi:10.1098/rstb.1995.0089

Shedlock DJ, Silvestri G, Weiner DB (2009) Monkeying around with HIV vaccines: using rhesus macaques to define 'gatekeepers' for clinical trials. Nat Rev Immunol 9(10):717–728. doi:10.1038/nri2636

Shen A, Zink MC, Mankowski JL, Chadwick K, Margolick JB, Carruth LM, Li M, Clements JE, Siliciano RF (2003) Resting CD4$^+$ T lymphocytes but not thymocytes provide a latent viral reservoir in a simian immunodeficiency virus-Macaca nemestrina model of human immunodeficiency virus type 1-infected patients on highly active antiretroviral therapy. J Virol 77 (8):4938–4949

Shen A, Yang HC, Zhou Y, Chase AJ, Boyer JD, Zhang H, Margolick JB, Zink MC, Clements JE, Siliciano RF (2007) Novel pathway for induction of latent virus from resting CD4(+) T cells in the simian immunodeficiency virus/macaque model of human immunodeficiency virus type 1 latency. J Virol 81(4):1660–1670

Shen L, Peterson S, Sedaghat AR, McMahon MA, Callender M, Zhang H, Zhou Y, Pitt E, Anderson KS, Acosta EP, Siliciano RF (2008) Dose-response curve slope sets class-specific limits on inhibitory potential of anti-HIV drugs. Nat Med 14(7):762–766. doi:10.1038/nm1777

Sheppard NC, Jones RB, Burwitz BJ, Nimityongskul FA, Newman LP, Buechler MB, Reed JS, Piaskowski SM, Weisgrau KL, Castrovinci PA, Wilson NA, Ostrowski MA, Park B, Nixon DF, Rakasz EG, Sacha JB (2014) Vaccination against endogenous retrotransposable element consensus sequences does not protect rhesus macaques from SIVsmE660 infection and replication. PLoS ONE 9(3):e92012. doi:10.1371/journal.pone.0092012

Shinohara K, Sakai K, Ando S, Ami Y, Yoshino N, Takahashi E, Someya K, Suzaki Y, Nakasone T, Sasaki Y, Kaizu M, Lu Y, Honda M (1999) A highly pathogenic simian/human immunodeficiency virus with genetic changes in cynomolgus monkey. J Gen Virol 80(Pt 5):1231–1240. doi:10.1099/0022-1317-80-5-1231

Shytaj IL, Norelli S, Chirullo B, Della Corte A, Collins M, Yalley-Ogunro J, Greenhouse J, Iraci N, Acosta EP, Barreca ML, Lewis MG, Savarino A (2012) A highly intensified ART regimen induces long-term viral suppression and restriction of the viral reservoir in a simian AIDS model. PLoS Pathog 8(6):e1002774

Silvestri G, Sodora DL, Koup RA, Paiardini M, O'Neil SP, McClure HM, Staprans SI, Feinberg MB (2003) Nonpathogenic SIV infection of sooty mangabeys is characterized by limited bystander immunopathology despite chronic high-level viremia. Immunity 18(3):441–452

Silvestri G, Fedanov A, Germon S, Kozyr N, Kaiser WJ, Garber DA, McClure H, Feinberg MB, Staprans SI (2005) Divergent host responses during primary simian immunodeficiency virus SIVsm infection of natural sooty mangabey and nonnatural rhesus macaque hosts. J Virol 79 (7):4043–4054. doi:10.1128/JVI.79.7.4043-4054.2005

Simonetti FR, Sobolewski MD, Fyne E, Shao W, Spindler J, Hattori J, Anderson EM, Watters SA, Hill S, Wu X, Wells D, Su L, Luke BT, Halvas EK, Besson G, Penrose KJ, Yang Z, Kwan RW, Van Waes C, Uldrick T, Citrin DE, Kovacs J, Polis MA, Rehm CA, Gorelick R, Piatak M, Keele BF, Kearney MF, Coffin JM, Hughes SH, Mellors JW, Maldarelli F (2016) Clonally expanded CD4$^+$ T cells can produce infectious HIV-1 in vivo. Proc Natl Acad Sci USA 113(7):1883–1888. doi:10.1073/pnas.1522675113

Snippenburg W, Nellen F, Smit C, Wensing A, Godfried MH, Mudrikova T (2017) Factors associated with time to achieve an undetectable HIV RNA viral load after start of antiretroviral treatment in HIV-1-infected pregnant women. J Virus Erad 3(1):34–39

Somsouk M, Dunham RM, Cohen M, Albright R, Abdel-Mohsen M, Liegler T, Lifson J, Piatak M, Gorelick R, Huang Y, Wu Y, Hsue PY, Martin JN, Deeks SG, McCune JM, Hunt PW (2014) The immunologic effects of mesalamine in treated HIV-infected individuals with incomplete CD4$^+$ T cell recovery: a randomized crossover trial. PLoS ONE 9(12): e116306. doi:10.1371/journal.pone.0116306

Song RJ, Chenine AL, Rasmussen RA, Ruprecht CR, Mirshahidi S, Grisson RD, Xu W, Whitney JB, Goins LM, Ong H, Li PL, Shai-Kobiler E, Wang T, McCann CM, Zhang H, Wood C, Kankasa C, Secor WE, McClure HM, Strobert E, Else JG, Ruprecht RM (2006) Molecularly cloned SHIV-1157ipd3N4: a highly replication-competent, mucosally transmissible R5 simian-human immunodeficiency virus encoding HIV clade C Env. J Virol 80 (17):8729–8738. doi:10.1128/JVI.00558-06

Steingrover R, Pogany K, Fernandez Garcia E, Jurriaans S, Brinkman K, Schuitemaker H, Miedema F, Lange JM, Prins JM (2008) HIV-1 viral rebound dynamics after a single treatment interruption depends on time of initiation of highly active antiretroviral therapy. Aids 22 (13):1583–1588

Stoddart CA, Galkina SA, Joshi P, Kosikova G, Moreno ME, Rivera JM, Sloan B, Reeve AB, Sarafianos SG, Murphey-Corb M, Parniak MA (2015) Oral administration of the nucleoside EFdA (4'-ethynyl-2-fluoro-2'-deoxyadenosine) provides rapid suppression of HIV viremia in humanized mice and favorable pharmacokinetic properties in mice and the rhesus macaque. Antimicrob Agents Chemother 59(7):4190–4198. doi:10.1128/AAC.05036-14

Stremlau M, Owens CM, Perron MJ, Kiessling M, Autissier P, Sodroski J (2004) The cytoplasmic body component TRIM5alpha restricts HIV-1 infection in Old World monkeys. Nature 427 (6977):848–853. doi:10.1038/nature02343

Stremlau M, Perron M, Lee M, Li Y, Song B, Javanbakht H, Diaz-Griffero F, Anderson DJ, Sundquist WI, Sodroski J (2006) Specific recognition and accelerated uncoating of retroviral capsids by the TRIM5alpha restriction factor. Proc Natl Acad Sci USA 103(14):5514–5519. doi:10.1073/pnas.0509996103

Sumpter B, Dunham R, Gordon S, Engram J, Hennessy M, Kinter A, Paiardini M, Cervasi B, Klatt N, McClure H, Milush JM, Staprans S, Sodora DL, Silvestri G (2007) Correlates of preserved CD4(+) T cell homeostasis during natural, nonpathogenic simian immunodeficiency virus infection of sooty mangabeys: implications for AIDS pathogenesis. J Immunol 178 (3):1680–1691

Tan RC, Harouse JM, Gettie A, Cheng-Mayer C (1999) In vivo adaptation of SHIV(SF162): chimeric virus expressing a NSI, CCR5-specific envelope protein. J Med Primatol 28(4–5):164–168

Tanner Z, Lachowsky N, Ding E, Samji H, Hull M, Cescon A, Patterson S, Chia J, Leslie A, Raboud J, Loutfy M, Cooper C, Klein M, Machouf N, Tsoukas C, Montaner J, Hogg RS, Canadian Observation Cohort C (2016) Predictors of viral suppression and rebound among HIV-positive men who have sex with men in a large multi-site Canadian cohort. BMC Infect Dis 16(1):590. doi:10.1186/s12879-016-1926-z

Tartaglia LJ, Chang HW, Lee BC, Abbink P, Ng'ang'a D, Boyd M, Lavine CL, Lim SY, Sanisetty S, Whitney JB, Seaman MS, Rolland M, Tovanabutra S, Ananworanich J, Robb ML, Kim JH, Michael NL, Barouch DH (2016) Production of mucosally transmissible SHIV challenge stocks from HIV-1 circulating recombinant form 01_AE env sequences. PLoS Pathog 12(2):e1005431. doi:10.1371/journal.ppat.1005431

Thippeshappa R, Polacino P, Yu Kimata MT, Siwak EB, Anderson D, Wang W, Sherwood L, Arora R, Wen M, Zhou P, Hu SL, Kimata JT (2011) Vif substitution enables persistent infection of pig-tailed macaques by human immunodeficiency virus type 1. J Virol 85(8):3767–3779

Trichel AM, Rajakumar PA, Murphey-Corb M (2002) Species-specific variation in SIV disease progression between Chinese and Indian subspecies of rhesus macaque. J Med Primatol 31(4–5):171–178

Tsai CC, Follis KE, Sabo A, Beck TW, Grant RF, Bischofberger N, Benveniste RE, Black R (1995) Prevention of SIV infection in macaques by (R)-9-(2-phosphonylmethoxypropyl) adenine. Science 270(5239):1197–1199

Tsai CC, Emau P, Follis KE, Beck TW, Benveniste RE, Bischofberger N, Lifson JD, Morton WR (1998) Effectiveness of postinoculation (R)-9-(2-phosphonylmethoxypropyl) adenine treatment for prevention of persistent simian immunodeficiency virus SIVmne infection depends critically on timing of initiation and duration of treatment. J Virol 72(5):4265–4273

Tsibris AM, Pal U, Schure AL, Veazey RS, Kunstman KJ, Henrich TJ, Klasse PJ, Wolinsky SM, Kuritzkes DR, Moore JP (2011) SHIV-162P3 infection of rhesus macaques given maraviroc gel vaginally does not involve resistant viruses. PLoS ONE 6(12):e28047. doi:10.1371/journal.pone.0028047

Uberla K, Stahl-Hennig C, Bottiger D, Matz-Rensing K, Kaup FJ, Li J, Haseltine WA, Fleckenstein B, Hunsmann G, Oberg B et al (1995) Animal model for the therapy of acquired immunodeficiency syndrome with reverse transcriptase inhibitors. Proc Natl Acad Sci USA 92 (18):8210–8214

Vaccari M, Boasso A, Fenizia C, Fuchs D, Hryniewicz A, Morgan T, Weiss D, Doster MN, Heraud JM, Shearer GM, Franchini G (2012) Fatal pancreatitis in simian immunodeficiency virus SIV(mac251)-infected macaques treated with 2',3'-dideoxyinosine and stavudine following cytotoxic-T-lymphocyte-associated antigen 4 and indoleamine 2,3-dioxygenase blockade. J Virol 86(1):108–113

Vagenas P, Aravantinou M, Williams VG, Jasny E, Piatak M Jr, Lifson JD, Salazar AM, Blanchard JL, Gettie A, Robbiani M (2010) A tonsillar PolyICLC/AT-2 SIV therapeutic vaccine maintains low viremia following antiretroviral therapy cessation. PLoS ONE 5(9): e12891. doi:10.1371/journal.pone.0012891

Van Rompay KK, Matthews TB, Higgins J, Canfield DR, Tarara RP, Wainberg MA, Schinazi RF, Pedersen NC, North TW (2002) Virulence and reduced fitness of simian immunodeficiency virus with the M184 V mutation in reverse transcriptase. J Virol 76(12):6083–6092

Van Rompay KK, Durand-Gasselin L, Brignolo LL, Ray AS, Abel K, Cihlar T, Spinner A, Jerome C, Moore J, Kearney BP, Marthas ML, Reiser H, Bischofberger N (2008) Chronic administration of tenofovir to rhesus macaques from infancy through adulthood and pregnancy: summary of pharmacokinetics and biological and virological effects. Antimicrob Agents Chemother 52(9):3144–3160

Veazey RS, DeMaria M, Chalifoux LV, Shvetz DE, Pauley DR, Knight HL, Rosenzweig M, Johnson RP, Desrosiers RC, Lackner AA (1998) Gastrointestinal tract as a major site of CD4+ T cell depletion and viral replication in SIV infection. Science 280(5362):427–431

Veazey RS, Klasse PJ, Schader SM, Hu Q, Ketas TJ, Lu M, Marx PA, Dufour J, Colonno RJ, Shattock RJ, Springer MS, Moore JP (2005a) Protection of macaques from vaginal SHIV challenge by vaginally delivered inhibitors of virus-cell fusion. Nature 438(7064):99–102. doi:10.1038/nature04055

Veazey RS, Springer MS, Marx PA, Dufour J, Klasse PJ, Moore JP (2005b) Protection of macaques from vaginal SHIV challenge by an orally delivered CCR5 inhibitor. Nat Med 11 (12):1293–1294. doi:10.1038/nm1321

Veazey RS, Ketas TJ, Dufour J, Moroney-Rasmussen T, Green LC, Klasse PJ, Moore JP (2010) Protection of rhesus macaques from vaginal infection by vaginally delivered maraviroc, an inhibitor of HIV-1 entry via the CCR5 co-receptor. J Infect Dis 202(5):739–744. doi:10.1086/655661

Vinikoor MJ, Cope A, Gay CL, Ferrari G, McGee KS, Kuruc JD, Lennox JL, Margolis DM, Hicks CB, Eron JJ (2013) Antiretroviral therapy initiated during acute HIV infection fails to prevent persistent T-cell activation. J Acquir Immune Defic Syndr 62(5):505–508. doi:10.1097/QAI.0b013e318285cd33

Virgen CA, Kratovac Z, Bieniasz PD, Hatziioannou T (2008) Independent genesis of chimeric TRIM5-cyclophilin proteins in two primate species. Proc Natl Acad Sci USA 105(9):3563–3568. doi:10.1073/pnas.0709258105

von Wyl V, Gianella S, Fischer M, Niederoest B, Kuster H, Battegay M, Bernasconi E, Cavassini M, Rauch A, Hirschel B, Vernazza P, Weber R, Joos B, Gunthard HF (2011) Early antiretroviral therapy during primary HIV-1 infection results in a transient reduction of the viral setpoint upon treatment interruption. PLoS ONE 6(11):e27463

Wagner TA, McLaughlin S, Garg K, Cheung CY, Larsen BB, Styrchak S, Huang HC, Edlefsen PT, Mullins JI, Frenkel LM (2014) HIV latency. Proliferation of cells with HIV integrated into cancer genes contributes to persistent infection. Science 345(6196):570–573. doi:10.1126/science.1256304

Wang F, Flanagan J, Su N, Wang LC, Bui S, Nielson A, Wu X, Vo HT, Ma XJ, Luo Y (2012) RNAscope: a novel in situ RNA analysis platform for formalin-fixed, paraffin-embedded tissues. J Mol Diagn 14(1):22–29

Whetter LE, Ojukwu IC, Novembre FJ, Dewhurst S (1999) Pathogenesis of simian immunodeficiency virus infection. J Gen Virol 80(Pt 7):1557–1568

Whitney JB, Hill AL, Sanisetty S, Penaloza-MacMaster P, Liu J, Shetty M, Parenteau L, Cabral C, Shields J, Blackmore S, Smith JY, Brinkman AL, Peter LE, Mathew SI, Smith KM, Borducchi EN, Rosenbloom DI, Lewis MG, Hattersley J, Li B, Hesselgesser J, Geleziunas R, Robb ML, Kim JH, Michael NL, Barouch DH (2014) Rapid seeding of the viral reservoir prior to SIV viraemia in rhesus monkeys. Nature 512(7512):74–77. doi:10.1038/nature13594

Yeh WW, Rao SS, Lim SY, Zhang J, Hraber PT, Brassard LM, Luedemann C, Todd JP, Dodson A, Shen L, Buzby AP, Whitney JB, Korber BT, Nabel GJ, Mascola JR, Letvin NL (2011) The TRIM5 gene modulates penile mucosal acquisition of simian immunodeficiency virus in rhesus monkeys. J Virol 85(19):10389–10398. doi:10.1128/JVI.00854-11

Zeldin RK, Petruschke RA (2004) Pharmacological and therapeutic properties of ritonavir-boosted protease inhibitor therapy in HIV-infected patients. J Antimicrob Chemother 53(1):4–9. doi:10.1093/jac/dkh029

Zink MC, Suryanarayana K, Mankowski JL, Shen A, Piatak M Jr, Spelman JP, Carter DL, Adams RJ, Lifson JD, Clements JE (1999) High viral load in the cerebrospinal fluid and brain correlates with severity of simian immunodeficiency virus encephalitis. J Virol 73(12):10480–10488

Zink MC, Brice AK, Kelly KM, Queen SE, Gama L, Li M, Adams RJ, Bartizal C, Varrone J, Rabi SA, Graham DR, Tarwater PM, Mankowski JL, Clements JE (2010) Simian immunodeficiency virus-infected macaques treated with highly active antiretroviral therapy have reduced central nervous system viral replication and inflammation but persistence of viral DNA. J Infect Dis 202(1):161–170

Zuber B, Bottiger D, Benthin R, ten Haaft P, Heeney J, Wahren B, Oberg B (2001) An in vivo model for HIV resistance development. AIDS Res Hum Retroviruses 17(7):631–635

# SIV Latency in Macrophages in the CNS

Lucio Gama, Celina Abreu, Erin N. Shirk, Suzanne E. Queen,
Sarah E. Beck, Kelly A. Metcalf Pate, Brandon T. Bullock,
M. Christine Zink, Joseph L. Mankowski and Janice E. Clements

## Contents

**Abstract** Lentiviruses infect myeloid cells, leading to acute infection followed by persistent/latent infections not cleared by the host immune system. HIV and SIV are lentiviruses that infect CD4+ lymphocytes in addition to myeloid cells in blood and tissues. HIV infection of myeloid cells in brain, lung, and heart causes tissue-specific diseases that are mostly observed during severe immunosuppression, when the number of circulating CD4+ T cells declines to exceeding low levels. Antiretroviral therapy (ART) controls viral replication but does not successfully eliminate latent virus, which leads to viral rebound once ART is interrupted. HIV latency in CD4+ lymphocytes is the main focus of research and concern when HIV

L. Gama · C. Abreu · E. N. Shirk · S. E. Queen · S. E. Beck · K. A. Metcalf Pate
B. T. Bullock · M. C. Zink · J. L. Mankowski · J. E. Clements
Department of Molecular and Comparative Pathobiology, Johns Hopkins University,
Baltimore, MD 21205, USA

J. L. Mankowski · J. E. Clements
Department of Neurology, Johns Hopkins University, Baltimore, MD 21205, USA

J. L. Mankowski · J. E. Clements (✉)
Department of Pathology, Johns Hopkins University, Baltimore, MD 21205, USA
e-mail: jclements@jhmi.edu

Current Topics in Microbiology and Immunology (2018) 417:111–130
DOI 10.1007/82_2018_89
© Springer International Publishing AG, part of Springer Nature 2018
Published Online: 17 May 2018

eradication efforts are considered. However, myeloid cells in tissues are long-lived and have not been routinely examined as a potential reservoir. Based on a quantitative viral outgrowth assay (QVOA) designed to evaluate latently infected CD4+ lymphocytes, a similar protocol was developed for the assessment of latently infected myeloid cells in blood and tissues. Using an SIV ART model, it was demonstrated that myeloid cells in blood and brain harbor latent SIV that can be reactivated and produce infectious virus in vitro, demonstrating that myeloid cells have the potential to be an additional latent reservoir of HIV that should be considered during HIV eradication strategies.

# 1   Introduction

AIDS emerged as a new disease in 1980 and was shown to be caused by a retrovirus, the human T-cell lymphotropic virus III (HTVL-III), thought to be similar to human viruses HTLV-I and HTLV-II (Masur et al. 1981; Gottlieb et al. 1981; Barre-Sinoussi et al. 1983; Popovic et al. 1984). AIDS pathogenesis included not only immunosuppression but also infection of tissues, in particular the brain, causing encephalitis and dementia in adults and children (Shaw et al. 1985; Epstein et al. 1984). Soon after the AIDS virus was isolated and molecular clones were constructed to further characterize the virus, molecular hybridization studies demonstrated that HTLV-III was actually more closely related to the ungulate lentiviruses than to the human deltaretroviruses HTLV-I and HTLV-II, and the virus was renamed human immunodeficiency virus (HIV) (Gonda et al. 1986; Gonda et al. 1985). HIV infection in vivo has many parallels to lentivirus pathogenesis causing not only primary immunodeficiency but also CNS- and lung-specific diseases. In contrast to most lentiviruses, the cellular tropism of HIV included not only macrophage lineage cells but also CD4+ lymphocytes (Gartner et al. 1986). While classic lentiviruses like visna virus do not infect lymphocytes, infection of macrophages does cause lymphocyte activation and lymphocytic proliferation in infected tissues such as brain, lung, and joints (Craig et al. 1997; Zink et al. 1987). Lentivirus infections are characterized by an acute phase followed by suppression of virus replication in blood and tissues that led to a state of undetectable virus in most animals (Gendelman et al. 1985). Despite lack of detectable viral RNA, cells from infected, suppressed animals can be activated to produce virus in vitro. This was also an early observation in studies of the ovine (visna virus) and caprine (caprine arthritis encephalitis virus, CAEV) viruses in which monocytes from infected animals without detectable viral RNA mature in vitro into macrophages with subsequent reactivation and detection of viral cellular RNA and virus in the culture supernatant (Narayan et al. 1983). Viral latency in myeloid lineage cells in lentiviruses in vivo is also a feature of nonprimate lentiviruses shared by SIV and HIV.

The pathogenesis of HIV during the early AIDS epidemic and before the development of antiretroviral therapy was characterized by immunodeficiency disease as a result of loss of circulating CD4+ T cells and by subsequent opportunistic infections. CNS neurologic disease accompanies these infections, and infectious virus is detected in the cerebrospinal fluid (CSF) (Shaw et al. 1985; Ho et al. 1985; Chiodi et al. 1988). HIV infection in brain of infected adults and children was shown to be responsible for the neurologic disease and was called AIDS Dementia Complex (ADC). This neurologic syndrome, ADC, was the cause of mortality in HIV-infected individuals. Although HIV enters the CNS during acute infection, CNS disease manifested mainly during later stages of infection, when individuals were immunosuppressed (Kennedy 1988).

Initially, the cause for the late-stage development of ADC was not clear; however, in later studies of the replication and regulation of HIV in macrophages, the cells infected in brain demonstrated differential regulation of HIV in myeloid cells as compared to lymphocytes (Honda et al. 1998; Weiden et al. 2000; Descombes and Schibler 1991). HIV transcription in macrophages is regulated by the transcription factor c/EBPβ and its isoforms (Honda et al. 1998; Weiden et al. 2000; Descombes and Schibler 1991; Henderson et al. 1995; Henderson et al. 1996; Henderson and Calame 1997), in contrast to transcriptional regulation of HIV by NF-kB in CD4+ lymphocytes (Kinoshita et al. 1997; Tong-Starksen et al. 1987). The differential expression of c/EBPβ isoforms is modulated in macrophages by IFNβ (Honda et al. 1998; Weiden et al. 2000; Descombes and Schibler 1991). The presence of this cytokine in brain causes the translation of a dominant negative form of c/EBPβ that down-regulates viral transcription and histone acetylation of the HIV LTR, resulting in transcriptional silencing of HIV in vitro and SIV in vitro and in vivo (Henderson et al. 1995; Henderson et al. 1996; Henderson and Calame 1997; Barber et al. 2006a, b). Thus, regulation of HIV transcriptional activation and suppression by IFNβ may be one mechanism for establishing HIV latency in macrophages in tissues. Viral latency is a state of reversibly nonproductive infection of individual cells and provides an important mechanism for viral persistence and escape from immune recognition and drug pressure.

In the era of ART, fully suppressed HIV-infected individuals usually control virus replication in blood controlling viral levels below 50 copies of HIV/ml. The occurrence of systemic immunosuppression and HIV-associated dementia has been greatly diminished by treatment. ART does not eliminate the viral provirus from tissues but suppresses virus that becomes latent, and the latent reservoir is recognized as a major barrier to curing HIV-1 infection. HIV research is mainly focused on the suppression of virus replication in CD4+ lymphocytes (CD4+ T) and on mechanisms of virus latency and the formation of long-lived CD4+ T reservoirs (Pierson et al. 2000). The dramatic decrease in CNS dementia suggests that the infected brain macrophages (microglia and perivascular macrophages) are no longer actively infected during ART. In contrast to the availability of CD4+ T cells in blood, however, brain macrophages cannot be directly studied in humans because of the difficulty of analysis in situ. To circumvent this hurdle, HIV studies performed in cerebrospinal fluid (CSF) and brain (postmortem) of HIV-infected individuals on ART demonstrated that HIV is present in brain despite undetectable virus in the plasma (Spudich et al. 2005; Anderson et al. 2017). The identification of HIV-neurocognitive disorders (HAND) and HIV RNA in the CSF in HIV-infected individuals on suppressive ART further demonstrates that HIV infection persists in brain in either a latent or persistent form (Spudich et al. 2005; Anderson et al. 2017). SIV-infected macaques on ART regimens, similar to those used in humans, provide the opportunity to study longitudinal progression of AIDS, CNS infection, disease pathogenesis, and viral latency of both CD4+ T and myeloid cells in blood, CSF, and tissues, including brain.

## 2   HIV Infection and SIV Infection in the CNS

Both HIV and SIV infect the CNS as early as the first week after infection, and both viruses are detectable in CSF as well as in the blood of infected individuals during acute, chronic, and late-stage disease. Infection of the CNS is caused by entry of infected CD4+ T cells and monocytes trafficking across the blood–brain barrier (BBB). HIV infection and SIV infection are then spread to perivascular macrophages that line the BBB and to microglia, the resident brain macrophages. Microglia are embryonically derived cells that self-renew rather than being replenished from circulating monocytes (Guilliams et al. 2014; Schulz et al. 2012). Infected microglial cells have been identified in HIV-infected humans and SIV-infected nonhuman primates (Cosenza et al. 2002; Zhou et al. 2009; Neuen-Jacob et al. 1993; Lamers et al. 2011; Brinkmann et al. 1993; Clements et al. 2002). However, the role of microglia infection in long-term HIV latency and persistence is controversial despite the detection of HIV DNA in postmortem brain of ART-suppressed individuals. In SIV infection, microglia isolated from both viremic and ART-suppressed macaques contain SIV DNA and RNA (see below for detailed studies).

## 3   New Insights on Macrophage Origin and Phenotypes

Depletion of $CD4^+$ T cells is the hallmark of HIV-1 infection, and most studies of pathogenesis and latency of HIV and SIV have focused on lymphocytes. Nonetheless, macrophages are a natural host cell for lentiviruses (Narayan et al. 1983; Gendelman et al. 1985, 1986; Peluso et al. 1985) and multiple lines of evidence point to the importance of macrophages during HIV infection: (1) The accessory protein Vpx (HIV-2) specifically enhances viral replication in macrophages (Westmoreland et al. 2014; Sharova et al. 2008), but not in $CD4^+$ T cells. Comparably, Vpr (HIV-1) recruits UNG2 into virions and modulates viral mutation rates in macrophages (Chen et al. 2004); (2) many HIV-1 strains replicate efficiently in macrophages, independent of the presence of Vpx (Gorry et al. 2005; Karita et al. 1997); (3) AIDS is characterized by dramatic depletion of $CD4^+$ T cells; however, despite the depletion of these cells high plasma viral load persists, suggesting that viral replication is occurring in cells other than CD4$^+$ lymphocytes. In the macaque models for SIV infection, experimental depletion of CD4$^+$ T cells results in an increase in viral load and selection in vivo of CD4-independent macrophage-tropic SIV phenotypes (Ortiz et al. 2011; Francella et al. 2013; Igarashi et al. 2003); (4) damage to both lung and brain (interstitial pneumonia and encephalitis) is directly associated with infections of macrophages (Mankowski et al. 1997, 1998); (5) finally, activation of monocytes and macrophages during cART suppression in HIV is associated with higher morbidity (Hearps et al. 2012; McArthur et al. 2010; Burdo et al. 2013).

In the last several years, advances in myeloid cell biology have shown that every tissue harbors distinct populations of macrophages: those arriving during embryogenesis (both yolk sac and fetal liver-derived) and postnatal bone marrow-derived blood monocytes (Schulz et al. 2012). A similar classification can be extended to humans based on transcriptomic and phenotypic profiling (Guilliams et al. 2014). Most resident tissue mononuclear phagocytes—including Kupffer cells in the liver, Langerhans cells in the skin, microglia in brain, and alveolar and peritoneal macrophages—originate from *Myb*-independent progenitor cells that migrated directly or indirectly from the yolk sac to their respective tissues during embryogenesis (Guilliams et al. 2014; Yona et al. 2013). They are predominantly maintained through self-renewal during steady state, independently of adult hematopoiesis (Hashimoto et al. 2013). These cells have only recently been thoroughly characterized as distinct from monocyte-derived macrophages, and little is known about their in vitro function (Yona et al. 2013). Resident tissue macrophages, infected with HIV or SIV, have the potential to divide and expand the viral reservoirs in tissues. In addition, HIV- and SIV-infected macrophages are not efficiently killed by CD8+T cells unlike infected CD4+ T cells (Rainho et al. 2015; Vojnov et al. 2012). Thus, resident tissue macrophages remain in tissues long term, are capable of self-renewal, are relatively resistant to the cytopathic effects of HIV infection compared to CD4+ T cells, and may serve as stable viral reservoirs.

Some tissue macrophages are directly derived from blood monocytes, which arise from common monoblasts in bone marrow. Especially during infection or inflammation, circulating monocytes infiltrate tissues via pro-inflammatory mediators, including chemokine gradients, and differentiate into cells with a broad range of functions, depending on the microenvironment (Tamoutounour et al. 2013; Gordon and Taylor 2005). Although morphologically similar, macrophages originating from monocytes have distinct transcriptomes, surface markers, and phenotypic profiles from those of embryonic origin (Schulz et al. 2012; Gautier et al. 2012). Macrophages with the same ontogeny can express different sets of transcripts and may respond differently to pathogens, depending on the tissue location (Okabe and Medzhitov 2014).

In addition to distinct ontogenies, in vitro studies have demonstrated that monocyte-derived macrophages (MDM) can dramatically change their phenotype, pattern of gene expression, and functionality under different culture conditions (Gordon 2003). Generally, macrophages matured in CSF-1 are the starting point for many murine and human experiments. Any alteration to these culture conditions will present specialized functional properties, referred to as polarization. Historically, post-differentiated MDMs treated with IFN-γ were defined as M1, or classically activated, and MDMs treated with IL-4 termed M2, or alternatively activated. The current perspective acknowledges a spectrum of polarization, with large shifts in gene expression based on stimuli used in culture, and suggests terminology specific to the activator, such as M(IFN-γ + LPS). MDMs activated by IFN-γ alone or in combination with microbial stimuli (LPS) or cytokines (TNF) express copious amounts of inflammatory and effector molecules (IL-6, IL-2, CCL2, CXCL10, iNOS, and ROS) and contribute to the induction and maintenance of $T_H1$ and $T_H17$ responses. They also have enhanced complement- and antibody-mediated phagocytosis with microbicidal capacity. Conversely, MDMs exposed to anti-inflammatory stimuli such as IL-4, IL-13, IL-10, TGF-beta, immune complexes, or glucocorticoids are associated with $T_H2$ responses, high levels of Fc receptors (CD16, CD32, CD64), and the resolution of inflammatory responses (Martinez and Gordon 2014; Martinez et al. 2006). These phenotypes are likely to be less distinctive in vivo since the microenvironments that macrophages inhabit are exposed to a broad and changing range of signaling molecules. Also, in vitro studies show that macrophages, even after full polarization, can rapidly change phenotypes when exposed to a novel stimulus (Davis et al. 2013).

In the context of HIV infection, polarization studies demonstrate the antiviral features of M1 MDM (Alfano et al. 2013). This inflammatory phenotype has poor surface expression of CD4 and DC-SIGN, which are important receptors for viral binding. They also inhibit intracellular steps of viral replication due to high levels of *APOBEC3A*, *tetherin*, and *TRIM22* (Cobos Jimenez et al. 2012; Cassetta et al. 2013). These cells present a transcription profile (Martinez et al. 2006), which, at least in theory, should support HIV RNA expression (Sirois et al. 2011; Sgarbanti et al. 2008). Thus, while cell activation is directly related to increased viral transcription in CD4$^+$ T lymphocytes, this is not the case in macrophages. Polarization has been mainly explored in bone marrow-derived MDMs in vitro, while in vivo macrophages are incredibly heterogeneous and likely exist along a continuum of the M1-M2 spectrum. During inflammatory and infectious processes, there is a major influx of monocytes into tissues (Cai et al. 2014; Ginhoux and Jung 2014), making it difficult to demonstrate that shifts in polarization are occurring in resident cells and not in recently arrived monocytes. Recent studies on brain macrophages show that microglia respond to cytokine stimulation similar to MDMs (Orihuela et al. 2016; Qin et al. 2015; Xu et al. 2017). While the concepts described by in vitro studies have been useful in tissue macrophages (Kobayashi et al. 2013; Wan et al. 2014; Redente et al. 2010), direct translation of these studies has not been comprehensively explored in vivo (Cherry et al. 2014).

# 4 SIV Macaque Models

SIV infection in macaques comprehensively reproduces the immunodeficiency symptoms observed in HIV-infected humans, with infection of CD4+ T cells and monocytes in blood, and of macrophages in tissues such as lymph nodes, bowel, brain, lung, spleen, and heart (Williams et al. 2016; Kumar et al. 2016). Antiretroviral drugs have been shown to fully suppress SIV replication in blood (Kumar et al. 2016) and, in limited studies, CSF (Zink et al. 2010; Gama et al. 2017; Avalos et al. 2017) to levels comparable to those in ART-suppressed HIV-infected individuals. SIV-infected macaques carry latently infected CD4+ T cells that harbor replication-competent virus, as shown by quantitative viral outgrowth assays (QVOA) (Dinoso et al. 2009) and by the rapid rebound of SIV in plasma when ART is discontinued (Whitney et al. 2014). The role of infection and latency in monocytes and tissue macrophages in ART-suppressed macaques has only been recently addressed. This is important to pursue because, in the era of ART and potential HIV cure approaches, fully characterizing all latently infected cells that may contribute to viral rebound after cessation of ART has become a priority. Initial trials of HIV eradication strategies have focused on viral load (VL) in plasma as an indication of HIV reactivation or change in the latent reservoir, although there is evidence from the "Boston patients" that virus rebound occurred not only in the blood but also in the brain, based on CNS symptoms prior to virus rebound and presence of HIV in CSF (Henrich et al. 2014). However, the mechanisms that drive latency in macrophages remain unclear and, probably, are distinct from those in CD4+ T cells. Also, new evidence indicates that many latent SIV genomes located in tissues may respond differently to latency reversing agents (LRA) (Gama et al. 2017).

Several well-characterized SIV macaque models have been used mainly to study the development of AIDS and the pathogenesis of infection using a variety of SIV viral strains and molecular clones. The most commonly used strains are cloned SIVmac239 and SIVmac251 strains, or viruses derived from these strains. Also, there are SIV models focused on the study of infection and disease progression in the CNS (Williams et al. 2016). Each model uses distinct mechanisms to achieve SIV encephalitis, which includes infecting macaques with naturally occurring neurotropic and immunosuppressive virus swarms, neurotropic virus adapted by in vivo passage of SIV, and nonneurotropic strains in association with CD8+ lymphocyte depletion (Williams et al. 2016). A recent review has compared these SIV models, concluding that all the models include monocyte/macrophage infection and activation, and increased number of macrophages in the brain of macaques that develop encephalitis (Williams et al. 2016).

This review focuses on studies using an SIV macaque model in which animals are inoculated with a viral strain swarm (SIVdelta/B670) that contains 22 SIV *env*-defined genotypes and a neurovirulent, molecular clone (SIV/17E) that consistently causes AIDS in 90 days with a high incidence of CNS infection and encephalitis (Clements et al. 2002; Zink and Clements 2002; Zink et al. 1999). This SIV model has been characterized longitudinally, demonstrating that SIV infection in brain occurred in the first week of infection (by 4 d p.i.) and that virus infection in brain was differentially regulated from the periphery (Clements et al. 2002; Witwer et al. 2009). Macrophages in brain, including resident microglia and perivascular macrophages, are the major target cell in the CNS; SIV infection and HIV infection of macrophages have been shown to be transcriptionally regulated by C/EBPβ isoforms (Honda et al. 1998; Weiden et al. 2000; Descombes and Schibler 1991; Henderson et al. 1995, 1996; Henderson and Calame 1997), which are regulated by innate immune responses as discussed previously in this review. The regulation of SIV transcription in brain macrophages provides a mechanism for silencing of the viral genome in macrophages and is likely to contribute to mechanism of SIV and HIV latency in tissue-resident macrophages, particularly in the CNS.

ART regimens in this dual-infection SIV model result in suppression of viral load to undetectable levels in the plasma and CSF (Dinoso et al. 2009). There is an extensive literature that addresses the frequency of HIV infection and latency in CD4+ T cells in ART-suppressed humans, but this same rigorous analysis had not been applied to the ART-suppressed SIV macaque models. Therefore, we developed an SIV rCD4+ QVOA analogous to the HIV quantitative viral outgrowth assay (QVOA) (Dinoso et al. 2009) and used it to measure the frequency of rCD4+ cells harboring replication-competent SIV latent genomes, not only plasma but also in spleen and multiple lymph nodes and spleen (Fig. 1) (Dinoso et al. 2009). These studies demonstrated that the frequencies of latently infected rCD4+ cells in blood, lymph nodes, and spleen are very similar to those in ART-suppressed HIV-infected individuals. In another study using the same macaque model, it was shown that SIV DNA persists in the brain despite undetectable levels of SIV cellular RNA in the CNS (Zink et al. 2010).

## 5  Evidence for a Functional Viral Reservoir in Brain Macrophages

Our well-characterized and consistent macaque model for AIDS and CNS disease was also used to evaluate the contribution of brain macrophages in SIV latency and reactivation during ART (Gama et al. 2017). In this study, SIV-infected macaques were fully suppressed with ART for over one year (<30 copies of SIV RNA per ml of plasma). To induce in vivo activation of latent reservoirs, we tested a combination of two synergistic LRAs: the protein kinase C (PKC) activator ingenol-B and the histone deacetylase (HDAC) inhibitor vorinostat. We had previously shown that the ingenol-B reactivated HIV-1 genomes in two different in vitro HIV-1 latency models as well as in CD4+ T isolated from HIV-infected individuals (Abreu et al. 2014). Our results show that LRA administration led to an increase in VL in cerebrospinal fluid (CSF) in one of two SIV-infected macaques. The increase in virus in the CSF was 10-fold higher than virus rebound in the plasma, and phylogenetic analyses of viruses demonstrated distinct genotypes in the plasma and CSF, suggesting compartmentalization of virus in the brain. These findings suggest that the CNS harbors latent SIV genomes despite long-term ART suppression and that these reservoirs can be activated with LRAs. Although a small number of animals were assessed, this study is the first in vivo demonstration that the brain represents a consequential viral reservoir (Gama et al. 2017).

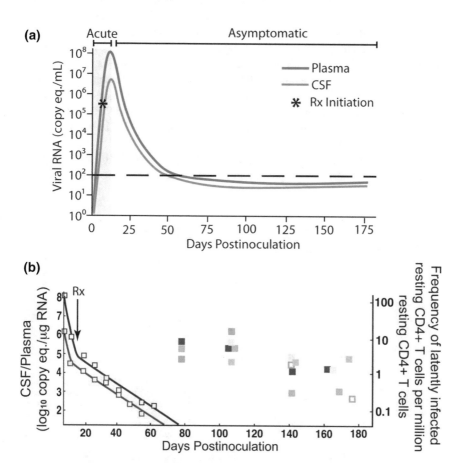

**Fig. 1  a** Viral RNA levels in plasma and CSF increase rapidly during the first 7–10 days prior to ART treatment. Within a few days after initiation of ART, plasma and CSF viral load decline. By approximately 60 days p.i., plasma and CSF viral load have declined to below the level of detection (<50 copy eq./ml) and viral loads remain low during ART. **b** The decline in plasma and CSF viral RNA occurred in two phases: an initial short-term rapid decline followed by a longer term slower decline similar to the two-phase decline seen in the plasma of HIV-infected individuals on HAART. At 80 days p.i., there were 8–10 latently infected resting CD4+ T cells per million resting CD4+ T cells in the blood. These numbers declined gradually to ∼ one latently infected resting CD4+ T cells per million by 175 days p.i. Abs., absolute; CSF, cerebrospinal fluid; p.i., post-inoculation; Rx, therapy

## 6  Relative Levels of Infection of CD4+ T Lymphocytes and Macrophages in SIV-Infected Macaques

The frequency of HIV or SIV infection of macrophages in tissues has been examined previously in a number of studies by measuring viral DNA in cells isolated from tissues (Avalos et al. 2016). However, this approach overestimates the number of productively infected CD4$^+$ T cells due to the presence of a large proportion of defective proviruses in vivo (Pollack et al. 2017). Thus, we developed a quantitative viral outgrowth assay similar to the CD4+ T-cell assay for HIV and for SIV to estimate the size of the potential latent reservoir of monocytes and macrophages (MΦ-QVOA) (Avalos et al. 2016) (Fig. 2 Macrophage QVOA). To validate this assay, we first examined the number of macrophages and CD4+ T cells in blood and tissues of viremic SIV-infected macaques (Avalos et al. 2016). To eliminate the potential contribution of CD4$^+$ T cells to the quantitation of infected macrophages, we also assessed the number of CD3$^+$ T cells in each assay by measuring *TCRβ* RNA.

The MΦ-QVOA utilized the expression of the integrin CD11b (Arnaout 1990) on monocytes and tissue macrophages and separated these myeloid cells from other

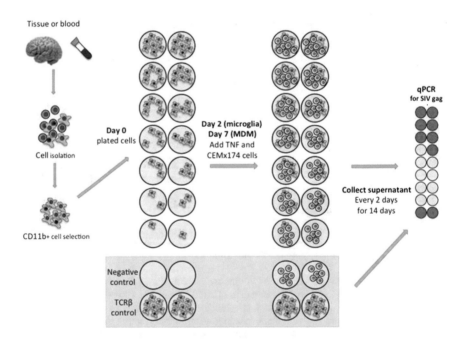

**Fig. 2** MΦ-QVOA. Monocytes from blood and macrophages from brain were collected from SIV-infected animals and purified by CD11b-specific bead selection. Macrophages expressing CD11b were plated in serial dilutions in triplicate wells. Cells were cultured with zidovudine (AZT) and darunavir (DRV). Nonadherent cells and the antiretrovirals were removed prior to activation with TNF and co-culture with CEMx174 cells (Avalos et al. 2016, 2017)

cell types by sorting with CD11b Miltenyi magnetic beads (Avalos et al. 2016). Like the CD4 + T-cell QVOA, the MΦ-QVOA involved a serial dilution of selected cells. Antiretroviral drugs were added to the culture to prevent virus spread from any CD4$^+$ T cells that might be in the culture during the first couple of days, while macrophage differentiated and matured in vitro. Unlike T cells, macrophages do not divide exponentially when activated in culture and strongly adhere to culture plates when grown in vitro. Cell supernatants were collected from the MΦ-QVOA wells after 12 days of cultivation (Fig. 2). Viral RNA was isolated from replicate wells and quantitated individually by qRT-PCR. The frequency of infectious virus per million (IUPM) was calculated using limiting dilution statistical analyses (Rosenbloom et al. 2015).

Quantitating macrophages with the QVOA from SIV-chronically and late-stage infected macaques demonstrated that the number of productively infected macrophages in a given tissue was surprisingly similar from macaque to macaque, whereas the number of productively infected macrophages varied widely across different tissues from the same SIV-infected macaque. The highest number of infected macrophages (424 IUPM) was measured in spleen, demonstrating that splenic macrophages are highly susceptible to SIV infection and harbor high levels of productive genomes (Fig. 3 IUPM CD4+ T cells, monocytes, macrophages). This suggests a role for tissue microenvironments in mediating virus infection of macrophages, since populations of macrophages that reside in each tissue may be differentially susceptible to SIV/HIV infection based on the cytokine profiles of the organs (Mulder et al. 2014).

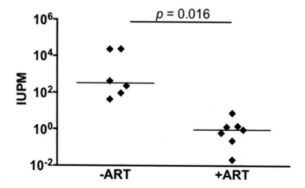

**Fig. 3** Quantitation of latently infected brain macrophages in ART-treated macaques by MΦ-QVOA. Quantitation of infected brain macrophages from ART-treated macaques (Avalos et al. 2017). Comparison between the numbers of SIV-infected brain macrophages isolated from animals that were not given ART (-ART) and the numbers isolated from animals that were treated with ART and with viral suppression <10 copies SIV RNA/ml plasma. The horizontal black line represents the median IUPM values. The MΦ QVOA results from SIV-infected animals with and without ART have been reported (Avalos et al. 2016, 2017). Significance was determined by Mann–Whitney nonparametric $t$ test; a $P$ of < 0.05 was considered significant

The number of infected brain macrophages, including both microglia and perivascular macrophages, was quantitated by MΦ-QVOA in SIV-infected macaques during both the chronic and late-stage diseases. Brain sections of these animals were examined for pathological changes associated with SIV encephalitis and were scored as none, mild, moderate, or severe disease. It was found that the brain of animals with mild-to-severe CNS disease contained the next highest level of infected cells (median 231 IUPM) compared to spleen. The two macaques with the most productively infected brain macrophages (Pm3 and Pm4 with 24,000 IUPM) had severe encephalitis and high levels of viral RNA in brain. The macaques without CNS disease had undetectable numbers of infected microglia/macrophages and little or no detectable viral RNA in the brain. Thus, the number of productively infected cells in the brain correlated with the severity of disease and the level of viral RNA detected in brain by qPCR. This study provided the first estimate of productively infected CD4$^+$ T cells and myeloid cells in SIV-infected tissues in vivo.

# 7 Quantitation of Latently Infected Brain Macrophages in ART-Suppressed SIV Macaques

ART has dramatically reduced the severe forms of HAND, but milder forms of neurologic impairment are still observed in HIV-infected individuals virally suppressed on ART. HAND is thought to be a result of chronic central nervous system (CNS) inflammation in the brain (Heaton et al. 2010, 2011; Rao et al. 2014; Rappaport and Volsky 2015). It is unclear whether inflammation is caused by incomplete penetrance of antiretroviral drugs into the CNS or the persistence of virus in brain macrophages (BrMΦ) in a latent state that reactivate causing sporadic inflammatory responses (Zayyad and Spudich 2015). Indeed, some HIV-infected individuals on ART have no detectable virus in the plasma but have measurable levels of HIV RNA in the CSF (Canestri et al. 2010; Eden et al. 2010). Also, HIV was detected after rebound in the CSF of the Boston patients, who had undetectable plasma HIV during ART interruption for several months (Henrich et al. 2014). There is a continuing debate on the sources of virus in the CSF and the cause of the chronic inflammation in brain that leads to HAND.

Using our SIV macaque model with SIV-infected macaques suppressed with four antiretroviral drugs for 100–500 days, we evaluated whether infected cells persist in brain despite ART. SIV-infected pigtailed macaques were virally suppressed with ART, and plasma and CSF VL were analyzed longitudinally to demonstrate viral suppression in the peripheral blood and the CNS. To assess whether virus persisted in brain macrophages (BrMΦ) in these long-term ART-suppressed macaques, we used MΦ-QVOA, qPCR, and in situ hybridization (ISH) to measure the frequency of infected cells and levels of viral RNA and DNA in brain. Viral RNA in brain tissue of suppressed macaques was undetectable, although viral DNA was observed in all animals. The MΦ-QVOA demonstrated that the majority of suppressed animals contained latently infected BrMΦ. We also showed that virus produced in the MΦ-QVOAs was replication competent, suggesting that latently infected BrMΦ are capable of re-establishing productive infection upon ART interruption. This study provides the first confirmation of replication-competent SIV in BrMΦ of ART-suppressed macaques and suggests that the highly debated question of viral latency in macrophages, at least in brain, has been addressed in SIV-infected macaques treated with ART.

In this study, we identified latently infected BrMΦ in brain samples containing fewer than 10 copies of SIV DNA per million cells. In animals suppressed for more than 500 days, the number of infected macrophages measured in the Mø-QVOA ranged from 3.6 to 15 in a hundred million cells, supporting the low level of DNA quantitated by qPCR. Thus, the quantitation of SIV DNA or RNA by PCR in brain tissue does not fully reflect the size of the latent functional reservoir, which is the main target in eradication strategies.

Most animals in the study harbored latently infected macrophages in regions of the brain that contained no detectable viral RNA. After 1.7 years of viral suppression, three macaques in the study showed no viral RNA in basal ganglia and parietal cortex. Nevertheless, all three macaques had replication-competent virus produced in the isolated BrMΦ. Of note, we detected viral RNA by ISH in the brain of one of the macaques in the study treated with a LRA. However, the RNA was detected in the occipital cortex, a brain section not used for the BrMΦ QVOA. These results corroborate findings, showing that SIV, and potentially HIV, infection in brain is highly focal (Gama et al. 2017) and can provide variable results depending on the brain region analyzed for each specific assay.

Also, the results from the MΦ-QVOA showing that a small number of replication-competent viruses are sporadically released in some latent BrMΦ indicate that parameters we used to define a positive QVOA well supernatants (with >50 SIV RNA copies/mL) underestimate the number of latently infected cells that produce replication-competent virus, at least in macrophages. Indeed, viruses collected from most QVOA assay supernatants were able to spread in healthy PBMC.

The demonstration that there is latent replication-competent virus in SIV-infected ART-suppressed macaque brain provides a mechanism for the ongoing macrophage activation observed both in the macaques and in the HIV individuals suppressed on ART. Recent studies have suggested that, while virus does not spread during ART suppression, there is ongoing stochastic activation of virus genomes in latently infected cells (Weinberger et al. 2005; Dar et al. 2014). Reactivation of virus without spread in the macrophage is likely to induce innate immune responses and cellular activation. Thus, productively infected latent macrophages in brain provide a mechanism for the ongoing inflammation of HIV in a fully suppressed individual. Also, it has been recently demonstrated that defective provirus expressed in rCD4s could be recognized by adaptive immune responses, shaping the proviral landscape (Pollack et al. 2017). It is possible that similar responses might happen with viral proteins generated from defective proviruses in BrMΦ.

# 8  Conclusions

The presence of a long-term functional reservoir of SIV in brain macrophages that parallels the biologic and pathologic features of infected individuals with HIV encephalitis suggests that the HIV in brain may be a formidable barrier to strategies to decrease or eliminate latent reservoirs. Further, the presence of low levels of viral DNA in brain of ART-suppressed macaques can contribute to virus spread in brain and potentially in the periphery during cessation of ART or eradication treatments. While the brain is protected by the blood–brain barrier and eradication approaches may not penetrate the brain, immune activation in the periphery could potentially activate virus in the CNS. On the other hand, the lack of CNS penetrance of such eradication therapies would potentially leave the CNS functional reservoir intact and undermine virus eradication. Strategies that include activation of virus in brain may have the effect of increasing inflammation and neuronal toxicity due to increased macrophage activation and production of cytokines, as we observed in a suppressed macaque treated with two cycles of LRAs (Gama et al. 2017). Our studies demonstrating the presence of a functional latent reservoir in brain macrophages have major implications for SIV eradication studies used to model treatment for HIV individuals. Examining recrudescence of virus in plasma but not CSF may overlook a source of virus that significantly contributes to the virus rebound.

**Acknowledgements**  These studies were funded by NIH awards R01NS089482, R01NS077869, P40OD0131117, R01NS055651, R56AI118753, R01AI127142, P01MH070306, P01AI131306, and the Johns Hopkins University Center for AIDS Research P30AI094189.

Anti-retroviral compounds for these studies were kindly donated by Gilead, ViiV Healthcare, Bristol-Meyers Squibb, Merck, Abbvie, Janssen, and Roche. These studies were supported by the excellent technical staff in the Retrovirus Lab at Johns Hopkins.

# References

Abreu CM, Price SL, Shirk EN, Cunha RD, Pianowski LF, Clements JE, Tanuri A, Gama L (2014) Dual role of novel ingenol derivatives from *Euphorbia tirucalli* in HIV replication: inhibition of de novo infection and activation of viral LTR. PLoS ONE 9:e97257

Alfano M, Graziano F, Genovese L, Poli G (2013) Macrophage polarization at the crossroad between HIV-1 infection and cancer development. Arterioscler Thromb Vasc Biol 33:1145–1152

Anderson AM, Munoz-Moreno JA, McClernon DR, Ellis RJ, Cookson D, Clifford DB, Collier C, Gelman BB, Marra CM, McArthur JC, McCutchan JA, Morgello S, Sacktor N, Simpson DM, Franklin DR, Heaton RK, Grant I, Letendre SL, Group C (2017) Prevalence and correlates of persistent HIV-1 RNA in cerebrospinal fluid during antiretroviral therapy. J Infect Dis 215:105–113

Arnaout MA (1990) Structure and function of the leukocyte adhesion molecules CD11/CD18. Blood 75:1037–1050

Avalos CR, Price SL, Forsyth ER, Pin JN, Shirk EN, Bullock BT, Queen SE, Li M, Gellerup D, O'Connor SL, Zink MC, Mankowski JL, Gama L, Clements JE (2016) Quantitation of productively infected monocytes and macrophages of simian immunodeficiency virus-infected macaques. J Virol 90:5643–5656

Avalos CR, Abreu CM, Queen SE, Li M, Price S, Shirk EN, Engle EL, Forsyth E, Bullock BT, Mac Gabhann F, Wietgrefe SW, Haase AT, Zink MC, Mankowski JL, Clements JE, Gama L (2017). Brain macrophages in simian immunodeficiency virus-infected, antiretroviral-suppressed macaques: a functional latent reservoir. MBio 8

Barber SA, Gama L, Dudaronek JM, Voelker T, Tarwater PM, Clements JE (2006a) Mechanism for the establishment of transcriptional HIV latency in the brain in a simian immunodeficiency virus-macaque model. J Infect Dis 193:963–970

Barber SA, Gama L, Li M, Voelker T, Anderson JE, Zink MC, Tarwater PM, Carruth LM, Clements JE (2006b) Longitudinal analysis of simian immunodeficiency virus (SIV) replication in the lungs: compartmentalized regulation of SIV. J Infect Dis 194:931–938

Barre-Sinoussi F, Chermann JC, Rey F, Nugeyre MT, Chamaret S, Gruest J, Dauguet C, Axler-Blin C, Vezinet-Brun F, Rouzioux C, Rozenbaum W, Montagnier L (1983) Isolation of a T-lymphotropic retrovirus from a patient at risk for acquired immune deficiency syndrome (AIDS). Science 220:868–871

Brinkmann R, Schwinn A, Muller J, Stahl-Hennig C, Coulibaly C, Hunsmann G, Czub S, Rethwilm A, Dorries R, ter Meulen V (1993) In vitro and in vivo infection of rhesus monkey microglial cells by simian immunodeficiency virus. Virology 195:561–568

Burdo TH, Lackner A, Williams KC (2013) Monocyte/macrophages and their role in HIV neuropathogenesis. Immunol Rev 254:102–113

Cai Y, Sugimoto C, Arainga M, Alvarez X, Didier ES, Kuroda MJ (2014) In vivo characterization of alveolar and interstitial lung macrophages in rhesus macaques: implications for understanding lung disease in humans. J Immunol 192:2821–2829

Canestri A, Lescure FX, Jaureguiberry S, Moulignier A, Amiel C, Marcelin AG, Peytavin G, Tubiana R, Pialoux G, Katlama C (2010) Discordance between cerebral spinal fluid and plasma HIV replication in patients with neurological symptoms who are receiving suppressive antiretroviral therapy. Clin Infect Dis 50:773–778

Cassetta L, Kajaste-Rudnitski A, Coradin T, Saba E, Della Chiara G, Barbagallo M, Graziano F, Alfano M, Cassol E, Vicenzi E, Poli G (2013) M1 polarization of human monocyte-derived macrophages restricts pre and postintegration steps of HIV-1 replication. AIDS 27:1847–1856

Chen R, Le Rouzic E, Kearney JA, Mansky LM, Benichou S (2004) Vpr-mediated incorporation of UNG2 into HIV-1 particles is required to modulate the virus mutation rate and for replication in macrophages. J Biol Chem 279:28419–28425

Cherry JD, Olschowka JA, O'Banion MK (2014) Neuroinflammation and M2 microglia: the good, the bad, and the inflamed. J Neuroinflammation 11:98

Chiodi F, Albert J, Olausson E, Norkrans G, Hagberg L, Sonnerborg A, Asjo B, Fenyo EM (1988) Isolation frequency of human immunodeficiency virus from cerebrospinal fluid and blood of patients with varying severity of HIV infection. AIDS Res Hum Retroviruses 4:351–358

Clements JE, Babas T, Mankowski JL, Suryanarayana K, Piatak M Jr, Tarwater PM, Lifson JD, Zink MC (2002a) The central nervous system as a reservoir for simian immunodeficiency virus (SIV): steady-state levels of SIV DNA in brain from acute through asymptomatic infection. J Infect Dis 186:905–913

Clements JE, Babas T, Mankowski JL, Suryanarayana K, Piatak M Jr, Tarwater PM, Lifson JD, Zink MC (2002b) The central nervous system as a reservoir for simian immunodeficiency virus (SIV): steady-state levels of SIV DNA in brain from acute through asymptomatic infection. J Infect Dis 186:905–913

Cobos Jimenez V, Booiman T, de Taeye SW, van Dort KA, Rits MA, Hamann J, Kootstra NA (2012) Differential expression of HIV-1 interfering factors in monocyte-derived macrophages stimulated with polarizing cytokines or interferons. Sci Rep 2:763

Cosenza MA, Zhao ML, Si Q, Lee SC (2002) Human brain parenchymal microglia express CD14 and CD45 and are productively infected by HIV-1 in HIV-1 encephalitis. Brain Pathol 12:442–455

Craig LE, Sheffer D, Meyer AL, Hauer D, Lechner F, Peterhans E, Adams RJ, Clements JE, Narayan O, Zink MC (1997) Pathogenesis of ovine lentiviral encephalitis: derivation of a neurovirulent strain by in vivo passage. J Neurovirol 3:417–427

Dar RD, Hosmane NN, Arkin MR, Siliciano RF, Weinberger LS (2014) Screening for noise in gene expression identifies drug synergies. Science 344:1392–1396

Davis MJ, Tsang TM, Qiu Y, Dayrit JK, Freij JB, Huffnagle GB, Olszewski MA (2013) Macrophage M1/M2 polarization dynamically adapts to changes in cytokine microenvironments in *Cryptococcus neoformans* infection. MBio 4:e00264–00213

Descombes P, Schibler U (1991) A liver-enriched transcriptional activator protein, LAP, and a transcriptional inhibitory protein, LIP, are translated from the same mRNA. Cell 67:569–579

Dinoso JB, Rabi SA, Blankson JN, Gama L, Mankowski JL, Siliciano RF, Zink MC, Clements JE (2009) A simian immunodeficiency virus-infected macaque model to study viral reservoirs that persist during highly active antiretroviral therapy. J Virol 83:9247–9257

Eden A, Fuchs D, Hagberg L, Nilsson S, Spudich S, Svennerholm B, Price RW, Gisslen M (2010) HIV-1 viral escape in cerebrospinal fluid of subjects on suppressive antiretroviral treatment. J Infect Dis 202:1819–1825

Epstein LG, Sharer LR, Cho ES, Myenhofer M, Navia B, Price RW (1984) HTLV-III/LAV-like retrovirus particles in the brains of patients with AIDS encephalopathy. AIDS Res 1:447–454

Francella N, Elliott ST, Yi Y, Gwyn SE, Ortiz AM, Li B, Silvestri G, Paiardini M, Derdeyn CA, Collman RG (2013) Decreased plasticity of coreceptor use by CD4-independent SIV Envs that emerge in vivo. Retrovirology 10:133

Gama L, Abreu CM, Shirk EN, Price SL, Li M, Laird GM, Pate KA, Wietgrefe SW, O'Connor SL, Pianowski L, Haase AT, Van Lint C, Siliciano RF, Clements JE, Group L-SS (2017) Reactivation of simian immunodeficiency virus reservoirs in the brain of virally suppressed macaques. AIDS 31:5–14

Gartner S, Markovits P, Markovitz DM, Kaplan MH, Gallo RC, Popovic M (1986) The role of mononuclear phagocytes in HTLV-III/LAV infection. Science 233:215–219

Gautier EL, Shay T, Miller J, Greter M, Jakubzick C, Ivanov S, Helft J, Chow A, Elpek KG, Gordonov S, Mazloom AR, Ma'ayan A, Chua WJ, Hansen TH, Turley SJ, Merad M, Randolph GJ, Immunological Genome C (2012) Gene-expression profiles and transcriptional regulatory pathways that underlie the identity and diversity of mouse tissue macrophages. Nat Immunol 13:1118–1128

Gendelman HE, Narayan O, Molineaux S, Clements JE, Ghotbi Z (1985a) Slow, persistent replication of lentiviruses: role of tissue macrophages and macrophage precursors in bone marrow. Proc Natl Acad Sci U S A 82:7086–7090

Gendelman HE, Moench TR, Narayan O, Griffin DE, Clements JE (1985b) A double labeling technique for performing immunocytochemistry and in situ hybridization in virus infected cell cultures and tissues. J Virol Methods 11:93–103

Gendelman HE, Narayan O, Kennedy-Stoskopf S, Kennedy PG, Ghotbi Z, Clements JE, Stanley J, Pezeshkpour G (1986) Tropism of sheep lentiviruses for monocytes: susceptibility to infection and virus gene expression increase during maturation of monocytes to macrophages. J Virol 58:67–74

Ginhoux F, Jung S (2014) Monocytes and macrophages: developmental pathways and tissue homeostasis. Nat Rev Immunol 14:392–404

Gonda MA, Wong-Staal F, Gallo RC, Clements JE, Narayan O, Gilden RV (1985) Sequence homology and morphologic similarity of HTLV-III and visna virus, a pathogenic lentivirus. Science 227:173–177

Gonda MA, Braun MJ, Clements JE, Pyper JM, Wong-Staal F, Gallo RC, Gilden RV (1986) Human T-cell lymphotropic virus type III shares sequence homology with a family of pathogenic lentiviruses. Proc Natl Acad Sci U S A 83:4007–4011

Gordon S (2003) Alternative activation of macrophages. Nat Rev Immunol 3:23–35

Gordon S, Taylor PR (2005) Monocyte and macrophage heterogeneity. Nat Rev Immunol 5:953–964

Gorry PR, Churchill M, Crowe SM, Cunningham AL, Gabuzda D (2005) Pathogenesis of macrophage tropic HIV-1. Curr HIV Res 3:53–60

Gottlieb MS, Schroff R, Schanker HM, Weisman JD, Fan PT, Wolf RA, Saxon A (1981) *Pneumocystis carinii* pneumonia and mucosal candidiasis in previously healthy homosexual men: evidence of a new acquired cellular immunodeficiency. N Engl J Med 305:1425–1431

Guilliams M, Ginhoux F, Jakubzick C, Naik SH, Onai N, Schraml BU, Segura E, Tussiwand R, Yona S (2014) Dendritic cells, monocytes and macrophages: a unified nomenclature based on ontogeny. Nat Rev Immunol 14:571–578

Hashimoto D, Chow A, Noizat C, Teo P, Beasley MB, Leboeuf M, Becker CD, See P, Price J, Lucas D, Greter M, Mortha A, Boyer SW, Forsberg EC, Tanaka M, van Rooijen N, Garcia-Sastre A, Stanley ER, Ginhoux F, Frenette PS, Merad M (2013) Tissue-resident macrophages self-maintain locally throughout adult life with minimal contribution from circulating monocytes. Immunity 38:792–804

Hearps AC, Martin GE, Angelovich TA, Cheng WJ, Maisa A, Landay AL, Jaworowski A, Crowe SM (2012) Aging is associated with chronic innate immune activation and dysregulation of monocyte phenotype and function. Aging Cell 11:867–875

Heaton RK, Clifford DB, Franklin DR Jr., Woods SP, Ake C, Vaida F, Ellis RJ, Letendre SL, Marcotte TD, Atkinson JH, Rivera-Mindt M, Vigil OR, Taylor MJ, Collier AC, Marra CM, Gelman BB, McArthur JC, Morgello S, Simpson DM, McCutchan JA, Abramson I, Gamst A, Fennema-Notestine C, Jernigan TL, Wong J, Grant I, Group C (2010) HIV-associated neurocognitive disorders persist in the era of potent antiretroviral therapy: CHARTER study. Neurology 75:2087–2096

Heaton RK, Franklin DR, Ellis RJ, McCutchan JA, Letendre SL, Leblanc S, Corkran SH, Duarte NA, Clifford DB, Woods SP, Collier AC, Marra CM, Morgello S, Mindt MR, Taylor MJ, Marcotte TD, Atkinson JH, Wolfson T, Gelman BB, McArthur JC, Simpson DM, Abramson I, Gamst A, Fennema-Notestine C, Jernigan TL, Wong J, Grant I, Group C, Group H (2011) HIV-associated neurocognitive disorders before and during the era of combination antiretroviral therapy: differences in rates, nature, and predictors. J Neurovirol 17:3–16

Henderson AJ, Calame KL (1997) CCAAT/enhancer binding protein (C/EBP) sites are required for HIV-1 replication in primary macrophages but not CD4(+) T cells. Proc Natl Acad Sci U S A 94:8714–8719

Henderson AJ, Zou X, Calame KL (1995) C/EBP proteins activate transcription from the human immunodeficiency virus type 1 long terminal repeat in macrophages/monocytes. J Virol 69:5337–5344

Henderson AJ, Connor RI, Calame KL (1996) C/EBP activators are required for HIV-1 replication and proviral induction in monocytic cell lines. Immunity 5:91–101

Henrich TJ, Hanhauser E, Marty FM, Sirignano MN, Keating S, Lee TH, Robles YP, Davis BT, Li JZ, Heisey A, Hill AL, Busch MP, Armand P, Soiffer RJ, Altfeld M, Kuritzkes DR (2014) Antiretroviral-free HIV-1 remission and viral rebound after allogeneic stem cell transplantation: report of 2 cases. Ann Intern Med 161:319–327

Ho DD, Rota TR, Schooley RT, Kaplan JC, Allan JD, Groopman JE, Resnick L, Felsenstein D, Andrews CA, Hirsch MS (1985) Isolation of HTLV-III from cerebrospinal fluid and neural tissues of patients with neurologic syndromes related to the acquired immunodeficiency syndrome. N Engl J Med 313:1493–1497

Honda Y, Rogers L, Nakata K, Zhao BY, Pine R, Nakai Y, Kurosu K, Rom WN, Weiden M (1998) Type I interferon induces inhibitory 16-kD CCAAT/ enhancer binding protein (C/EBP)beta, repressing the HIV-1 long terminal repeat in macrophages: pulmonary tuberculosis alters C/EBP expression, enhancing HIV-1 replication. J Exp Med 188:1255–1265

Igarashi T, Imamichi H, Brown CR, Hirsch VM, Martin MA (2003) The emergence and characterization of macrophage-tropic SIV/HIV chimeric viruses (SHIVs) present in CD4+ T cell-depleted rhesus monkeys. J Leukoc Biol 74:772–780

Karita E, Nkengasong JN, Willems B, Vanham G, Fransen K, Heyndrickx L, Janssens W, Piot P, van der Groen G (1997) Macrophage-tropism of HIV-1 isolates of different genetic subtypes. AIDS 11:1303–1304

Kennedy PG (1988) Neurological complications of human immunodeficiency virus infection. Postgrad Med J 64:180–187

Kinoshita S, Su L, Amano M, Timmerman LA, Kaneshima H, Nolan GP (1997) The T cell activation factor NF-ATc positively regulates HIV-1 replication and gene expression in T cells. Immunity 6:235–244

Kobayashi K, Imagama S, Ohgomori T, Hirano K, Uchimura K, Sakamoto K, Hirakawa A, Takeuchi H, Suzumura A, Ishiguro N, Kadomatsu K (2013) Minocycline selectively inhibits M1 polarization of microglia. Cell Death Dis 4:e525

Kumar N, Chahroudi A, Silvestri G (2016) Animal models to achieve an HIV cure. Curr Opin HIV AIDS 11:432–441

Lamers SL, Gray RR, Salemi M, Huysentruyt LC, McGrath MS (2011) HIV-1 phylogenetic analysis shows HIV-1 transits through the meninges to brain and peripheral tissues. Infect Genet Evol 11:31–37

Mankowski JL, Flaherty MT, Spelman JP, Hauer DA, Didier PJ, Amedee AM, Murphey-Corb M, Kirstein LM, Munoz A, Clements JE, Zink MC (1997) Pathogenesis of simian immunodeficiency virus encephalitis: viral determinants of neurovirulence. J Virol 71:6055–6060

Mankowski JL, Carter DL, Spelman JP, Nealen ML, Maughan KR, Kirstein LM, Didier PJ, Adams RJ, Murphey-Corb M, Zink MC (1998) Pathogenesis of simian immunodeficiency virus pneumonia: an immunopathological response to virus. Am J Pathol 153:1123–1130

Martinez FO, Gordon S (2014) The M1 and M2 paradigm of macrophage activation: time for reassessment. F1000Prime Rep 6:13

Martinez FO, Gordon S, Locati M, Mantovani A (2006) Transcriptional profiling of the human monocyte-to-macrophage differentiation and polarization: new molecules and patterns of gene expression. J Immunol 177:7303–7311

Masur H, Michelis MA, Greene JB, Onorato I, Stouwe RA, Holzman RS, Wormser G, Brettman L, Lange M, Murray HW, Cunningham-Rundles S (1981) An outbreak of community-acquired *Pneumocystis carinii* pneumonia: initial manifestation of cellular immune dysfunction. N Engl J Med 305:1431–1438

McArthur JC, Steiner J, Sacktor N, Nath A (2010) Human immunodeficiency virus-associated neurocognitive disorders: Mind the gap. Ann Neurol 67:699–714

Mulder R, Banete A, Basta S (2014) Spleen-derived macrophages are readily polarized into classically activated (M1) or alternatively activated (M2) states. Immunobiology 219:737–745

Narayan O, Kennedy-Stoskopf S, Sheffer D, Griffin DE, Clements JE (1983) Activation of caprine arthritis-encephalitis virus expression during maturation of monocytes to macrophages. Infect Immun 41:67–73

Neuen-Jacob E, Arendt G, Wendtland B, Jacob B, Schneeweis M, Wechsler W (1993) Frequency and topographical distribution of CD68-positive macrophages and HIV-1 core proteins in HIV-associated brain lesions. Clin Neuropathol 12:315–324

Okabe Y, Medzhitov R (2014) Tissue-specific signals control reversible program of localization and functional polarization of macrophages. Cell 157:832–844

Orihuela R, McPherson CA, Harry GJ (2016) Microglial M1/M2 polarization and metabolic states. Br J Pharmacol 173:649–665

Ortiz AM, Klatt NR, Li B, Yi Y, Tabb B, Hao XP, Sternberg L, Lawson B, Carnathan PM, Cramer EM, Engram JC, Little DM, Ryzhova E, Gonzalez-Scarano F, Paiardini M, Ansari AA, Ratcliffe S, Else JG, Brenchley JM, Collman RG, Estes JD, Derdeyn CA, Silvestri G (2011) Depletion of CD4(+) T cells abrogates post-peak decline of viremia in SIV-infected rhesus macaques. J Clin Invest 121:4433–4445

Peluso R, Haase A, Stowring L, Edwards M, Ventura P (1985) A Trojan Horse mechanism for the spread of visna virus in monocytes. Virology 147:231–236

Pierson T, McArthur J, Siliciano RF (2000) Reservoirs for HIV-1: mechanisms for viral persistence in the presence of antiviral immune responses and antiretroviral therapy. Annu Rev Immunol 18:665–708

Pollack RA, Jones RB, Pertea M, Bruner KM, Martin AR, Thomas AS, Capoferri AA, Beg SA, Huang SH, Karandish S, Hao H, Halper-Stromberg E, Yong PC, Kovacs C, Benko E,

Siliciano RF, Ho YC (2017) Defective HIV-1 proviruses are expressed and can be recognized by cytotoxic T lymphocytes, which shape the proviral landscape. Cell Host Microbe 21(494–506):e494

Popovic M, Sarngadharan MG, Read E, Gallo RC (1984) Detection, isolation, and continuous production of cytopathic retroviruses (HTLV-III) from patients with AIDS and pre-AIDS. Science 224:497–500

Qin Y, Sun X, Shao X, Cheng C, Feng J, Sun W, Gu D, Liu W, Xu F, Duan Y (2015) Macrophage-microglia networks drive M1 microglia polarization after mycobacterium infection. Inflammation 38:1609–1616

Rainho JN, Martins MA, Cunyat F, Watkins IT, Watkins DI, Stevenson M (2015) Nef is dispensable for resistance of simian immunodeficiency virus-infected macrophages to CD8+T cell killing. J Virol 89:10625–10636

Rao VR, Ruiz AP, Prasad VR (2014) Viral and cellular factors underlying neuropathogenesis in HIV associated neurocognitive disorders (HAND). AIDS Res Ther 11:13

Rappaport J, Volsky DJ (2015) Role of the macrophage in HIV-associated neurocognitive disorders and other comorbidities in patients on effective antiretroviral treatment. J Neurovirol 21:235–241

Redente EF, Higgins DM, Dwyer-Nield LD, Orme IM, Gonzalez-Juarrero M, Malkinson AM (2010) Differential polarization of alveolar macrophages and bone marrow-derived monocytes following chemically and pathogen-induced chronic lung inflammation. J Leukoc Biol 88:159–168

Rosenbloom DI, Elliott O, Hill AL, Henrich TJ, Siliciano JM, Siliciano RF (2015) Designing and interpreting limiting dilution assays: general principles and applications to the latent reservoir for human immunodeficiency virus-1. Open Forum Infect Dis 2:ofv123

Schulz C, Gomez Perdiguero E, Chorro L, Szabo-Rogers H, Cagnard N, Kierdorf K, Prinz M, Wu B, Jacobsen SE, Pollard JW, Frampton J, Liu KJ, Geissmann F (2012) A lineage of myeloid cells independent of Myb and hematopoietic stem cells. Science 336:86–90

Sgarbanti M, Remoli AL, Marsili G, Ridolfi B, Borsetti A, Perrotti E, Orsatti R, Ilari R, Sernicola L, Stellacci E, Ensoli B, Battistini A (2008) IRF-1 is required for full NF-kappaB transcriptional activity at the human immunodeficiency virus type 1 long terminal repeat enhancer. J Virol 82:3632–3641

Sharova N, Wu Y, Zhu X, Stranska R, Kaushik R, Sharkey M, Stevenson M (2008) Primate lentiviral Vpx commandeers DDB1 to counteract a macrophage restriction. PLoS Pathog 4:e1000057

Shaw GM, Harper ME, Hahn BH, Epstein LG, Gajdusek DC, Price RW, Navia BA, Petito CK, O'Hara CJ, Groopman JE et al (1985) HTLV-III infection in brains of children and adults with AIDS encephalopathy. Science 227:177–182

Sirois M, Robitaille L, Allary R, Shah M, Woelk CH, Estaquier J, Corbeil J (2011) TRAF6 and IRF7 control HIV replication in macrophages. PLoS ONE 6:e28125

Spudich SS, Nilsson AC, Lollo ND, Liegler TJ, Petropoulos CJ, Deeks SG, Paxinos EE, Price RW (2005) Cerebrospinal fluid HIV infection and pleocytosis: relation to systemic infection and antiretroviral treatment. BMC Infect Dis 5:98

Tamoutounour S, Guilliams M, Montanana Sanchis F, Liu H, Terhorst D, Malosse C, Pollet E, Ardouin L, Luche H, Sanchez C, Dalod M, Malissen B, Henri S (2013) Origins and functional specialization of macrophages and of conventional and monocyte-derived dendritic cells in mouse skin. Immunity 39:925–938

Tong-Starksen SE, Luciw PA, Peterlin BM (1987) Human immunodeficiency virus long terminal repeat responds to T-cell activation signals. Proc Natl Acad Sci U S A 84:6845–6849

Vojnov L, Martins MA, Bean AT, Veloso de Santana MG, Sacha JB, Wilson NA, Bonaldo MC, Galler R, Stevenson M, Watkins DI (2012) The majority of freshly sorted simian immunodeficiency virus (SIV)-specific CD8(+) T cells cannot suppress viral replication in SIV-infected macrophages. J Virol 86:4682–4687

Wan J, Benkdane M, Teixeira-Clerc F, Bonnafous S, Louvet A, Lafdil F, Pecker F, Tran A, Gual P, Mallat A, Lotersztajn S, Pavoine C (2014) M2 Kupffer cells promote M1 Kupffer cell apoptosis: a protective mechanism against alcoholic and nonalcoholic fatty liver disease. Hepatology 59:130–142

Weiden M, Tanaka N, Qiao Y, Zhao BY, Honda Y, Nakata K, Canova A, Levy DE, Rom WN, Pine R (2000) Differentiation of monocytes to macrophages switches the Mycobacterium tuberculosis effect on HIV-1 replication from stimulation to inhibition: modulation of interferon response and CCAAT/enhancer binding protein beta expression. J Immunol 165:2028–2039

Weinberger LS, Burnett JC, Toettcher JE, Arkin AP, Schaffer DV (2005) Stochastic gene expression in a lentiviral positive-feedback loop: HIV-1 Tat fluctuations drive phenotypic diversity. Cell 122:169–182

Westmoreland SV, Converse AP, Hrecka K, Hurley M, Knight H, Piatak M, Lifson J, Mansfield KG, Skowronski J, Desrosiers RC (2014) SIV vpx is essential for macrophage infection but not for development of AIDS. PLoS ONE 9:e84463

Whitney JB, Hill AL, Sanisetty S, Penaloza-MacMaster P, Liu J, Shetty M, Parenteau L, Cabral C, Shields J, Blackmore S, Smith JY, Brinkman AL, Peter LE, Mathew SI, Smith KM, Borducchi EN, Rosenbloom DI, Lewis MG, Hattersley J, Li B, Hesselgesser J, Geleziunas R, Robb ML, Kim JH, Michael NL, Barouch DH (2014) Rapid seeding of the viral reservoir prior to SIV viraemia in rhesus monkeys. Nature 512:74–77

Williams K, Lackner A, Mallard J (2016) Non-human primate models of SIV infection and CNS neuropathology. Curr Opin Virol 19:92–98

Witwer KW, Gama L, Li M, Bartizal CM, Queen SE, Varrone JJ, Brice AK, Graham DR, Tarwater PM, Mankowski JL, Zink MC, Clements JE (2009) Coordinated regulation of SIV replication and immune responses in the CNS. PLoS ONE 4:e8129

Xu H, Wang Z, Li J, Wu H, Peng Y, Fan L, Chen J, Gu C, Yan F, Wang L, Chen G (2017) The polarization states of microglia in TBI: a new paradigm for pharmacological intervention. Neural Plast 2017:5405104

Yona S, Kim KW, Wolf Y, Mildner A, Varol D, Breker M, Strauss-Ayali D, Viukov S, Guilliams M, Misharin A, Hume DA, Perlman H, Malissen B, Zelzer E, Jung S (2013) Fate mapping reveals origins and dynamics of monocytes and tissue macrophages under homeostasis. Immunity 38:79–91

Zayyad Z, Spudich S (2015) Neuropathogenesis of HIV: from initial neuroinvasion to HIV-associated neurocognitive disorder (HAND). Curr HIV/AIDS Rep 12:16–24

Zhou L, Rua R, Ng T, Vongrad V, Ho YS, Geczy C, Hsu K, Brew BJ, Saksena NK (2009) Evidence for predilection of macrophage infiltration patterns in the deeper midline and mesial temporal structures of the brain uniquely in patients with HIV-associated dementia. BMC Infect Dis 9:192

Zink MC, Clements JE (2002) A novel simian immunodeficiency virus model that provides insight into mechanisms of human immunodeficiency virus central nervous system disease. J Neurovirol 8(Suppl 2):42–48

Zink MC, Narayan O, Kennedy PG, Clements JE (1987) Pathogenesis of visna/maedi and caprine arthritis-encephalitis: new leads on the mechanism of restricted virus replication and persistent inflammation. Vet Immunol Immunopathol 15:167–180

Zink MC, Suryanarayana K, Mankowski JL, Shen A, Piatak M Jr, Spelman JP, Carter DL, Adams RJ, Lifson JD, Clements JE (1999) High viral load in the cerebrospinal fluid and brain correlates with severity of simian immunodeficiency virus encephalitis. J Virol 73:10480–10488

Zink MC, Brice AK, Kelly KM, Queen SE, Gama L, Li M, Adams RJ, Bartizal C, Varrone J, Rabi SA, Graham DR, Tarwater PM, Mankowski JL, Clements JE (2010) Simian immunodeficiency virus-infected macaques treated with highly active antiretroviral therapy have reduced central nervous system viral replication and inflammation but persistence of viral DNA. J Infect Dis 202:161–170

# Mathematical Models of HIV Latency

Alison L. Hill

**Abstract** Viral latency is a major barrier to curing HIV infection with antiretroviral therapy, and consequently, for eliminating the disease globally. The establishment, maintenance, and potential clearance of latent infection are complex dynamic processes and can be best understood and described with the help of mathematical models. Here we review the use of viral dynamics models for HIV, with a focus on applications to the latent reservoir. Such models have been used to explain the multiphasic decay of viral load during antiretroviral therapy, the early seeding of the latent reservoir during acute infection and the limited inflow during treatment, the dynamics of viral blips, and the phenomenon of posttreatment control. In addition, mathematical models have been used to predict the efficacy of potential HIV cure strategies, such as latency-reversing agents, early treatment initiation, or gene therapies, and to provide guidance for designing trials of these novel interventions.

## Contents

## 1  History of Mathematical Modeling for HIV

The first models for HIV infection within individual people were developed in the late 1980s, quickly after the discovery of the virus itself. These models were inspired by the bread-and-butter models of mathematical epidemiologists, such as

A. L. Hill (✉)
Program for Evolutionary Dynamics, Harvard University, Cambridge, MA 02138, USA
e-mail: alhill@fas.harvard.edu

Current Topics in Microbiology and Immunology (2018) 417:131–156
DOI 10.1007/82_2017_77

the SIS and SIR models introduced by Kendrick and McCormack in the early 1900s, which are used widely to describe the spread of infections between individuals in a population (Anderson and May 1991; Brauer 2009; Kermack and McKendrick 1927). In contrast, "viral dynamics" models were developed to describe the spread of virus between infected cells within a single individual's body (Nowak and May 2000). These models have since been used to describe many other human virus infections, such as Hepatitis B (Dahari et al. 2009; Nowak et al. 1996; Perelson and Ribeiro 2004) and C (Chatterjee et al. 2012; Neumann et al. 1998), influenza (Murillo et al. 2013), dengue (Clapham et al. 2014; Nuraini et al. 2009), and herpes simplex virus (Schiffer et al. 2016).

A description of the most basic version of the virus dynamic model is known in Box 1 and Fig. 1. Despite its apparent simplicity, this model has provided tremendous insight into HIV dynamics. It predicted a viral load time course similar to what was observed for acute HIV infection: an initial exponential increase, followed by a peak and then decline to steady state (Fig. 1b,c). The model demonstrated that it was not necessary to have the onset of an adaptive immune response to explain the post-peak decline: the fact that target cells turn over slower than infected cells and had a limited production rate were enough to explain this drop (Phillips 1996). When it became possible to identify patients early in infection and follow them frequently, longitudinal data on viral load and CD4$^+$ T cell count of sufficient quality to fit the model were generated, allowing for estimation of the parameters governing viral dynamics (Burg et al. 2009; Stafford et al. 2000). As a result, it was later determined that the post-peak drop in viral load was larger than could be explained by target cell limitation alone, and the effect of the adaptive immune response could be quantified.

Mathematical analysis of the viral dynamics model demonstrates that it displays threshold behavior, similar to epidemiological models (Diekmann et al. 1990; Nowak and May 2000). It is possible to define a single composite quantity that depends on all the parameters of the system, and the value of this quantity alone

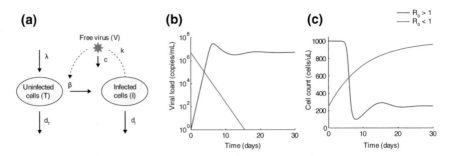

**Fig. 1** The basic viral dynamics model. **a** Flow diagram of processes that are tracked in the model, as described in Box 1. **b–c** Example trajectory of viral load (*V*) and uninfected target cells (CD4$^+$ T cells, *T*) when $R_0 > 1$ (red) and $R_0 < 1$ (blue). Graphs were generated by numerically integrating Eq. 2 with parameters $\lambda = 100$ cells/uL, $\beta = 10^{-7}$ or 0/day/(virus/mL), $k = 1000$ virus/cell, $d_T = 0.1$/day, $d_I = 1$/day, and $c = 25$/day

determines the qualitative behavior of the system. One such quantity is $R_0$, the basic reproductive ratio. This value describes the average number of secondary infected cells produced by a single infected cell over the course of its lifetime, in a population of otherwise susceptible target cells. When the basic reproductive ratio is greater than one $(R_0 > 1)$, an infection starting from a small size can grow and eventually establish a chronic state. When the basic reproductive ratio is less than one $(R_0 < 1)$, an infection starting from any size will decline and eventually be cleared. $R_0$ could be estimated from studies of acute infection, and its value falls anywhere in the range of 2–25 (averaging around 8) in patients (Ribeiro et al. 2010). This gave a benchmark for the required efficacy of drugs to inhibit infection. Similar to the vaccination threshold in epidemiology (Fine et al. 2011), the required drug efficacy $\varepsilon$, which describes the fraction of infections inhibited by the drug, must satisfy $\varepsilon > 1 - 1/R_0$ to achieve viral suppression within an individual.

---

**Box 1: The basic viral dynamics model**

The basic viral dynamics model (Fig. 1) tracks uninfected target cells of the virus ($T$), infected cells ($I$), and free virus ($V$). Target cells are produced at a rate $\lambda$ and die with a rate constant $d_T$. Infected cells are produced by contact between target cells and free virus at a rate $\beta$, and die at the rate $d_I$. Free virus is produced by infected cells at rate $k$ and is cleared at rate $c$. The model is generally formulated as a system of ordinary differential equations:

$$\dot{T} = \lambda - \beta TV - d_T T$$
$$\dot{I} = \beta TV - d_I I \tag{1}$$
$$\dot{V} = kI - cV$$

The *basic reproductive ratio*, which describes the average number of virions produced from in a single round of infection when one virion is introduced into an otherwise uninfected host, can be calculated for this model as

$$R_0 = \frac{\lambda \beta k}{d_T d_I c} \tag{2}$$

It represents a transcritical bifurcation in the system. When $R_0 < 1$ the only stable equilibrium is one with no virus or infected cells ($I = V = 0$). When inoculated from any size, the infection will decline towards eventual extinction. However, when $R_0 > 1$, infection levels will eventually reach a stable steady-state level from any nonzero initial condition.

---

The standard formulation of the viral dynamics model (Fig. 1 and Eq. 2) assumes that virus and cells are well mixed throughout the entire anatomical region of interest (for HIV, usually the plasma and lymph), and that they are present at high enough levels so that understanding the average (deterministic) behavior of the system is sufficient. However, the model can also be simulated stochastically to account for small population sizes, chance extinction, and events that occur rarely. There are multiple related stochastic formulations of the model which are equivalent to the same time-average deterministic representation but have different levels of fluctuations around the average. Generally, in the model, the target cells of the virus are assumed to be all $CD4^+$ T cells, though some versions of the model consider only activated cells in this class. Many other details of infection can be added to the model, such as a delay between infection and virion production, tracking anti-viral immune responses, viral latency, multiple-classes of infected cells, or the impact of drug treatments (Nowak and May 2000).

As described in the next section, the altered viral load trajectories observed once potent antiretroviral therapy is initiated allow for estimation of the rate of turnover of actively infected cells. When these observations were first made in the mid-1990s, they demonstrated that despite the slow-progressing clinical nature of chronic HIV infection, infected cells turned over on the order of a day, meaning that the rate of new infections and the virion production were extremely high (Ho et al. 1995; Wei et al. 1995). This fundamentally changed the view of HIV from a slow-moving to a dynamic infection.

When antiretroviral drugs were first used as monotherapy in patients, temporary declines in viremia were followed by resurgence to high (though slightly below pretreatment) levels. Along with genotypic testing, models demonstrated that the appearance of drug-resistant strains and subsequent competition with wild-type strains could explain these dynamics (Frost and McLean 1994; McLean and Nowak 1992). Parametrized viral dynamics models were able to show that preexisting resistance at the time of therapy initiation was more likely than newly emerging mutations once treatment was started (Bonhoeffer and Nowak 1997; Bonhoeffer et al. 1997). Importantly, these models, along with estimates for the mutation rate of HIV, suggested that at least three drugs would likely be needed to prevent rapid failure via drug resistance (Colgrove and Japour 1999; Ribeiro et al. 1998).

Adaptations of the basic viral dynamics model which took into account interactions with the immune system (both immune control of the virus and viral-mediated immune toxicity) were developed early on in the epidemic to understand the mechanism of pathogenesis and progression to AIDS. These models demonstrated the difficulty in explaining the long period of asymptomatic infection before the onset of AIDS and the slow decline of $CD4^+$ T cells (Anderson et al. 1998; Ribeiro 2002; Yates et al. 2007) with simple mechanisms. An early model proposed a "diversity threshold", whereby continual immune escape eventually leads to a critical level of antigenic diversity in the viral population which prevented control of the infection at a level where progressive immune destruction was held at bay (Nowak et al. 1991; Regoes et al. 1998). To date, the mechanism of HIV pathogenesis remains unclear and no single model is fully supported by observations.

## 2 Models to Interpret the Decay of Viremia During Antiretroviral Therapy

The viral dynamics model can be used to interpret changes in viremia when antiretroviral therapy begins. For a fully effective drug which prevents productive infection of new cells (such as entry inhibitors, reverse transcriptase inhibitors, and integrase inhibitors), treatment can be modeled as $\beta \to 0$ in the viral dynamics model. Viral load is expected to decay exponentially at a rate determined by the half-life of actively infected cells, after a short shoulder phase (Ho et al. 1995; Wei et al. 1995) (Fig. 1b). This shoulder phase is determined by the clearance rate of the free virus in the plasma, which is quite fast (free virus half-life of around an hour) (Ramratnam et al. 1999), and therefore creates only a short lag before exponential decay. In reality, the shoulder phase could be longer due to the delay required for drug absorption into the plasma and diffusion into the cell. This insight from models, along with frequently sampled viral load values from patients on the first antiretroviral drugs, resulted in an estimated infected cell lifespan of around 2.5 days (half-life of 1.6 days in Perelson et al. (1996) and 1.8 days in Wei et al. (Wei et al. 1995)). Later, when more potent drugs were developed and eventually used in combination, this estimate was adjusted to $\approx 1$ day (half-life of 0.7 days in Markowitz et al. (Markowitz et al. 2003)). As mentioned above, this finding consolidated the understanding of HIV as a very dynamic infection, since at steady-state viral loads, each infected cell that dies must be replaced by a newly infected one.

When more sensitive assays of viral load were developed and drug combinations, which prevented the rapid evolution of resistance, were used, it became apparent that viral load decline did not follow a simple exponential decay. Instead, it could best be fit by a multiphasic decay pattern. Models were developed to explain this pattern in terms of different populations of infected cells with different death rates, and used to generate hypotheses about the nature of the cells responsible for each phase of decay.

First, a second phase decay, thought to represent a population of virus-producing cells with a half-life of 14 days, was discovered (Andrade et al. 2013; Perelson et al. 1997). Based on this decay rate and an estimated total body infected cell population size of maximum $10^{12}$, it was estimated that it would take no more than 3 years to eradicate all virus from the body (Perelson et al. 1997). However, later studies using assays that could quantify viral loads down to 1 copper per mL of plasma identified the third phase of decay (half-life of 40–60 weeks, in Palmer et al. (2008), Andrade et al. (2013)), and perhaps a subsequent stable level of viremia (Palmer et al. 2008). Along with the discovery of latent infection in the form of provirus integrated into resting CD4$^+$ T cells (Chun et al. 1997; Finzi et al. 1997; Siliciano et al. 2003) and the observation of rebound upon ART interruption (Davey et al. 1999; Ruiz et al. 2000), these findings quelled hopes of a cure for HIV via ART.

To this day, the identity of the later phases of viral decay is not completely known. The second phase remains particularly elusive. It could represent a longer lived population of cells infected with HIV, such as monocytes, macrophages, partially activated T cells, or cells with unintegrated HIV DNA. The third/fourth phase is likely due to reactivation of infection in long-lived latently infected cells.

Mathematical models have provided insight into this issue, in particular by interpreting the changes in decay kinetics observed in clinical trials when integrase inhibitors are used instead of reverse transcriptase inhibitors. Spivak et al. (2010) have used this approach to suggest that macrophages are not a likely candidate for second phase decay, and Andrade et al. (2013) have similarly suggested that pre-integration latency is an unlikely cause for the third phase decay.

Models have also suggested alternative hypotheses for the mechanism behind the decelerating decay. Arnaout et al. (2000) suggested that as viral loads declined after ART initiation, so would antigenic stimulation of cytotoxic T cell (CTL) levels. Lower CTLs would result in less killing of productively infected cells, and hence an apparent decrease in viral load decay rate. Kim and Perelson (2006) hypothesized that viremia during the later phases was due to reactivation from latently infected memory $CD4^+$ T cells, and that infected cells specific to more common antigens would be reactivated early on. The latent pool would then be comprised of cells specific for rare antigens which appear infrequently, resulting in decelerating decay over time. Zhang and Perelson (2013) suggested that the third phase of viral load decline could be explained by the very slow release of virus bound to the surface of follicular dendritic cells, due to the complex multivalent binding kinetics.

---

**Box 2: Modeling HIV dynamics under antiretroviral therapy**

Antiretroviral therapy acts to block the production of newly infected cells, effectively reducing $\beta$ in the viral dynamics model (Fig. 1, Eq. 2). Assuming fully effective treatment (a good approximation for combination therapy or monotherapy with some newer highly potent drugs), $\beta \to 0$, and the basic viral dynamics model predicts (Perelson et al. 1996; Wei et al. 1995) that viral load decays following:

$$V(t) = V(0) \frac{(ce^{-d_I t} - d_I e^{-ct})}{c - d_I} \tag{3}$$

While this equation describes the first week of viral decay well (e.g., (Markowitz et al. 2003)), longer follow-up periods show that the decay slows over time. More realistic models that incorporate multiple populations of infected cells have been developed to help explain these dynamics (Fig. 2, adapted from Perelson et al. (1997)). A fraction $f$ of new infections of target cells ($T$, $CD4^+$ T cells) results in latency. Latently infected cells ($L$, generally representing resting memory $CD4^+$ T cells) do not produce virus and are extremely long lived. Occasionally (at rate $a$), these cells reactivate and produce virus. An alternative population of target cells (biological origin still debated) turns over more slowly and produces virus at a lower rate. The resulting equations are as follows:

$$\dot{T} = \lambda - \beta TV - d_T T$$
$$\dot{T}_2 = \lambda_2 - \beta_2 T_2 V - d_{T_2} T_2$$
$$\dot{I} = (1 - f)\beta TV - d_I I + aL$$
$$\dot{L} = f\beta TV - d_L L - aL \qquad (4)$$
$$\dot{I}_2 = \beta_2 T_2 V - d_{I_2} I_2$$
$$\dot{V} = kI + k_2 I_2 - cV$$

When fully effective therapy is given $(\beta, \beta_2 \to 0)$, viral load decays as

$$V(t) = V(0)[Ae^{-d_I t} + Be^{-(a+d_L)t} + Ce^{-d_{I_2} t} + (1 - A - B - C)e^{-ct}] \qquad (5)$$

where $T_0$ is the CD4$^+$ T cell level at the time of treatment start and

$$A = \frac{k\beta T_0}{(c - d_I)d_I} \left(1 - \frac{(d_I - d_L)f}{(d_I - d_L - a)}\right)$$

$$B = \frac{afk\beta T_0}{(a + d_L)(d_I - a - d_L)(c - a - d_L)}$$

$$C = \frac{c - \frac{k\beta T_0}{d_I}\left(1 - \frac{d_L f}{(a + d_L)}\right)}{c - d_{I_2}}$$

# 3   Mechanisms of Latent Reservoir Persistence

Plasma viral loads dramatically decrease during combination antiretroviral therapy, but ultrasensitive assays find that some residual viremia persists at very low levels ($\approx 1$ viral RNA copy per mL of plasma) indefinitely during treatment (Eriksson et al. 2013). Understanding the source of this persistent infection has been a major goal of HIV research for the past 20 years. The discovery of HIV latency has contributed to our understanding of the disease process but also introduced new questions that remain unanswered to this day. Throughout this period, mathematical models have been an important tool to test the many competing hypotheses about HIV persistence.

HIV integrates into host cell genomes as a part of its normal lifecycle, and CD4$^+$ T cells with resting, memory phenotypes can be isolated that contain proviral DNA (Chun et al. 1995). Many of these cells can be induced to produce infectious virus

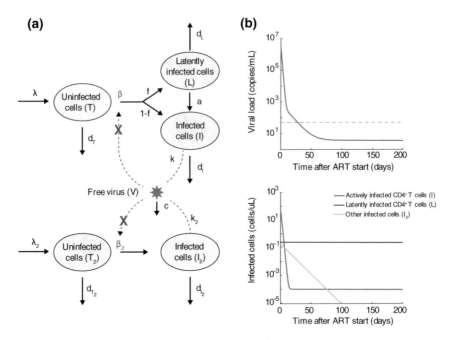

**Fig. 2** An augmented model of viral dynamics that takes into account multiple populations of infected cells. **a** Flow diagram of processes that occur in the model, as described in Box 2. Adapted from Perelson et al. (1997) **b** Multiphasic decay of viral load predicted by the model. **c** Decay of individual populations of infected cells predicted by the model. Graphs were generated by numerically integrating Eq. 5 with parameters approximately consistent with those estimated in Andrade et al. (2013). We assumed that other target cells ($I_2$) live 10× longer with or without infection, exist at 1000× lower level before infection, produce virus at 10× slower rate, and are infected at the same rate. 1 in 10,000 infections result in latency, and that latently infected cells ($L$) live 10,000 times longer than actively infected cells. Parameters are $\lambda = 100$ cells/uL, $\beta = 10^{-7}$ or 0/day/(virus/mL), $k = 1000$ virus/cell, $d_T = 0.1$/day, $d_I = 1$/day, $\lambda_2 = 0.01$ cells/uL, $\beta_2 = 10^{-7}$ or 0/day/(virus/mL), $k_2 = 100$ virus/cell, $d_{T_2} = 0.01$/day, $d_{I_2} = 0.1$/day, $c = 25$/day, $f = 10^{-4}$, $a = 4 \times 10^{-4}$/day, and $d_L = 10^{-4}$/day

when isolated, activated, and cultured ex vivo (Chun et al. 1997; Finzi et al. 1997; Siliciano et al. 2003). The term "HIV latency" refers to this process, and the pool of long-lived cells that contain provirus in a quiescent state is referred to as the "latent reservoir". Latently infected cells can be found early on in the infection and after therapy has begun (Ananworanich et al. 2015; Blankson et al. 2000; Chun et al. 1997; Finzi et al. 1997). One of the most important questions about the latent reservoir has been how quickly it decays during therapy—if it decays at all. Longitudinal samples of inducible virus in resting memory CD⁺ T cells can be analyzed similar to plasma viral loads, using models of exponential decay. Initial studies using single-exponential models beginning at timepoints immediately after ART initiation estimated a 6-month half-life of latent cells, implying 10 years of therapy would be needed to clear the estimated $10^6$ latently infected cells (Chun et al. 1997). Later work that with a broader range and density of timepoints demonstrated multiphase decay: This source of virus initially drops around 1000-fold once ART is started, with a half-life on the order of weeks, likely due to resting cells with unintegrated virus that did not complete the transition to latent infection (Blankson et al. 2000). Then, after about 6 months of treatment, a more stable underlying population is revealed at a frequency of about 1 in a million, which decays extremely slowly, with a median half-life of 44 months (Crooks et al. 2015; Finzi et al. 1999; Siliciano et al. 2003). This finding consolidated the realization that the duration of therapy needed for achieving a cure ($\approx$70 years) was too long to have any practical value, and that antiretroviral therapy would need to be continued for life. This was supported by observations of universal and rapid rebound in viremia to pretreatment levels when therapy was stopped (Davey et al. 1999; Ruiz et al. 2000).

Once it was established that both latent HIV and residual plasma viremia persist almost indefinitely despite therapy, the major question became the mechanism for this persistence. One hypothesis is that the latent infection is maintained by the same mechanisms as immunologic memory, and the occasional reactivation of latent cells is responsible for residual viremia (Blankson et al. 2002; Siliciano et al. 2003). Another theory is that ART is not completely effective in stopping viral replication ($R_0 > 1$) and that continual rounds of infection cause persistent viremia or persistent latent infection, or both (Cory et al. 2013; Martinez-Picado and Deeks 2016).

A large body of modeling work has investigated the relationship between the dynamics of latently infected cells and plasma viremia during ART (Conway and Coombs 2011; Conway and Perelson 2016; Jones and Perelson 2007; Kim and Perelson 2006; Rong and Perelson 2009b; Sedaghat et al. 2008). The most important insight from these studies is that as long as ART is marginally effective, the persistence of latent virus is dominantly influenced by the longevity of infected lymphocytes and the rate at which they reactivate, and not by any ongoing replication that continually seeds the reservoir. Models predict that low amounts of ongoing replication ($0 < R_0 < 1$) may influence the level of detectable plasma virus, but that these levels of replication are highly unlikely to allow for long-term sequence evolution of this virus (for example, the development of drug resistance) (Conway and Perelson 2016; Kim and Perelson 2006; Rong and Perelson 2009b; Sedaghat et al. 2008). Various clinical studies have shown that residual plasma virus levels during ART may fluctuate overtime, occasionally producing higher "blips" in viral RNA, and models suggest that the size, duration, and frequency of these blips can be explained by occasional antigen-driven reactivation of latent cells (Conway and Coombs 2011; Jones and Perelson 2007; Kim and Perelson 2006; Rong and Perelson 2009b).

Experimental work backs up many of these modeling predictions. Ex vivo quantifications of antiretroviral efficacy suggest a multi-log reduction in viral replication (Jilek et al. 2012; Shen et al. 2008). Long-term studies of low-level plasma virus sequences show a lack of evolution, including a lack of development of drug resistance (Joos et al. 2008; Kieffer et al. 2004). "Treatment intensification" studies add a fourth or fifth antiretroviral drug to existing combination therapy, and show no change in residual viremia levels, suggesting that viral replication is optimally suppressed (Dinoso et al. 2009; Gandhi et al. 2010). The size of the latent reservoir is highly correlated with the level of plasma viremia during ART (Archin et al. 2012). A study that tracked the movement of particular unique viral genetic sequences between plasma virus and resting memory $CD4^+$ T cells was able to put an upper limit on the number of new latent cells produced each day ($\approx$100), and show that this had a negligible impact on reservoir persistence (Sedaghat et al. 2007).

While most evidence from models and experiments points toward ART being highly effective and the latent reservoir persisting due to intrinsic memory T cell longevity, some findings appear to conflict with this theory. When treatment intensification studies were done by adding the integrase inhibitor raltegravir to patients taking three drugs from other classes, some studies tracked not only plasma viral RNA but also levels of a type of circularized viral DNA ("2-LTR circle") that forms in cells when the virus is unable to complete the integration step. Similar to other studies, treatment intensification did not influence plasma viremia, but 2-LTR levels showed a transient increase in some patients (Buzon et al. 2010; Hatano et al. 2013). Modeling work by Luo et al. (2013) showed that these kinetics could not be explained the dynamics of latent and actively infected cells under fully effective ART ($R_0 < 1$ everywhere), but instead were highly suggestive of at least a small compartment in the body where viral replication continued ($R_0 > 1$) despite ART. However, this effect was only observed in about 1 in 6 patients, model conclusions were sensitive to the (noisy) experimental data, and so the clinical significance is still being debated.

Even if the persistence of residual viremia can be explained completely by occasional reactivation of latent cells, and if the slow decay of the latent reservoir is purely a product of memory T cell maintenance, the dynamics of latency itself remain opaque. In the absence of any reseeding, the total decay rate of the reservoir is the net effect of cell death, cell reactivation, and cell proliferation. Various combinations of each of these processes could lead to the same observed net decay. For example, if latent cells lived on average 5.5 years, reactivated on average every 70 years, and never proliferated, then the reservoir would decay at a rate of 0.2/ years, equivalent to a 44-month half-life. The same net decay would occur if cells instead lived only 4 months and proliferated every 114 days. Some researchers initially assumed that latently infected $CD4^+$ T cells could not undergo proliferation without reactivating viral gene expression and succumbing to viral cytopathic effects or cytolytic immune responses. However, multiple lines of evidence now suggest that latently infected cells can indeed divide and expand without reacti-vating, including studies of cell-surface markers and cytokines indicative of pro-liferation in virally suppressed individuals (Chomont et al. 2009), repeated occurrences of an exact viral integration site or full-length genetic sequence (Cohn et al. 2015; Maldarelli et al. 2014; Simonetti et al. 2016), or ex vivo culture experiments (Hosmane et al. 2017). Even before this evidence existed, various mathematical models included the possibility of all three changes to reservoir size, but parameter values have simply been educated guesses and to date, the relative contribution of each process to reservoir stability remains unknown (Chomont et al. 2011; Kim and Perelson 2006; Rong and Perelson 2009a). The efficacy of hypothetical interventions to speed up the killing of latently infected cells or inhibit their proliferation can only be estimated if quantitative values can be determined for these processes.

# 4 Strategies to Clear the Reservoir and Cure HIV Infection

The persistence of latent HIV prevents combination antiretroviral therapy from being curative. If therapy is stopped, viral load rebounds in approximately 2 weeks and rapidly reaches high levels similar to acute infection (Davey et al. 1999; Ruiz et al. 2000). This rebound occurs in patients treated early in infection (Kaufmann et al. 2004; Rosenberg et al. 2000), in primate models of HIV (SIV-infected rhesus macaques) treated within a few days of infection (Whitney et al. 2014), and in individuals in whom the reservoir size is reduced by multiple orders of magnitude compared to typical infected people (Chun et al. 2010; Henrich et al. 2014; Luzuriaga et al. 2015). Consequently, it is now accepted that achieving a cure in most patients will require either a method to augment immunity or complete clearance of the latent reservoir (Ananworanich and Mellors 2015; Saez-Cirion et al. 2014).

The observed decay rate of the reservoir during ART suggests that 60+ years of therapy would likely be needed to clear all latently infected cells (around $10^6$ cells with a 44-month half-life), and even a 1000-fold smaller than average reservoir size would take $\approx$30 years to clear on its own. These calculations optimistically assume that the reservoir is homogeneous and the decay rate is constant, while in reality, decay may slow over time, for example, if cells specific to more common antigens reactivate and are cleared sooner (Kim and Perelson 2006). Even if some low-level, subcritical replication persists during ART that allows occasional entry of new cells into the latent pool, the body of modeling work cited in the previous section concludes that complete elimination of replication would have an insignificant impact on the rate of reservoir decay (Conway and Perelson 2016; Kim and Perelson 2006; Rong and Perelson 2009b; Sedaghat et al. 2008). Clearly, eradicating the latent reservoir will require the development of new interventions.

The important question posed to modelers in regards to disrupting HIV latency is "How much must the latent reservoir be reduced to delay or prevent viral rebound once ART is stopped?". Stochastic models of the latent reservoir and low-level infection are required to answer this question. Hill et al. (Hill et al. 2014) estimated used such a model to estimate that reservoir reductions of 2–3 logs are required to delay rebound for a few months, while >4-log reductions may be required for cure in most patients. They predicted that patients may rebound even after experiencing several years of treatment without detectable viremia, which was later observed

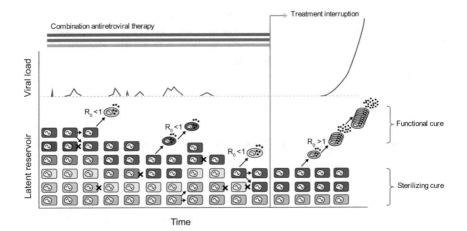

**Fig. 3** Schematic of latent reservoir dynamics. Long-lived resting memory CD4$^+$ T cells (squares) may contain integrated HIV provirus—these cells comprise the latent reservoir. The majority of evidence suggests that during combination antiretroviral therapy in most patients, viral replication is subcritical ($R_0 < 1$), and so the persistence of virus represents the maintenance of the latent reservoir. Latently infected cells may occasionally die (black X), proliferate, or reactivate. The relative contribution of slow death versus proliferation to stability of the reservoir is still being worked out. A large contribution of proliferation would result in the virus in the reservoir becoming less diverse over time. Reactivated latent cells (ovals) can produce virus which can contribute to observed levels of residual viremia. New infections may occasionally occur but continuous chains of replication and evolution cannot occur when $R_0 < 1$. When treatment is interrupted ($R_0 > 1$), reactivated cells can produce a virus that infects other cells, which, if it escapes stochastic extinction, can lead to eventual viral rebound. Research into curative interventions for HIV aims to prevent viral rebound by one of two mechanisms: either clearing enough latent virus or blocking reactivation to an extent that there is insufficient reactivation to seed rebound ("sterilizing cure"), or, by equipping the body with the ability to control the infection even in the face of reactivating virus ("functional cure"). Examples of potential sterilizing cures involve latency-reversing agents or bone marrow transplants, and examples of functional cures involve therapeutic vaccination or CCR5-knockout gene therapies

(Henrich et al. 2014; Luzuriaga et al. 2015). Additionally, their model demonstrated that large inter-patient variation in times to rebound is expected when reservoir sizes were reduced—simply due to the stochastic nature of reactivation from latency and infection dynamics at low population sizes. Overall, this work suggested that reservoir clearance must be nearly complete to have any significant clinical benefit and that designing and interpreting clinical trials for reservoir-reducing strategies would be complex. Pinkevych et al. (2015) developed a similar model to address the same questions, but their conclusions were much more optimistic: they suggested that a 50-fold reduction in the reservoir size could result in a 1-year remission (Fig. 3).

The main difference between these modeling approaches, and the main source of uncertainty in the predictions of any given model, are the parameter values. Hill et al. (2014) examined a large class of stochastic viral dynamics models analytically and numerically and determined that the predictions for rebound time depend only on a small number of composite parameters. Two of these parameters can be estimated from clinical data relatively easily: the half-life of the pool of latently infected cells is estimated from studies of LR decay (Crooks et al. 2015; Siliciano et al. 2003), and the viral growth rate can be estimated from studies of supervised cART interruption (Luo et al. 2012; Ruiz et al. 2000). A third parameter is the rate at which latently infected cells reactivate (or, framed another way, the number of cells that reactivate daily from typical reservoir sizes), and different methods to estimate this rate are the major source of discrepancy between different models. Many earlier models that tracked latent infection during ART but were not specifically designed to be used to model reservoir clearance estimated this rate based on the cellular activation required to explain the observed levels of residual viremia ($\approx$1 HIV RNA/ml), getting values on the order of 1000 cells/day in the whole body (Conway and Coombs 2011; Conway and Perelson 2015, 2016; Kim and Perelson 2006; Rong and Perelson 2009b; Rosenbloom et al. 2012; Sedaghat et al. 2008). Other work by Luo et al. (2012) estimated this rate based on the timing of viral rebound upon ART cessation ($\approx$ 0.03 viral RNA/mL/day or 60 cells/day). Hill et al. (2014) used this value but also needed to estimate a fourth composite parameter: the probability that a single activated cell manages to establish a growing infection. This quantity can only be indirectly estimated from in vitro (Bui et al. 2015; Singh et al. 2010) and population-genetic studies (Pennings 2012; Pennings et al. 2014; Rouzine and Coffin 1999) [reviewed in Rouzine et al. (2014)], and remains quite uncertain.

Pinkevych et al. (2015) used a simplified, approximate model in which viral growth occurs deterministically once a cell that is fated to establish a growing infection exits the reservoir. Consequently, they only estimated the product of the reservoir exit rate and the establishment probability. To do this, they assumed that all variations in rebound time between different patients were caused by different timings of viral reactivation. Since rebound times vary by up to a week between patients with (assumed) identical reservoir sizes, they concluded that the effective activation rate was low, and therefore that any reservoir reduction would significantly delay rebound. Hill et al. (2016b) have argued that varying rebound times are better explained by inter-patient variation in parameter values, and that the lack of variation in typical rebound times despite the 2–3 orders of magnitude variation in typical reservoir sizes is more consistent with their estimate of multiple effective reactivations per day versus rare reactivation ($\approx$4/day vs. $\approx$1/6 days). Additional experimental work will be needed to resolve this issue and narrow down estimates for these important rates.

Both Pinkevych and Hill have developed methods to use model predictions to interpret treatment interruption trials. They demonstrate that large trial sizes will be needed to estimate the reservoir reduction achieved by a particular therapy from the timing of rebound (Hill et al. 2016a; Pinkevych et al. 2015). They offer estimates for the reservoir size in the Boston patients (Henrich et al. 2014) and Mississippi baby (Luzuriaga et al. 2015) based on rebound delay and suggest ways to design the frequency of viral load sampling during interruption trials. These activities demonstrate how mathematical models can be used to design the most informative and cost-effective clinical trials and optimally interpret the findings in the context of prior knowledge.

These modeling conclusions are agnostic to how the latent reservoir is reduced, which could be accomplished by a variety of means. Much experimental work has focused on identifying drug candidates which can reactivate latent infection, with the hope that cells that are rapidly reactivated will then die to viral cytopathic effects of cytolytic immune responses, without reverting back to latency. Many such compounds, called "latency-reversing agents (LRA)" have been identified (see next chapter), but their efficacy so far in vitro (Bullen et al. 2014) or in vivo (Elliott et al. 2014, 2015; Rasmussen et al. 2014; Sgaard et al. 2015) has been relatively limited. Mathematical modelers have evaluated the expected dynamics of latent infection, active infection, and viremia under latency-reversing treatment (Hill et al. 2017; Petravic et al. 2017). Petravic et al. (2017) have outlined a method to relate the observed dynamics of cells expressing HIV during ART + LRA to an expectation for reservoir reduction and used it to interpret recent clinical trials of two LRAs (Elliott et al. 2014; Rasmussen et al. 2014). Hill et al. (2017) have suggested that under most realistic conditions, LRA, which significantly reduce the latent reservoir, should cause relatively large, detectable blips in plasma viremia. The shape and size of these blips may be used to estimate parameters related to LRA-activated cells and reservoir reduction. However, both sets of authors concluded that the expected levels of HIV transcription or viremia are dependent on the lifespan and burst size of LRA-reactivated cells, which may differ from naturally activated cells, complicating interpretation.

Other work has focused on understanding when—in the course of infection and treatment—is the ideal time to administer latency-reversing treatment. Petravic et al. (2014) suggest that because the turnover of the latent reservoir in resting CD4 T cells is higher when viral loads are higher, it would be most effective to administer latency-reversing drugs immediately following ART initiation. This is in contrast to the common practice in existing trials of latency-reversing agents, where only patients who have been ART for many years with fully suppressed viral loads are included.

Instead of trying to reduce the size of the latent reservoir once it's formed, other work has focused on preventing seeding of the latent reservoir in the first place. The question is then how early ART must be initiated to result in small enough reservoir sizes to delay or prevent viral rebound when ART is later stopped. Archin et al. (2012) used paired measurements of viral load and CD4 count pretreatment along with reservoir size posttreatment to estimate the rate at which the latent reservoir is seeded. They compared the observed data to a mathematical model in which some fraction of newly infected cells become latent, as opposed to actively producing virus. In agreement with predictions of this model, the reservoir size was highly correlated with the area-under-the-curve of the product of viral load and CD4 count during acute infection, and they were able to estimate the fraction of cells that became latent. This work suggested that the vast majority of reservoir seeding happens during peak viremia in acute infection, and very early ART would be needed to substantially limit seeding. In a primate model of infection, there was a similarly strong relationship between pretreatment viral burden and reservoir size that was well described by simple models, and even ART initiated 3 days after infection did not prevent viral rebound when treatment was stopped (Whitney et al. 2014). Similarly, case studies of immediate postpartum treatment of neonates infected during gestation showed that reservoir sizes could be limited to extremely small levels and rebound delayed for months to years, but not be prevented altogether (Luzuriaga et al. 2015). Therefore, it is now clear that for the majority of patients even extremely early ART initiation is unlikely to make a cure possible.

There is a subset of patients in whom early treatment seems to have a disproportionately beneficial effect. Multiple cohort studies have retrospectively identified a group of "posttreatment controllers", who had high viremia during acute infection, but do not rebound, or rebound only to very low levels, following ART interruption (Chret et al. 2015; Etemad et al. 2016; Goujard et al. 2012; Lodi et al. 2012; Maenza et al. 2015; Sez-Cirin et al. 2013). The mechanisms of this effect are still unknown. Most models of HIV infection do not display this behavior (Nowak and May 2000). Early work by Wodarz et al. (2000a, b) suggested that the dynamics of memory $CD8^+$ T cells (i.e., cytotoxic T lymphocytes, CTL)—in particular their dependence on CD4 help and their persistence during ART—could lead to control after ART interruption, a phenomenon that had been observed in the SIV system as early as 2000 by Lifson et al. (2000, 2001). More recently, Conway and Perelson (2015) have explored the potential basis of posttreatment control using a viral dynamics models that track latent and active infection, and cytolytic immune responses which expand in the presence of viral antigen but may also become exhausted from antigen exposure. They found that this model yields bistability: Even for identical parameter values, different initial conditions can make the difference between the establishment of a high set-point viral load, or control of viral load to a low level. This implies that the size of the latent reservoir at the time of ART interruption (the "initial condition") could determine in which state the infection ends up. Their work suggests that early treatment initiation either (a) in a subset of patients with relatively effective immune responses, or, (b) in combination with an immune-boosting intervention, could be used to induce indefinite posttreatment control. Such a curative intervention would be termed a "functional cure"—meaning that latent cells would remain and occasionally reactivate but high-level rebound would be avoided—as opposed to a "sterilizing cure" in which latent cells would be cleared. However, this work currently remains speculative, because the observation of bistability in models is sensitive to the model structure and to values of parameters, and, because we do not know the immunological basis of posttreatment control and do not know how to identify patients who may benefit from it.

The conclusion of this body of modeling work is that clearing the latent reservoir to a low enough level to significantly delay or prevent viral rebound upon ART cessation is likely impossible in most patients. Instead, curative interventions will likely need instead to limit the ability of the reactivated virus to restart full-blown infection. One approach is to administer immunotherapy or therapeutic vaccination along with ART or after ART cessation, to reduce the fitness of reactivated virus. Recent studies in SIV-infected primates have highlighted the potential promise of this approach (Borducchi et al. 2016; Byrareddy et al. 2016; Lim et al. 2017). In these studies, viral rebound was delayed, reached lower peak levels, and was eventually controlled (suppressed below the detection limit) in a subset of animals, and modeling work suggested that the dominant mechanism for this control was based on the immunologic effect, and not the reservoir reduction.

## 5 HIV Latency and Viral Evolution

The rapid evolution of HIV is known to play a major role in pathogenesis and is also responsible for viral adaptation to humans after crossing over from primates approximately 100 years ago. Researchers have used mathematical models to investigate the role that latent infection plays in the rate and direction of HIV evolution within and between hosts. Other work has theorized about whether latency is itself an evolved trait of HIV, as opposed to simply a byproduct of the lifecycle of the lymphocytes the virus infects.

The long-lived nature of latently infected cells compared to those actively producing virus means that the latent reservoir serves as an archive of viral strains present at earlier time points in infection. Most viruses produced in the bloodstream are from cells that were recently infected, and so will contain the most recently evolved strains, whereas the provirus contained in latently infected cells—which may occasionally reactivate and produce virions—may contain a more ancestral virus that is more genetically similar to the transmitted strain. Genetic sequencing of plasma viral RNA and proviral DNA in cells that remain after long-term therapy has confirmed this hypothesis, showing that archived strains mean that wild-type virus can persist despite years of antiretroviral therapy and drug-resistant strains can persist after therapy has stopped (Kieffer et al. 2004; No et al. 2005; Persaud et al. 2000; Ruff et al. 2002; Wind-Rotolo et al. 2009).

Mathematical models of viral evolution which take into account this strain archiving have suggested that it has multiple effects on infection dynamics. Within an individual patient, the rate of intra-host evolution may be decreased relative to the case without latency (Doekes et al. 2017; Kelly 1996), drug-resistant mutations may appear in the absence of ongoing replication from preexisting strains the reservoir (Rosenbloom et al. 2012), and recombination could occur between strains which dominated the viral population at very different times during infection (Immonen et al. 2015). At a population level, strain-archiving resulting in selective transmission of ancestral virus may explain the paradoxically slow rate of between-host HIV evolution (Alizon and Fraser 2013; Lythgoe and Fraser 2012); by allowing transmission of strains similar to those which initiated the infection despite years of diversification during chronic infection, and could theoretically change the evolutionarily optimal set-point viral load (Doekes et al. 2017). The relative magnitude of these effects is more uncertain, as it depends strongly on the fraction of infections that become latent, the turnover of the latent reservoir, and the effective population size of active infection, which are all difficult to quantify.

Given that the presence of viral latency alters HIV dynamics, a natural question is whether latency itself is an adaptive trait that the virus has evolved to increase some aspect of fitness. While most experimental evidence is consistent with latency as a byproduct of the natural lifecycle of activated T cells (Murray et al. 2016), some cell-culture work on latency also supports the idea that viral gene expression can be switched off by a virally encoded, transcription-factor mediated feedback loop (Razooky et al. 2015; Weinberger et al. 2005). Inspired by that work, a mathematical modeling study showed that if latency enhances trafficking of infected cells across the musoca after inoculation, viral latency could increase transmission rates and be evolutionarily favored (Rouzine et al. 2015). However, some predictions of this model—such as the high fraction of cells reverting to latency during early infection and the potential for a long delay between inoculation and establishment of infection—continue to be debated.

# References

Alizon S, Fraser C (2013) Within-host and between-host evolutionary rates across the HIV-1 genome. Retrovirology 10:49. doi:10.1186/1742-4690-10-49, URL:https://doi.org/10.1186/1742-4690-10-49

Ananworanich J, Mellors JW (2015) A cure for HIV: what will it take? Curr Opin HIV AIDS 10 (1):1–3. doi:10.1097/COH.0000000000000125

Ananworanich J, Dubé K, Chomont N (2015) How does the timing of antiretroviral therapy initiation in acute infection affect HIV reservoirs? Curr Opin HIV AIDS 10(1):18–28. doi:10.1097/COH. 0000000000000122, URL:http://www.ncbi.nlm.nih.gov/pmc/articles/PMC4271317/

Anderson RM, May RM (1991) Infectious diseases of humans: dynamics and control. Oxford University Press, USA

Anderson RW, Ascher MS, Sheppard HW (1998) Direct HIV cytopathicity cannot account for CD4 decline in AIDS in the presence of homeostasis: a worst-case dynamic analysis. J Acquir Immune Defic Syndr 17(3):245–252. URL:http://www.ncbi.nlm.nih.gov/pubmed/9495225

Andrade A, Rosenkranz SL, Cillo AR, Lu D, Daar ES, Jacobson JM, Lederman M, Acosta EP, Campbell T, Feinberg J, Flexner C, Mellors JW, Kuritzkes DR, Team ftACTGA (2013) Three distinct phases of HIV-1 RNA decay in treatment-naive patients receiving raltegravir-based antiretroviral therapy: ACTG A5248. J Infect Dis 208(6):884–891. doi:10.1093/infdis/jit272, URL:http://jid.oxfordjournals.org/content/208/6/884

Archin NM, Vaidya NK, Kuruc JD, Liberty AL, Wiegand A, Kearney MF, Cohen MS, Coffin JM, Bosch RJ, Gay CL, Eron JJ, Margolis DM, Perelson AS (2012) Immediate antiviral therapy appears to restrict resting $CD4^+$ cell HIV-1 infection without accelerating the decay of latent infection. Proc Natl Acad Sci USA 109(24):9523–9528. doi:10.1073/pnas.1120248109

Arnaout RA, Nowak MA, Wodarz D (2000) HIV-1 dynamics revisited: biphasic decay by cytotoxic T lymphocyte killing? Proc Roy Soc B Biol Sci 267(1450):1347–1354. URL:http://www.ncbi.nlm.nih.gov/pmc/articles/PMC1690670/

Blankson JN, Finzi D, Pierson TC, Sabundayo BP, Chadwick K, Margolick JB, Quinn TC, Siliciano RF (2000) Biphasic decay of latently infected $CD4^+$ T cells in acute human immunodeficiency virus type 1 infection. J Infect Dis 182(6):1636–1642. URL:http://jid.oxfordjournals.org/content/182/6/1636.short

Blankson J, Persaud D, Siliciano R (2002) The challenge of viral reservoirs in HIV-1 infection. Annu Rev Med 53(1):557–593

Bonhoeffer S, Nowak MA (1997) Pre-existence and emergence of drug resistance in HIV-1 infection. Proc Roy Soc B Biol Sci 264(1382):631–637

Bonhoeffer S, May RM, Shaw GM, Nowak MA (1997) Virus dynamics and drug therapy. Proc Natl Acad Sci USA 94:6971–6976

Borducchi EN, Cabral C, Stephenson KE, Liu J, Abbink P, Ng'ang'a D, Nkolola JP, Brinkman AL, Peter L, Lee BC, Jimenez J, Jetton D, Mondesir J, Mojta S, Chandrashekar A, Molloy K, Alter G, Gerold JM, Hill AL, Lewis MG, Pau MG, Schuitemaker H, Hesselgesser J, Geleziunas R, Kim JH, Robb ML, Michael NL, Barouch DH (2016) Ad26/MVA therapeutic vaccination with TLR7 stimulation in SIV-infected rhesus monkeys. Nature 540(7632):284–287. doi:10.1038/nature20583, URL:http://www.nature.com.ezp-prod1.hul.harvard.edu/nature/journal/v540/n7632/full/nature20583.html

Brauer F (2009) Mathematical epidemiology is not an oxymoron. BMC Publ Health 9(Suppl 1): S2. doi:10.1186/1471-2458-9-S1-S2, URL:http://www.biomedcentral.com/1471-2458/9/S1/S2/abstract

Bui JK, Mellors JW, Cillo AR (2015) HIV-1 virion production from single inducible proviruses following T-cell activation ex vivo. J Virol. doi:10.1128/JVI.02520-15, URL:http://jvi.asm.org.ezp-prod1.hul.harvard.edu/content/early/2015/11/05/JVI.02520-15

Bullen CK, Laird GM, Durand CM, Siliciano JD, Siliciano RF (2014) New ex vivo approaches distinguish effective and ineffective single agents for reversing HIV-1 latency in vivo. Nat Med 20(4):425–429. doi:10.1038/nm.3489, URL:http://www.nature.com/nm/journal/v20/n4/abs/nm.3489.html

Burg D, Rong L, Neumann AU, Dahari H (2009) Mathematical modeling of viral kinetics under immune control during primary HIV-1 infection. J Theor Biol 259(4):751–759. doi:10.1016/j.jtbi.2009.04.010, URL:http://www.sciencedirect.com/science/article/pii/S0022519309001702

Buzon MM, Massanella M, Llibre JM, Esteve A, Dahl V, Puertas MC, Gatell JM, Domingo P, Paredes R, Sharkey M, Palmer S, Stevenson M, Clotet B, Blanco J, Martinez-Picado J (2010) HIV-1 replication and immune dynamics are affected by raltegravir intensification of HAART-suppressed subjects. Nat Med 16(4):460–465. doi:10.1038/nm.2111, URL:http://dx.doi.org/10.1038/nm.2111

Byrareddy SN, Arthos J, Cicala C, Villinger F, Ortiz KT, Little D, Sidell N, Kane MA, Yu J, Jones JW, Santangelo PJ, Zurla C, McKinnon LR, Arnold KB, Woody CE, Walter L, Roos C, Noll A, Ryk DV, Jelicic K, Cimbro R, Gumber S, Reid MD, Adsay V, Amancha PK, Mayne AE, Parslow TG, Fauci AS, Ansari AA (2016) Sustained virologic control in SIV+ macaques after antiretroviral and 47 antibody therapy. Science 354(6309):197–202. doi:10.1126/science.aag1276, URL:http://science.sciencemag.org/content/354/6309/197

Chatterjee A, Guedj J, Perelson AS (2012) Mathematical modelling of HCV infection: what can it teach us in the era of direct-acting antiviral agents? Antiviral Ther 17(6 Pt B):1171–1182. doi:10.3851/IMP2428, URL:http://europepmc.org/articles/PMC3641583

Chomont N, El-Far M, Ancuta P, Trautmann L, Procopio FA, Yassine-Diab B, Boucher G, Boulassel MR, Ghattas G, Brenchley JM, Schacker TW, Hill BJ, Douek DC, Routy JP, Haddad EK, Sekaly RP (2009) HIV reservoir size and persistence are driven by T cell survival and homeostatic proliferation. Nat Med 15(8):893–900. doi:10.1038/nm.1972, URL:http://dx.doi.org/10.1038/nm.1972

Chomont N, DaFonseca S, Vandergeeten C, Ancuta P, Sékaly RP (2011) Maintenance of CD4$^+$ T-cell memory and HIV persistence: keeping memory, keeping HIV. Curr Opin in HIV and AIDS 6(1):30–36. doi:10.1097/COH.0b013e3283413775

Chret A, Bacchus-Souffan C, Avettand-Fenol V, Mlard A, Nembot G, Blanc C, Samri A, Sez-Cirin A, Hocqueloux L, Lascoux-Combe C, Allavena C, Goujard C, Valantin MA, Leplatois A, Meyer L, Rouzioux C, Autran B, OPTIPRIM ANRS-147 Study Group (2015) Combined ART started during acute HIV infection protects central memory CD4$^+$ T cells and can induce remission. The Journal of Antimicrobial Chemotherapy 70(7):2108–2120. doi:10.1093/jac/dkv084

Chun TW, Finzi D, Margolick J, Chadwick K, Schwartz D, Siliciano RF (1995) In vivo fate of HIV-1-infected T cells: quantitative analysis of the transition to stable latency. Nat Med 1 (12):1284–1290

Chun TW, Carruth L, Finzi D, Shen X, DiGiuseppe JA, Taylor H, Hermankova M, Chadwick K, Margolick J, Quinn TC et al (1997) Quantification of latent tissue reservoirs and total body viral load in HIV-1 infection. Nature 387(6629):183–188

Chun TW, Justement JS, Murray D, Hallahan CW, Maenza J, Collier AC, Sheth PM, Kaul R, Ostrowski M, Moir S, Kovacs C, Fauci AS (2010) Rebound of plasma viremia following cessation of antiretroviral therapy despite profoundly low levels of HIV reservoir: implications for eradication. AIDS 24(18):2803–2808. doi:10.1097/QAD.0b013e328340a239, URL:http://www.ncbi.nlm.nih.gov/pmc/articles/PMC3154092/

Clapham HE, Tricou V, Van Vinh Chau N, Simmons CP, Ferguson NM (2014) Within-host viral dynamics of dengue serotype 1 infection. J Roy Soc Interface 11(96). doi:10.1098/rsif.2014.0094, URL:http://www.ncbi.nlm.nih.gov/pmc/articles/PMC4032531/

Cohn LB, Silva IT, Oliveira TY, Rosales RA, Parrish EH, Learn GH, Hahn BH, Czartoski JL, McElrath MJ, Lehmann C, Klein F, Caskey M, Walker BD, Siliciano JD, Siliciano RF, Jankovic M, Nussenzweig MC (2015) HIV-1 integration landscape during latent and active infection. Cell 160(3):420–432. doi:10.1016/j.cell.2015.01.020

Colgrove R, Japour A (1999) A combinatorial ledge: reverse transcriptase fidelity, total body viral burden, and the implications of multiple-drug HIV therapy for the evolution of antiviral resistance. Antiviral Res 41(1):45–56. doi:10.1016/S0166-3542(98)00062-X, URL:http://www.sciencedirect.com/science/article/pii/S016635429800062X

Conway JM, Coombs D (2011) A stochastic model of latently infected cell reactivation and viral blip generation in treated HIV patients. PLoS Comput Biol 7(4):e1002033. doi:10.1371/journal.pcbi.1002033, URL:http://dx.doi.org/10.1371/journal.pcbi.1002033

Conway JM, Perelson AS (2015) Post-treatment control of HIV infection. Proc Nat Acad Sci 112 (17):5467–5472. doi:10.1073/pnas.1419162112, URL:http://www.pnas.org/content/112/17/5467

Conway JM, Perelson AS (2016) Residual viremia in treated HIV+ individuals. PLoS Comput Biol 12(1):e1004677. doi:10.1371/journal.pcbi.1004677

Cory TJ, Schacker TW, Stevenson M, Fletcher CV (2013) Overcoming pharmacologic sanctuaries. Current Opin HIV AIDS 8(3):190–195

Crooks AM, Bateson R, Cope AB, Dahl NP, Griggs MK, Kuruc JD, Gay CL, Eron JJ, Margolis DM, Bosch RJ, Archin NM (2015) Precise quantitation of the latent HIV-1 reservoir: implications for eradication strategies. J Infect Dis. doi:10.1093/infdis/jiv218, URL:http://jid.oxfordjournals.org/content/early/2015/04/15/infdis.jiv218

Dahari H, Shudo E, Ribeiro RM, Perelson AS (2009) Modeling complex decay profiles of hepatitis B virus during antiviral therapy. Hepatology 49(1):32–38. doi:10.1002/hep.22586, URL:http://onlinelibrary.wiley.com/doi/10.1002/hep.22586/abstract

Davey RT Jr, Bhat N, Yoder C, Chun TW, Metcalf JA, Dewar R, Natarajan V, Lempicki RA, Adelsberger JW, Miller KD, Kovacs JA, Polis MA, Walker RE, Falloon J, Masur H, Gee D, Baseler M, Dimitrov DS, Fauci AS, Lane HC (1999) HIV-1 and T cell dynamics after interruption of highly active antiretroviral therapy (HAART) in patients with a history of sustained viral suppression. Proc Natl Acad Sci USA 96(26):15109–15114

Diekmann O, Heesterbeek JaP, Metz JaJ (1990) On the definition and the computation of the basic reproduction ratio R0 in models for infectious diseases in heterogeneous populations. J Math Biol 28(4):365–382. doi:10.1007/BF00178324, URL:https://link-springer-com.ezp-prod1.hul.harvard.edu/article/10.1007/BF00178324

Dinoso JB, Kim SY, Wiegand AM, Palmer SE, Gange SJ, Cranmer L, O'Shea A, Callender M, Spivak A, Brennan T et al (2009) Treatment intensification does not reduce residual HIV-1 viremia in patients on highly active antiretroviral therapy. Proc Natl Acad Sci 106(23):9403

Doekes HM, Fraser C, Lythgoe KA (2017) Effect of the latent reservoir on the evolution of HIV at the within- and between-host levels. PLOS Comput Biol 13(1):e1005228. doi:10.1371/journal.pcbi.1005228, URL:http://journals.plos.org/ploscompbiol/article?id=10.1371/journal.pcbi.1005228

Elliott JH, Wightman F, Solomon A, Ghneim K, Ahlers J, Cameron MJ, Smith MZ, Spelman T, McMahon J, Velayudham P, Brown G, Roney J, Watson J, Prince MH, Hoy JF, Chomont N, Fromentin R, Procopio FA, Zeidan J, Palmer S, Odevall L, Johnstone RW, Martin BP, Sinclair E, Deeks SG, Hazuda DJ, Cameron PU, Sékaly RP, Lewin SR (2014) Activation of HIV Transcription with short-course vorinostat in HIV-infected patients on suppressive antiretroviral therapy. PLoS Pathog 10(11):e1004473. doi:10.1371/journal.ppat.1004473, URL:http://dx.doi.org/10.1371/journal.ppat.1004473

Elliott JH, McMahon JH, Chang CC, Lee SA, Hartogensis W, Bumpus N, Savic R, Roney J, Hoh R, Solomon A, Piatak M, Gorelick RJ, Lifson J, Bacchetti P, Deeks SG, Lewin SR (2015) Short-term administration of disulfiram for reversal of latent HIV infection: a phase 2 dose-escalation study. The Lancet HIV 2(12):e520–e529. doi:10.1016/S2352-3018(15)00226-X, URL:http://www.sciencedirect.com/science/article/pii/S235230181500226X

Eriksson S, Graf EH, Dahl V, Strain MC, Yukl SA, Lysenko ES, Bosch RJ, Lai J, Chioma S, Emad F, Abdel-Mohsen M, Hoh R, Hecht F, Hunt P, Somsouk M, Wong J, Johnston R, Siliciano RF, Richman DD, O'Doherty U, Palmer S, Deeks SG, Siliciano JD (2013) Comparative analysis of measures of viral reservoirs in HIV-1 eradication studies. PLoS Pathog 9(2):e1003174. doi:10.1371/journal.ppat.1003174, URL:http://dx.doi.org/10.1371/journal.ppat.1003174

Etemad B, Sun X, Lederman MM, Gottlieb R, Aga E, Bosch RJ, Jacobson JM, Gandhi RT, Yu X, Li JZ (2016) Viral and immune characteristics of HIV post-treatment controllers in ACTG studies. In: CROI, Boston. URL:http://www.croiconference.org/sessions/viral-and-immune-characteristics-hiv-post-treatment-controllers-actg-studies

Fine P, Eames K, Heymann DL (2011) "Herd immunity": a rough guide. Clinical infectious diseases: an official publication of the infectious diseases society of America 52(7):911–916. doi:10.1093/cid/cir007

Finzi D, Hermankova M, Pierson T, Carruth LM, Buck C, Chaisson RE, Quinn TC, Chadwick K, Margolick J, Brookmeyer R, Gallant J, Markowitz M, Ho DD, Richman DD, Siliciano RF (1997) Identification of a reservoir for HIV-1 in patients on highly active antiretroviral therapy. Science 278(5341):1295–1300. doi:10.1126/science.278.5341.1295, URL:http://www.sciencemag.org/content/278/5341/1295.abstract

Finzi D, Blankson J, Siliciano JD, Margolick JB, Chadwick K, Pierson T, Smith K, Lisziewicz J, Lori F, Flexner C, Quinn TC, Chaisson RE, Rosenberg E, Walker B, Gange S, Gallant J, Siliciano RF (1999) Latent infection of CD4+ T cells provides a mechanism for lifelong persistence of HIV-1, even in patients on effective combination therapy. Nat Med 5(5):512–517. doi:10.1038/8394

Frost SD, McLean AR (1994) Quasispecies dynamics and the emergence of drug resistance during zidovudine therapy of HIV infection. Aids 8(3):323–332

Gandhi RT, Zheng L, Bosch RJ, Chan ES, Margolis DM, Read S, Kallungal B, Palmer S, Medvik K, Lederman MM, Alatrakchi N, Jacobson JM, Wiegand A, Kearney M, Coffin JM, Mellors JW, Eron JJ, on behalf of the AIDS Clinical Trials Group A5244 team (2010) The effect of raltegravir intensification on low-level residual viremia in HIV-infected patients on antiretroviral therapy: a randomized controlled trial. PLoS Med 7(8):e1000321. doi:10.1371/journal.pmed.1000321, URL:http://dx.doi.org/10.1371/journal.pmed.1000321

Goujard C, Girault I, Rouzioux C, Lécuroux C, Deveau C, Chaix ML, Jacomet C, Talamali A, Delfraissy JF, Venet A, Meyer L, Sinet M, ANRS CO6 PRIMO Study Group (2012) HIV-1 control after transient antiretroviral treatment initiated in primary infection: role of patient characteristics and effect of therapy. Antiviral Ther 17(6):1001–1009. doi:10.3851/IMP2273

Hatano H, Strain MC, Scherzer R, Bacchetti P, Wentworth D, Hoh RA, Martin JN, McCune JM, Neaton J, Tracy R, Richman DD, Deeks SG (2013) Increase in 2-LTR circles and decrease in D-dimer after raltegravir intensification in treated HIV-infected patients: a randomized, placebo-controlled trial. J Infect Dis 208(9):1436–1442

Henrich TJ, Hanhauser E, Marty FM, Sirignano MN, Keating S, Lee TH, Robles YP, Davis BT, Li JZ, Heisey A, Hill AL, Busch MP, Armand P, Soiffer RJ, Altfeld M, Kuritzkes DR (2014) Antiretroviral-free HIV-1 remission and viral rebound after allogeneic stem cell transplantation: report of 2 cases. Ann Intern Med 161(5):319–327. doi:10.7326/M14-1027, URL:http://dx.doi.org/10.7326/M14-1027

Hill AL, Rosenbloom DIS, Fu F, Nowak MA, Siliciano RF (2014) Predicting the outcomes of treatment to eradicate the latent reservoir for HIV-1. Proc Nat Acad Sci 111(37):13475–13480. doi:10.1073/pnas.1406663111, URL:http://www.pnas.org/content/111/37/13475

Hill AL, Rosenbloom DIS, Goldstein E, Hanhauser E, Kuritzkes DR, Siliciano RF, Henrich TJ (2016a) Real-time predictions of reservoir size and rebound time during antiretroviral therapy interruption trials for HIV. PLOS Pathog 12(4):e1005535. doi:10.1371/journal.ppat.1005535, URL:http://journals.plos.org/plospathogens/article?id=10.1371/journal.ppat.1005535

Hill AL, Rosenbloom DIS, Siliciano JD, Siliciano RF (2016b) Insufficient evidence for rare activation of latent HIV in the absence of reservoir-reducing interventions. PLOS Pathog 12(8): e1005679. doi:10.1371/journal.ppat.1005679, URL:http://journals.plos.org/plospathogens/article?id=10.1371/journal.ppat.1005679

Hill AL, Rosenbloom DIS, Siliciano R (2017) Estimating the efficacy of HIV latency reversing agents from residual viremia measurements. In preparation

Ho DD, Neumann AU, Perelson AS, Chen W, Leonard JM, Markowitz M et al (1995) Rapid turnover of plasma virions and CD4 lymphocytes in HIV-1 infection. Nature 373(6510):123–126

Hosmane NN, Kwon KJ, Bruner KM, Capoferri AA, Beg S, Rosenbloom DIS, Keele BF, Ho YC, Siliciano JD, Siliciano RF (2017) Proliferation of latently infected CD4(+) T cells carrying replication-competent HIV-1: potential role in latent reservoir dynamics. J Exp Med 214 (4):959–972. doi:10.1084/jem.20170193

Immonen TT, Conway JM, Romero-Severson EO, Perelson AS, Leitner T (2015) Recombination enhances HIV-1 envelope diversity by facilitating the survival of latent genomic fragments in the plasma virus population. PLOS Comput Biol 11(12):e1004625. doi:10.1371/journal.pcbi.1004625, URL:http://journals.plos.org/ploscompbiol/article?id=10.1371/journal.pcbi.1004625

Jilek BL, Zarr M, Sampah ME, Rabi SA, Bullen CK, Lai J, Shen L, Siliciano RF (2012) A quantitative basis for antiretroviral therapy for HIV-1 infection. Nat Med 18(3):446–451. doi:10.1038/nm.2649, URL:http://www.nature.com/nm/journal/v18/n3/full/nm.2649.html

Jones LE, Perelson AS (2007) Transient viremia, plasma viral load, and reservoir replenishment in HIV-infected patients on antiretroviral therapy. J Acquir Immune Defic Syndr (1999) 45 (5):483–493. doi:10.1097/QAI.0b013e3180654836, URL:http://www.ncbi.nlm.nih.gov/pmc/articles/PMC2584971/

Joos B, Fischer M, Kuster H, Pillai SK, Wong JK, Böni J, Hirschel B, Weber R, Trkola A, Günthard HF, Swiss HIV Cohort Study (2008) HIV rebounds from latently infected cells, rather than from continuing low-level replication. Proc Nat Acad Sci U S A 105(43):16725–16730. doi:10.1073/pnas.0804192105

Kaufmann DE, Lichterfeld M, Altfeld M, Addo MM, Johnston MN, Lee PK, Wagner BS, Kalife ET, Strick D, Rosenberg ES, Walker BD (2004) Limited durability of viral control following treated acute HIV infection. PLOS Med 1(2):e36. doi:10.1371/journal.pmed.0010036, URL:http://journals.plos.org/plosmedicine/article?id=10.1371/journal.pmed.0010036

Kelly JK (1996) Replication rate and evolution in the human immunodeficiency virus. J Theor Biol 180(4):359–364. doi:10.1006/jtbi.1996.0108, URL:http://www.sciencedirect.com/science/article/pii/S0022519396901083

Kermack WO, McKendrick AG (1927) A contribution to the mathematical theory of epidemics. Proc Roy Soc Lond A Math Phys Eng Sci 115(772):700–721. doi:10.1098/rspa.1927.0118, URL:http://rspa.royalsocietypublishing.org/content/115/772/700

Kieffer TL, Finucane MM, Nettles RE, Quinn TC, Broman KW, Ray SC, Persaud D, Siliciano RF (2004) Genotypic analysis of HIV-1 drug resistance at the limit of detection: virus production without evolution in treated adults with undetectable HIV loads. J Infect Dis 189(8):1452–1465. doi:10.1086/382488

Kim H, Perelson AS (2006) Viral and latent reservoir persistence in HIV-1-infected patients on therapy. PLoS Comput Biol 2(10):e135. doi:10.1371/journal.pcbi.0020135, URL:http://dx. plos.org/10.1371/journal.pcbi.0020135

Lifson JD, Rossio JL, Arnaout R, Li L, Parks TL, Schneider DK, Kiser RF, Coalter VJ, Walsh G, Imming RJ (2000) Containment of simian immunodeficiency virus infection: cellular immune responses and protection from rechallenge following transient postinoculation antiretroviral treatment. J Virol 74(6):2584–2593

Lifson JD, Rossio JL, Piatak M, Parks T, Li L, Kiser R, Coalter V, Fisher B, Flynn BM, Czajak S (2001) Role of CD8+ lymphocytes in control of simian immunodeficiency virus infection and resistance to rechallenge after transient early antiretroviral treatment. J Virol 75(21): 10187–10199

Lim SY, Osuna CE, Hesselgesser J, Hill AL, Miller MD, Geleziunas R, Lee W, Whitney JB (2017) TLR7 agonist treatment of SIV+ monkeys on ART can lead to complete viral remission. URL:http://www.croiconference.org/sessions/tlr7-agonist-treatment-siv-monkeys-art-can-lead-complete-viral-remission, #338LB

Lodi S, Meyer L, Kelleher AD, Rosinska M, Ghosn J, Sannes M, Porter K (2012) Immunovirologic control 24 months after interruption of antiretroviral therapy initiated close to HIV seroconversion. Arch Intern Med 172(16):1252–1255. doi:10.1001/archinternmed. 2012.2719

Luo R, Piovoso MJ, Martinez-Picado J, Zurakowski R (2012) HIV model parameter estimates from interruption trial data including drug efficacy and reservoir dynamics. PLoS One 7(7):e40198. doi:10.1371/journal.pone.0040198, URL:http://dx.doi.org/10.1371/journal.pone.0040198

Luo R, Cardozo EF, Piovoso MJ, Wu H, Buzon MJ, Martinez-Picado J, Zurakowski R (2013) Modelling HIV-1 2-LTR dynamics following raltegravir intensification. J Roy Soc Interface 10 (84). doi:10.1098/rsif.2013.0186, URL:http://rsif.royalsocietypublishing.org/content/10/84/ 20130186

Luzuriaga K, Gay H, Ziemniak C, Sanborn KB, Somasundaran M, Rainwater-Lovett K, Mellors JW, Rosenbloom D, Persaud D (2015) Viremic relapse after HIV-1 remission in a perinatally infected child. New Engl J Med 372(8):786–788. doi:10.1056/NEJMc1413931

Lythgoe KA, Fraser C (2012) New insights into the evolutionary rate of HIV-1 at the within-host and epidemiological levels. Proc Roy Soc B Biol Sci. doi:10.1098/rspb.2012.0595, URL:http:// www.ncbi.nlm.nih.gov/pubmed/22593106

Maenza J, Tapia K, Holte S, Stekler JD, Stevens CE, Mullins JI, Collier AC (2015) How often does treatment of primary HIV lead to post-treatment control? Antiviral Ther. doi:10.3851/ IMP2963

Maldarelli F, Wu X, Su L, Simonetti FR, Shao W, Hill S, Spindler J, Ferris AL, Mellors JW, Kearney MF, Coffin JM, Hughes SH (2014) Specific HIV integration sites are linked to clonal expansion and persistence of infected cells. Science 345(6193):179–183. doi:10.1126/science. 1254194, URL:http://www.sciencemag.org/content/345/6193/179

Markowitz M, Louie M, Hurley A, Sun E, Di Mascio M, Perelson AS, Ho DD (2003) A novel antiviral intervention results in more accurate assessment of human immunodeficiency virus type 1 replication dynamics and T-cell decay in vivo. J Virol 77(8):5037–5038. doi:10.1128/ JVI.77.8.5037-5038.2003

Martinez-Picado J, Deeks SG (2016) Persistent HIV-1 replication during antiretroviral therapy. Curr Opin HIV AIDS 11(4):417–423. doi:10.1097/COH.0000000000000287

McLean AR, Nowak MA (1992) Competition between zidovudine-sensitive and zidovudine-resistant strains of HIV. AIDS 6(1):71

Murillo LN, Murillo MS, Perelson AS (2013) Towards multiscale modeling of influenza infection. J Theor Biol 332:267–290. doi:10.1016/j.jtbi.2013.03.024, URL:http://www.sciencedirect. com/science/article/pii/S0022519313001501

Murray AJ, Kwon KJ, Farber DL, Siliciano RF (2016) The latent reservoir for HIV-1: how immunologic memory and clonal expansion contribute to HIV-1 persistence. J Immun 197 (2):407–417. doi:10.4049/jimmunol.1600343, URL:http://www.jimmunol.ezp-prod1.hul. harvard.edu/content/197/2/407

Neumann AU, Lam NP, Dahari H, Gretch DR, Wiley TE, Layden TJ, Perelson AS (1998) Hepatitis C viral dynamics in vivo and the antiviral efficacy of interferon- therapy. Science 282 (5386):103–107. doi:10.1126/science.282.5386.103, URL:http://www.sciencemag.org/content/282/5386/103

No A, Plum J, Verhofstede C (2005) The latent HIV-1 reservoir in patients undergoing HAART: an archive of pre-HAART drug resistance. J Antimicrob Chemother 55(4):410–412. doi:10.1093/jac/dki038, URL:http://jac.oxfordjournals.org/content/55/4/410.abstract

Nowak MA, May RMC (2000) Virus dynamics: mathematical principles of immunology and virology. Oxford University Press, USA

Nowak MA, Anderson RM, McLean AR, Wolfs TF, Goudsmit J, May RM (1991) Antigenic diversity thresholds and the development of AIDS. Science 254(5034):963–969

Nowak MA, Bonhoeffer S, Hill AM, Boehme R, Thomas HC, McDade H (1996) Viral dynamics in hepatitis B virus infection. Proc Natl Acad Sci 93:4398–4402

Nuraini N, Tasman H, Soewono E, Sidarto KA (2009) A with-in host dengue infection model with immune response. Math Comput Model 49(56):1148–1155. doi:10.1016/j.mcm.2008.06.016, URL:http://www.sciencedirect.com/science/article/pii/S0895717708002732

Palmer S, Maldarelli F, Wiegand A, Bernstein B, Hanna GJ, Brun SC, Kempf DJ, Mellors JW, Coffin JM, King MS (2008) Low-level viremia persists for at least 7 years in patients on suppressive antiretroviral therapy. Proc Nat Acad Sci U S A 105(10):3879–3884. doi:10.1073/pnas.0800050105, URL:http://www.ncbi.nlm.nih.gov/pubmed/18332425

Pennings PS (2012) Standing genetic variation and the evolution of drug resistance in HIV. PLoS Comput Biol 8(6):e1002527. doi:10.1371/journal.pcbi.1002527, URL:http://dx.doi.org/10.1371/journal.pcbi.1002527

Pennings PS, Kryazhimskiy S, Wakeley J (2014) Loss and recovery of genetic diversity in adapting populations of HIV. PLoS Genet 10(1):e1004000. doi:10.1371/journal.pgen.1004000, URL:http://dx.doi.org/10.1371/journal.pgen.1004000

Perelson AS, Ribeiro RM (2004) Hepatitis B virus kinetics and mathematical modeling. Semin Liver Dis 24(S 1):11–16. doi:10.1055/s-2004-828673, URL:https://www.thieme-connect.com/products/ejournals/html/10.1055/s-2004-828673

Perelson AS, Neumann AU, Markowitz M, Leonard JM, Ho DD (1996) HIV-1 dynamics in vivo: virion clearance rate, infected cell life-span, and viral generation time. Science 271 (5255):1582–1586. URL:http://www.ncbi.nlm.nih.gov/pubmed/8599114

Perelson AS, Essunger P, Cao Y, Vesanen M, Hurley A, Saksela K, Markowitz M, Ho DD (1997) Decay characteristics of HIV-1-infected compartments during combination therapy. Nature 387 (6629):188–191. doi:10.1038/387188a0, URL:http://dx.doi.org/10.1038/387188a0

Persaud D, Pierson T, Ruff C, Finzi D, Chadwick KR, Margolick JB, Ruff A, Hutton N, Ray S, Siliciano RF (2000) A stable latent reservoir for HIV-1 in resting CD4$^+$ T lymphocytes in infected children. J Clin Invest 105(7):995–1003. URL:http://www.ncbi.nlm.nih.gov/pmc/articles/PMC377486/

Petravic J, Martyushev A, Reece JC, Kent SJ, Davenport MP (2014) Modeling the timing of antilatency drug administration during HIV treatment. J Virol 88(24):14050–14056. doi:10.1128/JVI.01701-14

Petravic J, Rasmussen TA, Lewin SR, Kent SJ, Davenport MP (2017) Relationship between measures of HIV reactivation and the decline of latent reservoir under latency-reversing agents. J Virol. doi:10.1128/JVI.02092-16

Phillips AN (1996) Reduction of HIV concentration during acute infection: independence from a specific immune response. Science 271(5248):497

Pinkevych M, Cromer D, Tolstrup M, Grimm AJ, Cooper DA, Lewin SR, Søgaard OS, Rasmussen TA, Kent SJ, Kelleher AD, Davenport MP (2015) HIV reactivation from latency after treatment interruption occurs on average every 5–8 days—implications for HIV remission. PLoS Pathog 11(7):e1005000. doi:10.1371/journal.ppat.1005000, URL:http://dx.doi.org/10.1371/journal.ppat.1005000

Ramratnam B, Bonhoeffer S, Binley J, Hurley A, Zhang L, Mittler JE, Markowitz M, Moore JP, Perelson AS, Ho DD (1999) Rapid production and clearance of HIV-1 and hepatitis C virus assessed by large volume plasma apheresis. Lancet 354(9192):1782–1785. doi:10.1016/S0140-6736(99)02035-8, URL:http://www.sciencedirect.com/science/article/pii/S0140673699020358

Rasmussen TA, Tolstrup M, Brinkmann CR, Olesen R, Erikstrup C, Solomon A, Winckelmann A, Palmer S, Dinarello C, Buzon M, others (2014) Panobinostat, a histone deacetylase inhibitor, for latent-virus reactivation in HIV-infected patients on suppressive antiretroviral therapy: a phase 1/2, single group, clinical trial. Lancet HIV 1(1):e13–e21. URL:http://www.science direct.com/science/article/pii/S2352301814700141

Razooky BS, Pai A, Aull K, Rouzine IM, Weinberger LS (2015) A hardwired HIV latency program. Cell 160(5):990–1001. doi:10.1016/j.cell.2015.02.009, URL:http://www.cell.com. ezp-prod1.hul.harvard.edu/article/S0092867415001749/abstract

Regoes RR, Wodarz D, Nowak MA (1998) Virus dynamics: the effect of target cell limitation and immune responses on virus evolution. J Theor Biol 191(4):451–462

Ribeiro RM (2002) In vivo dynamics of T cell activation, proliferation, and death in HIV-1 infection: why are $CD4^+$ but not $CD8^+$ T cells depleted? Proc Nat Acad Sci 99(24):15572–15577. doi:10. 1073/pnas.242358099, URL:http://www.ncbi.nlm.nih.gov.ezp-prod1.hul.harvard.edu/pmc/articles/PMC137758/

Ribeiro RM, Bonhoeffer S, Nowak MA (1998) The frequency of resistant mutant virus before antiviral therapy. AIDS 12(5):461

Ribeiro RM, Qin L, Chavez LL, Li D, Self SG, Perelson AS (2010) Estimation of the initial viral growth rate and basic reproductive number during acute HIV-1 infection. J Virol 84(12):6096–6102. doi:10.1128/JVI.00127-10, URL:http://jvi.asm.org/content/84/12/6096.abstract

Rong L, Perelson AS (2009a) Asymmetric division of activated latently infected cells may explain the decay kinetics of the HIV-1 latent reservoir and intermittent viral blips. Math Biosci 217 (1):77–87. doi:10.1016/j.mbs.2008.10.006

Rong L, Perelson AS (2009b) Modeling latently infected cell activation: viral and latent reservoir persistence, and viral blips in HIV-infected patients on potent therapy. PLoS Comput Biol 5 (10):e1000533. doi:10.1371/journal.pcbi.1000533, URL:http://dx.doi.org/10.1371/journal. pcbi.1000533

Rosenberg ES, Altfeld M, Poon SH, Phillips MN, Wilkes BM, Eldridge RL, Robbins GK, D'Aquila RT, Goulder PJR, Walker BD (2000) Immune control of HIV-1 after early treatment of acute infection. Nature 407(6803):523–526. doi:10.1038/35035103, URL:http://dx.doi.org/ 10.1038/35035103

Rosenbloom DIS, Hill AL, Rabi SA, Siliciano RF, Nowak MA (2012) Antiretroviral dynamics determines HIV evolution and predicts therapy outcome. Nat Med 18(9):1378–1385. doi:10.1038/ nm.2892, URL:http://www.nature.com.ezp-prod1.hul.harvard.edu/nm/journal/v18/n9/full/nm. 2892.html

Rouzine I, Coffin J (1999) Linkage disequilibrium test implies a large effective population number for HIV in vivo. Proc Nat Acad Sci 96(19):10758

Rouzine IM, Razooky BS, Weinberger LS (2014) Stochastic variability in HIV affects viral eradication. Proc Nat Acad Sci 111(37):13251–13252. doi:10.1073/pnas.1413362111, URL: http://www.pnas.org/content/111/37/13251

Rouzine I, Weinberger A, Weinberger L (2015) An evolutionary role for HIV latency in enhancing viral transmission. Cell 160(5):1002–1012. doi:10.1016/j.cell.2015.02.017, URL:http://www. sciencedirect.com/science/article/pii/S009286741500183X

Ruff CT, Ray SC, Kwon P, Zinn R, Pendleton A, Hutton N, Ashworth R, Gange S, Quinn TC, Siliciano RF, Persaud D (2002) Persistence of wild-type virus and lack of temporal structure in the latent reservoir for human immunodeficiency virus type 1 in pediatric patients with extensive antiretroviral exposure. J Virol 76(18):9481–9492. doi:10.1128/JVI.76.18.9481-9492.2002, URL:http://jvi.asm.org/content/76/18/9481

Ruiz L, Martinez-Picado J, Romeu J, Paredes R, Zayat MK, Marfil S, Negredo E, Sirera G, Tural C, Clotet B (2000) Structured treatment interruption in chronically HIV-1 infected patients after long-term viral suppression. AIDS 14(4):397

Saez-Cirion A, Jacquelin B, Barré-Sinoussi F, Müller-Trutwin M (2014) Immune responses during spontaneous control of HIV and AIDS: what is the hope for a cure? Philos Trans Roy Soc Lond Series B Biol Sci 369(1645):20130436. doi:10.1098/rstb.2013.0436

Schiffer JT, Swan DA, Magaret A, Corey L, Wald A, Ossig J, Ruebsamen-Schaeff H, Stoelben S, Timmler B, Zimmermann H, Melhem MR, Van Wart SA, Rubino CM, Birkmann A (2016) Mathematical modeling of herpes simplex virus-2 suppression with pritelivir predicts trial outcomes. Sci Transl Med 8(324):324ra15, DOI 10.1126/scitranslmed.aad6654

Sedaghat AR, Siliciano JD, Brennan TP, Wilke CO, Siliciano RF (2007) Limits on replenishment of the resting $CD4^+$ T cell reservoir for HIV in patients on HAART. PLoS Pathog 3(8):e122. doi:10.1371/journal.ppat.0030122, URL:http://dx.plos.org/10.1371/journal.ppat.0030122

Sedaghat A, Siliciano R, Wilke C (2008) Low-level HIV-1 replication and the dynamics of the resting $CD4^+$ T cell reservoir for HIV-1 in the setting of HAART. BMC Infect Dis 8(1):2. doi:10.1186/1471-2334-8-2, URL:http://www.biomedcentral.com/1471-2334/8/2

Sez-Cirin A, Bacchus C, Hocqueloux L, Avettand-Fenoel V, Girault I, Lecroux C, Potard V, Versmisse P, Melard A, Prazuck T, Descours B, Guergnon J, Viard JP, Boufassa F, Lambotte O, Goujard C, Meyer L, Costagliola D, Venet A, Pancino G, Autran B, Rouzioux C (2013) Post-treatment HIV-1 controllers with a long-term virological remission after the interruption of early initiated antiretroviral therapy ANRS VISCONTI study. PLoS Pathog 9 (3):e1003211. doi:10.1371/journal.ppat.1003211

Sgaard OS, Graversen ME, Leth S, Olesen R, Brinkmann CR, Nissen SK, Kjaer AS, Schleimann MH, Denton PW, Hey-Cunningham WJ, Koelsch KK, Pantaleo G, Krogsgaard K, Sommerfelt M, Fromentin R, Chomont N, Rasmussen TA, Østergaard L, Tolstrup M (2015) The depsipeptide romidepsin reverses HIV-1 latency in vivo. PLoS Pathog 11(9):e1005142. doi:10.1371/journal.ppat.1005142, URL:http://dx.doi.org/10.1371/journal. ppat.1005142

Shen L, Peterson S, Sedaghat AR, McMahon MA, Callender M, Zhang H, Zhou Y, Pitt E, Anderson KS, Acosta EP et al (2008) Dose-response curve slope sets class-specific limits on inhibitory potential of anti-HIV drugs. Nat Med 14(7):762–766

Siliciano JD, Kajdas J, Finzi D, Quinn TC, Chadwick K, Margolick JB, Kovacs C, Gange SJ, Siliciano RF (2003) Long-term follow-up studies confirm the stability of the latent reservoir for HIV-1 in resting $CD4^+$ T cells. Nat Med 9(6):727–728. doi:10.1038/nm880, URL:http://dx.doi. org/10.1038/nm880

Simonetti FR, Sobolewski MD, Fyne E, Shao W, Spindler J, Hattori J, Anderson EM, Watters SA, Hill S, Wu X, Wells D, Su L, Luke BT, Halvas EK, Besson G, Penrose KJ, Yang Z, Kwan RW, Waes CV, Uldrick T, Citrin DE, Kovacs J, Polis MA, Rehm CA, Gorelick R, Piatak M, Keele BF, Kearney MF, Coffin JM, Hughes SH, Mellors JW, Maldarelli F (2016) Clonally expanded $CD4^+$ T cells can produce infectious HIV-1 in vivo. Proc Nat Acad Sci 113 (7):1883–1888. doi:10.1073/pnas.1522675113, URL:http://www.pnas.org/content/113/7/1883

Singh A, Razooky B, Cox CD, Simpson ML, Weinberger LS (2010) Transcriptional bursting from the HIV-1 promoter is a significant source of stochastic noise in HIV-1 gene expression. Biophys J 98(8):L32–L34. doi:10.1016/j.bpj.2010.03.001

Spivak A, Rabi A, McMahon MA, Shan L, Sedaghat AR, Wilke CO, Siliciano R (2010) Dynamic constraints on the second phase compartment of HIV-infected cells. AIDS Res Hum Retroviruses. doi:10.1089/AID.2010.0199, URL:http://www.ncbi.nlm.nih.gov/pubmed/ 21105850

Stafford MA, Corey L, Cao Y, Daar ES, Ho DD, Perelson AS (2000) Modeling plasma virus concentration during primary HIV infection. J Theor Biol 203(3):285–301. doi:10.1006/jtbi. 2000.1076, URL:http://www.sciencedirect.com/science/article/pii/S0022519300910762

Wei X, Ghosh SK, Taylor ME, Johnson VA, Emini EA, Deutsch P, Lifson JD, Bonhoeffer S, Nowak MA, Hahn BH (1995) Viral dynamics in human immunodeficiency virus type 1 infection. Nature 373(6510):117–122

Weinberger LS, Burnett JC, Toettcher JE, Arkin AP, Schaffer DV (2005) Stochastic gene expression in a lentiviral positive-feedback loop: HIV-1 Tat fluctuations drive phenotypic diversity. Cell 122(2):169–182. doi:10.1016/j.cell.2005.06.006, URL:http://www.science direct.com/science/article/B6WSN-4GRH1R2-7/2/856752b80260bf0e4ebecec7af038327

Whitney JB, Hill AL, Sanisetty S, Penaloza-MacMaster P, Liu J, Shetty M, Parenteau L, Cabral C, Shields J, Blackmore S, Smith JY, Brinkman AL, Peter LE, Mathew SI, Smith KM, Borducchi EN, Rosenbloom DIS, Lewis MG, Hattersley J, Li B, Hesselgesser J, Geleziunas R, Robb ML, Kim JH, Michael NL, Barouch DH (2014) Rapid seeding of the viral reservoir prior to SIV viraemia in rhesus monkeys. Nature 512(7512):74–77. doi:10.1038/nature13594, URL: http://www.nature.com/nature/journal/vaop/ncurrent/full/nature13594.html

Wind-Rotolo M, Durand C, Cranmer L, Reid A, Martinson N, Doherty M, Jilek BL, Kagaayi J, Kizza A, Pillay V, Laeyendecker O, Reynolds SJ, Eshleman SH, Lau B, Ray SC, Siliciano JD, Quinn TC, Siliciano RF (2009) Identification of nevirapine-resistant HIV-1 in the latent reservoir after single-dose nevirapine to prevent mother-to-child transmission of HIV-1. J Infect Dis 199(9):1301. doi:10.1086/597759, URL: https://www.ncbi.nlm.nih.gov/pmc/articles/PMC2703715/report=abstract

Wodarz D, Arnaout RA, Nowak MA, Lifson JD (2000a) Transient antiretroviral treatment during acute simian immunodeficiency virus infection facilitates long-term control of the virus. Philos Trans Roy Soc B Biol Sci 355(1400):1021–1029. URL:http://www.ncbi.nlm.nih.gov/pmc/articles/PMC1692816/

Wodarz D, Page KM, Arnaout RA, Thomsen AR, Lifson JD, Nowak MA (2000b) A new theory of cytotoxic T-lymphocyte memory: implications for HIV treatment. Philos Trans Roy Soc B Biol Sci 355(1395):329–343

Yates A, Stark J, Klein N, Antia R, Callard R (2007) Understanding the slow depletion of memory CD4+ T cells in HIV infection. PLoS Med 4(5):e177. doi:10.1371/journal.pmed.0040177, URL:http://dx.doi.org/10.1371/journal.pmed.0040177

Zhang J, Perelson AS (2013) Contribution of follicular dendritic cells to persistent HIV viremia. J Virol 87(14):7893–7901. doi:10.1128/JVI.00556-13

# Residual Immune Activation and Latency

Elena Bruzzesi and Irini Sereti

## Contents

**Abstract** The introduction of combination antiretroviral therapy (cART) in the 1990s has dramatically changed the course of HIV infection, decreasing the risk for both AIDS- and non-AIDS-related events. Cancers, cardiovascular disease (CVD), liver and kidney disease, neurological disorders and frailty have become of great importance lately in the clinical management as they represent the principal cause of death in people living with HIV who receive cART (Kirk et al. in Clin Infect Dis 45(1):103–10, 2007; Strategies for Management of Antiretroviral Therapy Study et al. N Engl J Med 355(22):2283–2296, 2006; Ances et al. J Infect Dis 201 (3):336–340, 2010; Desquilbet et al. J Gerontol A Biol Sci Med Sci 62(11): 1279–1286, 2007; Lifson et al. HIV Clin Trials 9(3):177–185, 2008). Despite the undeniable achievements of cART, we are now faced with its limitations: a considerable proportion of individuals, referred as to immunological non-responders, fails to reconstitute the immune system despite optimal treatment and viral suppression (Kelley et al. Clin Infect Dis 48(6):787–794, 2009; Robbins et al. Clin

E. Bruzzesi · I. Sereti (✉)
Laboratory of Immunoregulation, National Institutes of Allergy
and Infectious Diseases, National Institutes of Health, Bethesda, MD, USA
e-mail: ISereti@niaid.nih.gov

E. Bruzzesi · I. Sereti
Department of Infectious Diseases, IRCCS, San Raffaele
Scientific Institute, Milan, Italy

Current Topics in Microbiology and Immunology (2018) 417:157–180
DOI 10.1007/82_2018_118
Published Online: 14 August 2018

Infect Dis 48(3):350–361, 2009) and remains at high risk for opportunistic infections and non-AIDS-related events (Strategies for Management of Antiretroviral Therapy Study et al. N Engl J Med 355(22):2283–2296, 2006). Moreover, the generalized state of immune activation and inflammation, linked to serious non-AIDS events, persists despite successful HIV suppression with cART. Finally, the current strategies have so far failed to eradicate the virus, and inflammation appears a driving force in viral persistence. In the light of all this, it is of fundamental importance to investigate the pathophysiological processes that link incomplete immune recovery, immune activation and HIV persistence to design targeted therapies that could impact on the three.

# 1 Immune Activation

HIV disease is not merely a disease of viral replication and immunodeficiency, but also an inflammatory process from the acute to the chronic phase of infection. The creation of an inflammatory environment soon after infection favours not only dissemination of the virus, but also the very early formation of a reservoir of latently infected T cells (Hunt et al. 2011).

Although CD4 T cell counts and HIV viremia are important prognostic factors in the untreated population, it was clear from the beginning of the epidemic that immune activation and inflammation played a pivotal role in disease prognosis. Giorgi et al. proposed T cell activation, measured as either the frequency of CD38 expression or the co-expression of CD38 and HLA-DR in CD8+ T cells, as an independent predictor of mortality in the pre-cART era (Giorgi et al. 1993; Giorgi et al. 1999). Subsequent studies showed that both immune activation and the decline of CD4+ T cells were independent predictors of disease progression and robust prognostic factors in ART-suppressed patients (Hazenberg et al. 2003; Deeks et al. 2004). Early in the HIV epidemic, disease progression was found to strongly correlate with serum concentration of markers of immune activation, namely ß2-microglobulin and urine neopterin, which were also used for early diagnosis in children (Grieco et al. 1984; Fuchs et al. 1984; Chan et al. 1990). The association with HIV-driven inflammation regardless of CD4 and viral load levels has been confirmed more recently in the SMART trial (Kuller et al. 2008).

Immune activation is characterized by the acquisition of an activated phenotype by cells of both innate and adaptive immunity. Consistently, in both acute and chronic phases of HIV infection, elevated levels of serum cytokines, chemokines, pro-inflammatory soluble mediators, acute phase proteins, microbial sensors and coagulation factors are detected: their production starts very early during acute infection with a massive release of cytokines (IFN-alpha and IL-15, followed by IL-10, TNF-alpha, IP-10, IFN-gamma and IL-6), referred as to cytokine storm (Stacey et al. 2009). Some of these cytokines are associated with T cell activation, CD4+ depletion and disease progression (Boulware et al. 2011): IL-6, a pro-inflammatory cytokine produced mainly by monocytes and macrophages, is increased in HIV-infected individuals compared to matched uninfected controls (Neuhaus et al. 2010) and has been associated with all-cause mortality in the SMART study (Kuller et al. 2008). Similar findings were seen with hs-CRP, D-dimer and sCD14 suggesting that inflammation, coagulation, and innate immune activation are intertwined (Long et al. 2016) and predict mortality in persons living with HIV treated with cART (Kuller et al. 2008; Neuhaus et al. 2010).

In addition, plasmacytoid DCs (pDC) produce interferon alpha (IFN-α) upon activation (Chew 2014), which is released early after infection together with IL-15 and remains elevated through the chronic stages (O'Brien et al. 2011; Hardy et al. 2013). During acute infection, it induces the expression of interferon-stimulated genes (ISGs), facilitating clearance of intracellular pathogens (Ploquin et al. 2016). Conversely, in the chronic stages it is detrimentally involved in T cell activation and depletion, as demonstrated in a cohort of HCV-infected patients under IFN treatment (Boasso et al. 2008), as well as in disease progression (von Sydow et al. 1991). Furthermore, activated monocytes (CD14+CD16+, CD14dimCD16+) are important in the maintenance of a pro-inflammatory environment as they are persistently elevated, despite cART (Burdo et al. 2011). In particular, the CD14 molecule exposed on their membrane is a co-receptor for lipopolysaccharide (LPS), a component of the membrane of gram-negative bacteria; its soluble form (sCD14) has been found to be predictive of disease progression and mortality, accounting for a sixfold higher risk of death (Sandler et al. 2011) and to be associated with coronary artery calcification (Longenecker et al. 2014) and other inflammatory diseases, including hepatitis, paediatric inflammatory lung diseases and rheumatoid arthritis (Sandler, N.G., et al., Host response to translocated microbial products predicts outcomes of patients with HBV or HCV infection (Gastroenterology 2011; Marcos et al. 2010; Bas et al. 2004). In addition, circulating CD16+ monocytes were found to be an independent indicator of coronary artery calcium progression in HIV-infected individuals, demonstrating the role of immune activation in the spectrum of non-AIDS-related disorders (Baker et al. 2014).

Immune activation has a multifactorial origin, and the most important drivers are residual viral replication, gut translocation, co-infections and immune dysregulation. The pro-inflammatory environment does not entirely derive from viral replication as markers of immune activation decrease but do not revert to normal, once viremia has been suppressed (Sereti et al. 2017). Moreover, persistent viral replication alone is not necessary nor sufficient to explain the decline of CD4+ T cells, since the percentage of the HIV-infected CD4+ T cells is too low (0.01–1%) to account uniquely for the loss of CD4+ T cells, at least in the periphery (Haase et al. 1996; Lassen et al. 2004). Therefore, mechanisms other than increased destruction have been investigated, namely decreased production by low thymic output or inefficient hematopoiesis, redistribution from blood to tissues, altered function linked to tissue fibrosis, and immune activation, as shown in special groups of patients: in HIV-2 infected individuals, whose disease course is characterized by low levels of viremia and preserved immune function long term, a causal relationship between immune activation and CD4+ T cell depletion has been found (Sousa et al. 2002). Similarly, in elite controllers that spontaneously maintain viral load under detection but can progressively develop CD4+ depletion and eventually AIDS, the level of T cell activation and inflammatory markers are higher compared to healthy controls and to suppressed patients (Hunt et al. 2008; Li 2015). Consistently, long-term non-progressors (LTNP), who are by definition asymptomatic individuals with stable CD4 count of $\geq 500$ cells/mm$^3$ for at least seven or more years of HIV infection in the absence of antiretroviral treatment, show a decreased level of immune activation (Choudhary et al. 2007).

In HIV-1 infection, CD4+ T cell depletion occurs primarily in the mucosal-associated lymphoid tissues (MALT) where CD4+ T cells are more concentrated, affecting mainly the Th17 population (Schuetz et al. 2014), both in the acute and the chronic phases of the disease (Mehandru et al. 2004; Brenchley et al. 2004), and despite optimal treatment (Guadalupe et al. 2003). In fact, CD4+ T cells that reside in the GI tract are activated with higher frequency compared to the ones in peripheral blood and in lymph nodes (LN) (Kim et al. 1997), as most of them express CCR5 (Veazey et al. 2000; Anton et al. 2000), the chemokine co-receptor used by HIV to enter into the cells especially in the first phases of infection. T cell depletion, increase in mucosal T cell activation and cytokine production and the direct response of mucosal epithelial cells to HIV-1 can all be ascribed for the epithelial damage, characterized by the apoptosis of enterocytes and the disruption of tight junctions (Nazli et al. 2010). As a result of the loss of defence and of the imbalance between depleting Th17 and expanding regulatory T cells (Treg), gut permeability increases (Chung et al. 2014) and microbial translocation ensues (Brenchley et al. 2006). Gut translocation is defined as the non-physiological passage of GI flora or microbial products across the epithelium to the circulation, while normally, microbes and their products are phagocytized at the level of lamina propria or at the mesenteric lymph nodes, without reaching extra-intestinal sites. As a consequence, microbial translocation can be quantified by measuring the microbial by-products in plasma, such as LPS and 16S ribosomal DNA (16S rDNA). LPS is a component of the cell wall of gram-negative bacteria and triggers an innate immune response upon interaction with the LPS-binding protein (LBP). The LBP presents the LPS to Toll-like receptor 4, to CD14 and sCD14, leading to the activation of the NF-kB pathway and the production of the cytokines IL-6, IL-1b, and TNF, and type I interferons (Gioannini and Weiss 2007; Klatt et al. 2006). Moreover, it elicits IgM, IgA and IgG responses specific for both the core antigen and for the endotoxin core (EndoCAb). Therefore, while LPS and 16S rDNA represent the direct markers of gut translocation, sCD14, LBP and EndoCAb can be used as indirect indicators. In this regard, Brenchley et al. for the first time correlated LPS levels not only with the mucosal immune dysfunction, but also with chronic activation and, eventually, HIV disease progression (Brenchley et al. 2006), a finding that was later confirmed for 16S rDNA (Jiang et al. 2009). Non-human primate models have further confirmed this relationship: on the one hand, rhesus macaques (RM) develop CD4+ T cell depletion and AIDS upon simian immunodeficiency virus (SIV) infection; on the other hand, sooty mangabeys (SM) and African green monkeys (AGM) are considered natural hosts of SIV virus, as they do not develop immunodeficiency despite high levels of SIV viremia. Albeit the opposite disease course, both the progressors (RM) and the non-progressors (AGM and SM) display a considerable level of immune activation during acute infection. This suggested that the main difference is not the presence versus absence of immune activation as it was previously thought, but rather the duration and the timing of the innate immune response (Bosinger et al. 2012). Indeed, the innate system is activated early and transiently in non-pathogenic hosts, while it is delayed and protracted in pathogenic hosts (Benecke et al. 2012). Such delay could result in late activation of the adaptive immunity that in turn will be less selective against self-antigens and non-harmful pathogens (Jacquelin et al. 2009). Furthermore, another important difference is the absence of microbial translocation in the non-progressors, despite an early and severe CD4+ T cell depletion, suggesting that the CD4+ T cell loss is not sufficient, although possibly necessary, to explain the translocation (Vinton et al. 2011; Sundaravaradan et al. 2013; Chahroudi et al. 2012). Interestingly, the translocation can favour HIV replication in the early phases of the infection, as in SIV-infected macaques the translocation occurs before the peak of the viremia and since, after the administration of a GI permeabilizing compound, the dextran sulphate sodium (DSS), two 'elite controller' SIV-infected macaques showed a prolonged period of viremia (Ericsen et al. 2016).

Intestinal inflammation, T cell activation and microbial translocation have also been related to a disequilibrium in the gut microbiome, in that the diversity is decreased (Mutlu et al. 2014) and the microbiota composition altered, with a shift from Firmicutes to Proteobacteria at the phylum level and from Bacteroidetes to Prevotella at the genus one (Dillon et al. 2014).

Epithelial barrier dysfunctions have also been related to the asymptomatic reactivation of Cytomegalovirus (CMV) (Steininger et al. 2006), as CMV disrupts the tight junctions of polarized enterocytes and induces the production of IL-6, which promotes chronic immune activation and gut permeability (Maidji et al. 2017). Also, the presence of HCV can potentially affect the acute phase of HIV infection acting on gut permeability (Sajadi et al. 2012), as microbial translocation markers increase with a more advanced liver disease and regress upon HCV treatment (Chew 2014; Sandberg et al. 2010; Giron-Gonzalez et al. 2007; Kushner et al. 2013). Therefore, co-infections, particularly from herpes and hepatitis viruses, have been proposed as important contributors to the chronic immune activation. Indeed, the association between CMV and immune activation has been demonstrated both in the general population and with greater significance in patients co-infected with HIV (Effros et al. 1996). CMV is a herpesvirus with a high prevalence that increases with age, affecting 70–90% of the adult population, with variability related to socioeconomic status (George et al. 2015). In the immunocompetent host, the primary infection gives little or no symptoms, while in the immunosuppressed, the virus possibly sheds in genital secretions, saliva, urine, blood, stool and breast milk. It then establishes its latency in myeloid progenitor cells and is able to reactivate at any time, especially concomitantly to immunodepression. In seropositive patients, shedding of CMV has been correlated with

CD4+ T cell activation and proliferation, as well as with a higher amount of proviral HIV DNA in blood (Gianella et al. 2014). A low CD4/CD8 ratio is an indicator of baseline immune activation in primary HIV infection and represents a good predictor of CD4+ T cell reconstitution and non-AIDS-related morbidity and mortality (Lu et al. 2015), making it an easily accessible marker that should be taken into consideration in the clinical management (Serrano-Villar et al. 2014). In the co-infected patients, the CD4/CD8 ratio is diminished due to the HIV-driven loss of CD4+ T cells, the CMV-driven expansion of CD8+ T cells and the failure by the depleted pool of CD4+ T cells at helping the CD8+ T cells, which become even more susceptible to CMV attack (Freeman et al. 2016). Strikingly, also in the general population, it represents an independent risk factor of mortality and a marker of immunosenescence (Sauter et al. 2016; Hadrup et al. 2006), a term that indicates a decline in immune system efficiency and precision occurring with age involving both innate and adaptive response. The associated pathologies are infectious, inflammation-related (i.e. CVD and neurocognitive disorders), autoimmune and neoplastic. In particular, the relationship between ageing and inflammation has been extensively described in the literature, as the release of pro-inflammatory cytokines and coagulation factors is involved in the development of age-related disease, such as atherosclerosis, CVDs, DMT2, cerebrovascular diseases, cancer, frailty, liver and kidney diseases, osteoporosis and osteopenia (De Martinis et al. 2006). Indeed, age-related diseases are the most common cause of morbidity and mortality in virologically suppressed and reconstituted HIV patients, as only 10% of deaths in the developed world are caused by AIDS-defining illnesses (Lifson et al. 2008; Deeks 2011).

In fact, ageing and HIV infection share several features, such as elevated markers of inflammation, increased T cell activation, reduced responsiveness to vaccines (Deeks 2011), thymic dysfunction, thymic involution and compromised regenerative potential in the hematopoietic stem cell compartment. A reduced thymic output can be caused either by the infection of thymocytes and/or by thymic involution: the former correlates with immune activation, independently from viremia or CD4 T cell counts, while the latter with poor immune reconstitution (Dion et al. 2004; Lelievre et al. 2012; Kolte et al. 2002). The fact that other T, B and NK cells are all depleted during HIV infection might suggest that the damage occurs upstream of thymopoiesis, at the level of hematopoiesis. Hematopoietic progenitor CD34+ cells are in fact able to express CCR5, CD4+ and CXCR4 receptors (Moir et al. 2008), and their amount decreases with disease progression (Sauce et al. 2011), despite elite control status or antiretroviral therapy. Other common characteristics are a reduced T cell repertoire, particularly in the naïve T cell subset, a low CD4/CD8 ratio and, most importantly, T cell senescence. In the context of HIV infection, it has been proposed that the chronic state of immune activation could lead to premature T cell dysfunctions ascribed to telomere shortening in terminally differentiated CD8+ T cells, loss of the co-stimulatory molecule CD28 and higher expression of CD57 (Effros et al. 1996). Interestingly, the CD8+ T cell population that has lost CD28 upon proliferation and differentiation after antigen contact is expanded in CMV and/or in HIV seropositive individuals. On the other hand, CD57, a marker of proliferative history, is always promoted by CMV and to a lower extent by HIV (Lee et al. 2014), being expressed predominantly by the CMV-specific CD28-CD57+ T cell memory cells. These cells are intrinsically more resistant to apoptosis due to a reduced telomerase activity (Dowd et al. 2013), and they can produce IFN-gamma, IL-1beta, IL-6 and TNF-alpha (Fulop et al. 2013), increasing the inflammatory state, and subsequently the rate of reactivation in the central nervous system (CNS), in the endothelial and smooth muscle tissues (Revello and Gerna 2010; Jarvis and Nelson 2007). Therefore, in both HIV and non-HIV positive individuals, CMV constitutes an independent risk factor for atherosclerotic diseases and CVD (Melnick 1993), neurocognitive diseases (Lichtner et al. 2015), frailty, and for faster progression towards AIDS in co-infected individuals (Kovacs et al. 1999). Similarly, in individuals co-infected with HIV and HCV, immune activation is increased, the liver disease progresses to a more advanced stage and there is increased incidence of extra-hepatic diseases (immune-rheumatologic, lymphoproliferative disorders, cardiovascular, renal, metabolic and central nervous system manifestations) (Gonzalez et al. 2009).

Finally, pro-inflammatory cytokines are elevated and T cells are activated at a higher frequency also in patients co-infected with TB and HIV, as compared to mono-infected individuals (Sullivan et al. 2015). The immune reconstitution inflammatory syndrome (IRIS), an aberrant inflammatory response which takes place after the initiation of ART, is an interesting model to study immune activation in HIV co-infected patients. In the paradoxical form, the immune response following the recovery in CD4+ T cells targets a known and treated pathogen; in the unmasking form, the infectious agent becomes manifest after the initiation of ART. In both cases, the peculiar feature is the inflammation-mediated tissue damage. In fact, immune activation markers are both predictors and indicators of this pathology: T cell activation (HLA-DR+CD38+), but also monocyte activation and the elevation of certain cytokines (CRP, TNF, IL-6), sTF and sCD14 are all predictors of IRIS development (Andrade et al. 2014). In addition, many cytokines remain elevated during the IRIS episode (Sereti et al. 2010). In more recent studies, the role of markers of immune activation as indicators of severity has been discarded, being attributed more to demographic and behavioural factors (Bastard et al. 2015; Wallet et al. 2015). Though the best indicator of disease progression has not been unanimously determined and despite the huge variety of findings, the role of immune activation in the pathogenesis of HIV infection has been confirmed at multiple levels. In particular, T cell activation markers and pathways related to the adaptive immunity are important predictors mainly in resource-limited settings, where infections are still the major cause of death as in the pre-ART era. On the other hand, innate immune activation and pro-inflammatory markers are more predictive of mortality in the developed world, where the incidence of non-AIDS events predominate (Tenorio et al. 2014).

## 2   Viral Persistence and Latency

HIV-1 reservoir has been defined by Eisele and Siliciano as an infected cell population that allows persistence of replication-competent HIV in patients on optimal cART regimens on a timescale of years (Eisele and Siliciano 2012). Its maintenance is currently the most important barrier to HIV eradication, as current therapies are only able to block HIV replication.

Two non-mutually exclusive mechanisms are potentially responsible for the low-level viremia detected during prolonged cART: ongoing cycles of HIV replication in certain anatomical sites with infection of new host cells, and survival and proliferation of infected long-lived cells that may produce the virus, either chronically or episodically in response to cell activation. Persistent immune dysfunction also fail at controlling both the ongoing replication and the reactivation from latently infected cells.

Already in 1998, it was clear that the reservoir is established very early in infection, as the pool of latently infected CD4+ T cells is formed also in individuals initiating HAART as early as 10 days after the onset of symptoms of primary HIV-1 infection (Chun et al. 1998) and in the non-human primate model starting ART at 3 days post-infection (Whitney et al. 2014). Still, the timing of ART administration is one of the primary determinants of cell-associated HIV DNA levels over time, and early ART initiation limits the size of the viral reservoir (Buzon et al. 2014).

Ongoing HIV replication during cART is controversial and is supported by the detection of free virions in plasma, particularly in the light of their short half-life although their presence is not necessarily a consequence of active viral replication (Ho et al. 1995; Palmer et al. 2003). Conversely, the lack of impact of treatment intensification on residual viremia (Vallejo et al. 2012), the absence of viral evolution (Kearney et al. 2014) and the detection of identical sequences call in question the role of active viral production in HIV persistence, even though mathematical extrapolations from drug-intensification studies have supported the presence of ongoing replication (Luo et al. 2013) and some deep-sequencing analyses have showed residual evolution (Lorenzo-Redondo et al. 2016).

Nevertheless, low-level viral replication has been demonstrated to occur especially in sites with suboptimal drug penetration or with inefficient control by the host immune system; lymphoid tissues, GI tract, CNS and the genital tract have been indicated as the main anatomical compartments for viral maintenance (Eisele and Siliciano 2012; Hatano et al. 2013; Yukl et al. 2010; Eden et al. 2010). In particular, in lymph nodes where the frequency of infection per cell is higher and the drug concentration is lower than in blood, residual replication has been documented (Lorenzo-Redondo et al. 2016). HIV remains detectable despite suppressed plasma viremia also in the cerebrospinal fluid (CSF) (Eden et al. 2010). This phenomenon, known as CSF viral escape, can be attributed to poor penetrance of ART through the blood–brain barrier, drug resistance, ongoing replication but also to latency, as genetic analyses suggest CNS compartmentalization in macrophages, microglia and astrocytes (Dahl et al. 2014; Churchill et al. 2006; Crowe et al. 2003).

Because of the failure of the intensification studies and the lack of diversity, chronically infected cells rather than new cycles of infection are probably the source of residual plasma viremia in the treated population.

Latency is a rare event, and the frequency of latently infected cells is extremely low in individuals under cART, typically around 0.1–10 infectious units per million resting CD4+ T cells, and only a small fraction of them carries an integrated provirus which is competent to replication (Eisele and Siliciano 2012; Chun et al. 1997). Resting CD4+ T cells are so far recognized as the main cellular reservoir, and latency can be established both pre-integration or post-integration: upon direct infection of resting but permissive cells, HIV-1 is able to enter latency immediately (Chan et al. 2016). Still, this pre-integration latency is thought to be short lived (Thierry et al. 2015). On the other hand, Siliciano and Chun proposed that upon infection, quiescent CD4+ T cells proliferate and differentiate into effector T cells; the majority of them die, but a small proportion survives and reverts to the resting state, persisting as memory cells and putatively constituting the latent source of infection (Siliciano and Greene 2011).

Resting memory CD4+ T cells can be classified according to their differentiation and memory status in central memory T cells (Tcm), transition memory T cells (Ttm) and effector memory T cells (Tem). Tcm cells (CD45RA−CCR7+CD27+) are long-lived and self-renewing, with low rate of proliferation; they reside in the lymph node, as they express CCR7. Upon antigen-stimulation, Tcm cells undergo proliferation and differentiation into Ttm cells (CD45RA−CCR7−CD27+), which represent a longer-term reservoir as they undergo low but continuous cycles of replication. Finally, Ttm cells differentiate into Tem cells (CD45RA − CCR7 −CD27−), a highly proliferating population involved in T cell homeostasis that contributes only marginally to the reservoir pool. The phenotype of the predominant cellular reservoir reflects the timing of treatment administration as early ART initiation limits the size of viral reservoir, harboured mainly in Tcm pool, whereas in individuals that start cART later in the course of infection, HIV genome is found more frequently in Ttm cells, which are able to proliferate at low level, ensuring the persistence of a genetically stable HIV pool (Chomont et al. 2009). Also, T-memory stem cells have been found to carry identical HIV sequences at a higher frequency as compared to other memory T cell subsets, and they become higher in percentage to the total reservoir over time (Buzon et al. 2014). At present, no consensus has been reached about the involvement of CD34+ hematopoietic progenitor cells (Nixon et al. 2013; Josefsson et al. 2012).

The CD4 compartment can also be subdivided according to function and homing capacities, including Th1, Th2, Th17, Treg and Tfh cells. In particular, as HIV DNA is more frequently found in Tregs than in non-Tregs subsets, they have been proposed as a potential cellular reservoir as they are less susceptible to death than conventional T cells and are clonally expanded both in blood and in tissues (Tran et al. 2008). The most important cellular reservoirs, namely follicular dendritic cells (FDC) and follicular T helper cells (Tfh), do not reside in blood but in the lymphoid tissue (Zhang and Perelson 2013; Miles and Connick 2016).

Non-conventional cellular HIV reservoirs have also been described, including CD8+ T cells and cells from the myeloid lineage especially macrophages that are long-lived cells, promoting HIV persistence in the spleen and gut, but also in the lungs and CNS (Coleman and Wu 2009; Falk and Stutte 1990; Brown and Mattapallil 2014). Similarly, in the CNS, microglial cells and astrocytes are permissive to HIV infection; these are long-lived cells, with low turnover that can sustain both low-level replication and latency (Thompson et al. 2011).

The persistence and the expansion of latently infected cells are influenced by immune mechanisms, namely homeostatic proliferation, antigenic stimulation and the generalized state of immune activation, and by pre- and post-integration mechanisms. In particular, HIV persistence in latently infected memory CD4+ T cell subsets is promoted by interleukin-7 (IL-7) signalling, the cytokine responsible for naïve T cells survival and for the homeostatic proliferation of Ttm cells (Chomont et al. 2009). These cells proliferate also following antigenic stimulation, resulting in the reactivation of the virus and potentially explaining the residual level of viremia (Wagner et al. 2013). Of note, activated CD4+ T cells have been shown to release viral particles even in the absence of stimuli (Chun et al. 2005). Further, evidence supporting the clonal expansion of HIV-infected cells was reported in studies of patients undergoing long-term therapy: in chronically infected patients, genetic diversity is impressive, while in patients under ART for a prolonged period, identical copies were repeatedly retrieved (Kearney et al. 2014). As identical copies of a single HIV hypermutant were detected in treated patients, were conserved during the years and were defective, they could not be the result of ongoing cycles of infection, but rather they must have been retrieved from cells carrying an integrated provirus that have undergone division (Josefsson et al. 2013).

Post-integration latency occurs when a provirus fails to effectively express its genome and is reversibly silenced after integration into the host cell genome. A key determinant for the fate of the HIV provirus is the site of integration, where transcriptional interference, epigenetic control and post-transcriptional processes (i.e. inhibition of nuclear RNA export and inhibition of HIV-1 translation by microRNAs) influence its expression. In general, integration site targets are non-specific, but the virus has site preference for intragenic regions of actively transcribed genes. Putative post-integration molecular mechanisms involve the site of viral integration into the host genome; the integration near genes related to cell growth and T cell development, as BACH2 and MKL2, and the interference with the promoter activity through insertional mutagenesis (Shan et al. 2011; Han et al. 2008; Maldarelli et al. 2014); the epigenetic control of the HIV-1 promoter with histone post-translational modifications; the absence of inducible host transcription factors, such as NF-kappaB; the presence of transcriptional repressors (i.e. TRIM22, COMMD1 and PML); and the insufficient concentration of the viral transactivator Tat (Munch et al. 2007). In particular, transcriptional silencing of the integrated provirus through histone acetylation is performed by histone deacetylases (HDAC), whose inhibition is sufficient to reactivate a portion of latent HIV. As a consequence, HDAC inhibitors have been used as latency-reversing agents in the 'shock and kill' or 'kick and kill' strategy for eradication of the pool of latently infected cells, as the forced viral transcription and the resulting expression of viral proteins would lead to virus- or host-induced cell death of the reactivated cells (Deeks 2012). Despite the strong rationale for this strategy, the reversal of the latent status is not sufficient to trigger death of the infected cells and other mechanisms are currently being investigated (Shan et al. 2012). Interestingly, one of the most promising therapeutical strategies involves the use of Tat-dependent transcriptional inhibitors such as the analogue of Cortistatin A, which aims at reducing the latent reservoir promoting a state of 'deep latency' (Mousseau and Valente 2016).

# 3 Viral Reservoir, Immune Activation and Therapeutic Strategies

Immune activation and viral reservoir are inter-dependent mechanisms: the inflammatory environment and T cell activation can contribute to the maintenance of HIV reservoir mainly by providing target cells for de novo HIV infection; simultaneously, the persistence of latently infected cells and ongoing replication lead to immune activation.

To date, it is still controversial to what extent clonally expanded cells are represented in the population of replication-competent provirus carrying cells, which are ultimately the relevant reservoir. It was previously believed that CD4+ T cell clones are not likely to contain replication-competent HIV-1 because the expression of viral proteins would have been cytotoxic and, accordingly, only a small proportion (11.3%) of integrated proviruses are intact (Ho et al. 2013). Therefore, Cohn et al. have proposed that clonally expanded CD4+ T cells contain only defective proviruses, while intact proviruses are found only in unexpanded cells (Cohn et al. 2015). Simonetti et al. in 2016 identified for the first time a highly expanded clone of HIV-infected CD4+ T cells that were able to produce infectious virus (Simonetti et al. 2016). Therefore, not only is the clonally expanded latently infected cells able to reactivate the infection but it is possible that they maintain the capacity to generate viral RNA and proteins, contributing to the elevated levels of immune activation. Moreover, it has been demonstrated that the unintegrated provirus can also be transcriptionally active, as it can integrate upon cell activation resulting in expression of viral particles eliciting a cytotoxic T cell response. Because of this, it may be a potentially extinguishable reservoir through the 'shock and kill' strategy. Hence, the role of unintegrated provirus in latency, virus production and immune response activation should be further investigated as the unintegrated genome is not targeted by current therapies (Chan et al. 2016).

Viral production occurs at low level in some compartments, such as the lymphoid tissues which provide the perfect environment for both residual replication and survival, proliferation and infection of latent cells. The lymphoid tissue is characterized by suboptimal penetration of drug, significant enrichment of cells permissive to HIV infection, a pro-inflammatory environment that fuels viral production, but also by a tissue architecture that allows for cell-to-cell contacts (Kulpa et al. 2013; Folkvord et al. 2005; Lafeuillade et al. 2001).

Interestingly, in the lymphoid tissues, more than 95% of CD4+ T cells are abortively infected and die by pyroptosis, a highly inflammatory type of programmed cell death, which results in the release of inflammatory signals. As a consequence, latent cells that reside in lymphoid tissues reactivate and more target cells are attracted at this site, establishing a vicious cycle and resulting in tissue fibrosis (Doitsh et al. 2010, 2014). Notably, TGF-beta upregulation and tissue fibrosis have been documented in elite controllers and not in natural SIV hosts (Sauce et al. 2011; Estes et al. 2008): lymphoid fibrosis seems to be sufficient to drive CD4+ T cell depletion, as it is also associated with poorer immune reconstitution (Schacker et al. 2006; Schacker et al. 2005). It starts only seven days after infection driven by TGF-$\beta$ production (Estes 2013) with the deposition of collagen which impacts on the maturation and activity of both T and B cells, altering the architectural structure of the B cell pathway to the primary follicle and of the naïve T cells pathway to the paracortical T cell zone. There, naïve T cells normally interact with APC and with IL-7, the cytokine responsible for the survival of naïve T cells and the homeostatic proliferation of Ttm cells that regulates viral release (Chomont et al. 2009).

The lymphoid tissues are also the sites where important cellular reservoirs reside, namely FDC and Tfh cells. Tfh cells expand in the germinal centre, where they are protected from cytotoxic CD8+ T cell killing and where they interact with antigen-specific B cells, promoting maturation and differentiation of follicular B cells into plasma cells and memory B cells. As a result of lymphoid tissue hyperplasia, plasma cells are also expanded, resulting in the hypergammaglobulinemia commonly seen in HIV patients (Klatt et al. 2011; Lindqvist et al. 2012; Pantaleo et al. 1993).

Another characteristic of the lymphoid tissue is the presence of cell-to-cell contacts: if the contact occurs between a HIV-infected CD4+ T cell and a target CD4+ T cell, it is called virological synapse and promotes viral dissemination. On the other hand, cell-to-cell contacts that involve a CD4+ T cell and an APC, called immunological synapse, ensuring viral dormancy and persistence. For example, FDCs, which store the virus at a concentration 40–50 fold higher than cells in peripheral blood, are able to transmit the virus to target cells through immunological synapses with an infected T cell (Haase et al. 1996; Yu et al. 2008). The immunological synapse promotes the differentiation of naïve T cells into effector helper cells following the interaction between the MHC class II on the APC and the TCR on the T cells. TCR stimulation induces the upregulation of inhibitory signals in the CD4+ T cell, like programmed cell death protein 1 (PD-1) and cytotoxic T lymphocyte antigen-4 (CTLA-4) receptors, resulting in the blockade of T cell activation and transforming a productive infection into a latent one (Kulpa et al. 2013). Notably, PD-1 and CTLA-4 are very reliable markers of immune exhaustion and of immunosenescence, correlating with immune activation, immune reconstitution and viral reservoirs (Hatano et al. 2013).

Finally, the lymphoid tissue is a pro-inflammatory environment, where cytokines like IL-2, TNF (tumour necrosis factor), IL-6, and IL-18, and chemokines CC-chemokine ligand 19 (CCL19) and CCL21 render resting memory T cells more permissive to infection (Cameron et al. 2010; Chun et al. 1998). Latently infected cells persist due to homeostatic proliferation driven by IL-7 and IL-15 and secondary to proliferation driven by inflammatory cytokines such as IL-1 and IL-6 (Gianella et al. 2013), as well as by antigen-driven proliferation, as discussed above. Along with other cytokines (IL-2, IL-15, IL-21), IL-7 induces the expression of PD-1 on both CD4+ and CD8+ T cells which correlates with the reservoir size (Kinter et al. 2008; Vandergeeten et al. 2013). IL-7 administration led to an increase of CD4+ T cells and an improvement of T cell homeostasis (Levy et al. 2012) and the implication of potential expansion of central memory cells with provirus are unclear.

Residual inflammation and HIV persistence intertwine, and most of the markers of immune activation correlate with the size of viral reservoir (Chomont et al. 2009; Hatano et al. 2013; Cockerham et al. 2014). Important indicators of both are PD-1 and CTLA-4 and the CD4/CD8 ratio, which is inversely proportional to the HIV-1 proviral DNA (Boulassel et al. 2012). However, the best clinical predictor of reservoir size after prolonged therapy is the CD4+ T cell nadir, independently from viral load, duration of ART and CD4 cell counts (Boulassel et al. 2012).

Despite the fact that a relationship of causality between immune activation and HIV persistence remains undefined, early ART initiation and intensification strategies have demonstrated to decrease both, indirectly confirming the association. In raltegravir intensification studies, a decrease in the presence of low-grade viral replication, measured as increase of 2-LTR circles, was correlated with a decrease in immune activation levels, quantified as CD38+HLA-DR+CD8+T cells and D-dimer. However, it did not affect plasma HIV RNA measured by a sensitive single copy assay (Buzon et al. 2010; Hatano et al. 2013). Similarly, early ART initiation, investigated in the Viro-Immunological Sustained Control after Treatment Interruption (VISCONTI) trial, led to a reduction in immune activation and limitation in reservoir formation (Saez-Cirion et al. 2013), but without any impact on the proviral composition according to Bruner et al. (2016). Consistently, the initiation of ART in early acute infection led to the normalization of some inflammatory biomarkers and of D-dimer, but markers of enterocyte turnover, tissue fibrosis and monocyte activation remained elevated (Sereti et al. 2017). Despite the evidence that early ART initiation is beneficial, viral eradication is not likely achievable with ART alone. It has been estimated that the cellular reservoir's half-life is between 40 and 44 months, requiring more than 70 years of intensive therapy for its elimination (Siliciano et al. 2003). Finding the link between immune activation and viral reservoir establishment and persistence can represent a starting point for new possible therapeutic strategies for remission.

In order to reduce the size of the reservoir, several strategies have been attempted: from the shock and kill, to the 'deep latency' strategy, from the blockade of immune checkpoints to the interference in cell cycle progression. As immune activation and viral persistence intertwine and are inter-dependent with one fuelling the other, it is hardly feasible that a strategy aiming at reducing only one of the two will be effective. In this regard, sirolimus, an immunosuppressant which blocks the mammalian target of rapamycin (mTOR), was shown to decrease the amount of HIV DNA in post-transplant patients (Stock et al. 2010). Panobinostat, a potent HDAC inhibitor used in the shock and kill strategy turned out to effectively reduce the reservoir size, but a close correlation with innate immune factors was found, indicating a direct role of innate immune cells in the reduction of the reservoir (Olesen et al. 2015). Treg cells are also considered optimal therapeutic targets, as they intrinsically have a dual activity: on one hand, an increased frequency of Treg cells results in a greater suppression of immune activation; on the other, the immune response to opportunistic infection and the virus itself may be dampened contributing to the formation of the reservoir (Chevalier and Weiss 2013). Another promising therapeutic target is CD32a, a low-affinity receptor for the immunoglobulin G Fc fragment, which was found to be a specific marker expressed exclusively by latently infected T cells that harbour replication-competent proviruses and not by non-infected nor by productively infected T cells. As it is expressed also by monocytes and B cells and is involved in antigen uptake and modulates signalling, it may also represent another link between immune activation and reservoir persistence (Descours et al. 2017; Pillai 2017).

Fibrosis, triggered by inflammatory cytokines such as IL-6 and TGF-β, is another important feature of chronic treated HIV infection that may be contributing to the maintenance of immune activation and reservoirs. As angiotensin II (AT2) induces fibrosis by increasing levels of TGF-β1, angiotensin-converting enzyme (ACE) inhibitors have been considered in addition to standard ART, and lisinopril was shown to decrease immune activation (Baker et al. 2012). Interestingly, also AT2 blockers, such as losartan and telmisartan, are currently under study to test whether they decrease inflammation, fibrosis and the spread of the HIV reservoir. TNF inhibitors have also been used to counteract tissue fibrosis in seropositive patients with autoimmune conditions: adalimumab, an anti-TNF-alpha monoclonal antibody, reduced lymphoid tissue fibrosis and limited CD4+ T cells depletion in SIV-infected primates, without affecting plasma SIV RNA or T cell immune activation (Tabb et al. 2013); the use of thalidomide and LMP-420, weak TNF inhibitor, was associated with a decreased viral load (Marriott et al. 1997; Haraguchi et al. 2006). On the other hand, TNF-alpha has been used to reactivate HIV from the latent cells, as it interferes with NF-kb signalling that promotes T cell survival at the moment of differentiation of the effector T cells into memory T cells, with unsatisfying results though (Reuse et al. 2009).

Administration of pre- and probiotics has also decreased lymphoid tissue fibrosis in the GI, with a positive impact on colonic CD4+ reconstitution and APC number in SIV-infected macaques (Klatt et al. 2013) and on inflammation, coagulation and immune activation with beneficial outcomes (Pandrea et al. 2016). Microbial translocation has been targeted directly with rifaximin in primates with a reduction of inflammatory indices, and by probiotics administration acting on the microbial composition (Marchetti et al. 2013; Gonzalez-Hernandez et al. 2012). Shutting down inflammation and decreasing immune activation can indirectly impact on reservoir persistence, as a correlation between LPS and sigmoid proviral reservoir has been found (Chege et al. 2011). Furthermore, as already tested in SIV-infected macaques, the use of IL-21, a cytokine responsible for the maintenance and functionality of Th17 cells, is effective on immune reconstitution, microbial translocation and on viral reservoir (Micci et al. 2015). On the other hand, induction of microbial translocation or the mimicry of microbial antigenemia will be possibly used to reactivate the latent reservoir.

Co-infections are associated with cytokine upregulation and T cell activation and may be involved in the seeding of HIV reservoir (Cockerham et al. 2014; Vandergeeten et al. 2012). In particular, CMV induces reservoir formation and maintenance, as shedding of CMV has been correlated with CD4+ T cell activation and proliferation, and with higher levels of proviral HIV DNA in blood (Gianella et al. 2014). In fact, CMV favours HIV persistence indirectly by increasing immune activation, but it also directly increases HIV transcription and promotes lymphocyte clonal expansion (Gianella et al. 2014). It also integrates into CD34+ cell genome and if CMV-infected CD34+ cells are prone to differentiation, HIV-infected CD34 + cells tend to remain in an early progenitor state. Interestingly, in cases of co-infection, HIV-1 is able to revert the differentiation, favouring the maintenance of both (Cheung 2017). A positive correlation with HIV DNA titres has also been found with EBV DNA, suggesting that EBV could be involved in the formation of the viral reservoir (Gianella et al. 2016; Scaggiante et al. 2016). Therefore, therapeutic strategies targeting reactivation of herpesviruses could be a possible strategy to reduce the reservoir. Hunt et al. demonstrated that valganciclovir therapy reduced T cell activation by suppressing CMV replication, rather than by affecting HIV replication, since during the 4 weeks after treatment discontinuation, CMV DNA remained suppressed, CD8+ T cells activation and CRP level were significantly lower than in the placebo group, while HIV RNA did not show any significant change (Hunt et al. 2011). Moreover, ganciclovir reduced also the risk for atherosclerosis in transplanted patients, while a related study in an HIV cohort is currently ongoing (Valantine et al. 1999; Berquist 2017).

In summary, immune activation and viral persistence should always be considered together in the management of HIV infection. Targeting immune activation may not only limit seeding and reactivation of the HIV viral reservoir but will also result in decreased inflammation and incidence of non-AIDS-related events, further improving the possibility of success of future remission strategies.

**Acknowledgements** The work of the authors was supported by the Intramural Research Program of NIAID/NIH.

# References

Ances BM et al (2010) HIV infection and aging independently affect brain function as measured by functional magnetic resonance imaging. J Infect Dis 201(3):336–340

Andrade BB et al (2014) Mycobacterial antigen driven activation of CD14 ++CD16− monocytes is a predictor of tuberculosis-associated immune reconstitution inflammatory syndrome. PLoS Pathog 10(10):e1004433

Anton PA et al (2000) Enhanced levels of functional HIV-1 co-receptors on human mucosal T cells demonstrated using intestinal biopsy tissue. AIDS 14(12):1761–1765

Baker JV et al (2012) Angiotensin converting enzyme inhibitor and HMG-CoA reductase inhibitor as adjunct treatment for persons with HIV infection: a feasibility randomized trial. PLoS ONE 7(10):e46894

Baker JV et al (2014) Immunologic predictors of coronary artery calcium progression in a contemporary HIV cohort. AIDS 28(6):831–840

Bas S et al (2004) CD14 is an acute-phase protein. J Immunol 172(7):4470–4479

Bastard JP et al (2015) Increased systemic immune activation and inflammatory profile of long-term HIV-infected ART-controlled patients is related to personal factors, but not to markers of HIV infection severity. J Antimicrob Chemother 70(6):1816–1824

Benecke A, Gale M Jr, Katze MG (2012) Dynamics of innate immunity are key to chronic immune activation in AIDS. Curr Opin HIV AIDS 7(1):79–85

Berquist V, Hoy JF, Trevillyan JM (2017) Contribution of common infections to cardiovascular risk in HIV positive individuals. AIDS Rev 19(2)

Boasso A et al (2008) HIV-induced type I interferon and tryptophan catabolism drive T cell dysfunction despite phenotypic activation. PLoS ONE 3(8):e2961

Bosinger SE et al (2012) Systems biology of natural simian immunodeficiency virus infections. Curr Opin HIV AIDS 7(1):71–78

Boulassel MR et al (2012) CD4 T cell nadir independently predicts the magnitude of the HIV reservoir after prolonged suppressive antiretroviral therapy. J Clin Virol 53(1):29–32

Boulware DR et al (2011) Higher levels of CRP, D-dimer, IL-6, and hyaluronic acid before initiation of antiretroviral therapy (ART) are associated with increased risk of AIDS or death. J Infect Dis 203(11):1637–1646

Brenchley JM et al (2004) CD4+ T cell depletion during all stages of HIV disease occurs predominantly in the gastrointestinal tract. J Exp Med 200(6):749–759

Brenchley JM et al (2006a) Microbial translocation is a cause of systemic immune activation in chronic HIV infection. Nat Med 12(12):1365–1371

Brenchley JM, Price DA, Douek DC (2006b) HIV disease: fallout from a mucosal catastrophe? Nat Immunol 7(3):235–239

Brown D, Mattapallil JJ (2014) Gastrointestinal tract and the mucosal macrophage reservoir in HIV infection. Clin Vaccine Immunol 21(11):1469–1473

Bruner KM et al (2016) Defective proviruses rapidly accumulate during acute HIV-1 infection. Nat Med 22(9):1043–1049

Burdo TH et al (2011) Soluble CD163 made by monocyte/macrophages is a novel marker of HIV activity in early and chronic infection prior to and after anti-retroviral therapy. J Infect Dis 204(1):154–163

Buzon MJ et al (2010) HIV-1 replication and immune dynamics are affected by raltegravir intensification of HAART-suppressed subjects. Nat Med 16(4):460–465

Buzon MJ et al (2014a) Long-term antiretroviral treatment initiated at primary HIV-1 infection affects the size, composition, and decay kinetics of the reservoir of HIV-1-infected CD4 T cells. J Virol 88(17):10056–10065

Buzon MJ et al (2014b) HIV-1 persistence in CD4+ T cells with stem cell-like properties. Nat Med 20(2):139–142

Cameron PU et al (2010) Establishment of HIV-1 latency in resting CD4+ T cells depends on chemokine-induced changes in the actin cytoskeleton. Proc Natl Acad Sci U S A 107(39): 16934–16939

Chahroudi A et al (2012) Natural SIV hosts: showing AIDS the door. Science 335(6073): 1188–1193

Chan MM et al (1990) Beta 2-microglobulin and neopterin: predictive markers for human immunodeficiency virus type 1 infection in children? J Clin Microbiol 28(10):2215–2219

Chan CN et al (2016) HIV-1 latency and virus production from unintegrated genomes following direct infection of resting CD4 T cells. Retrovirology 13:1

Chege D et al (2011) Sigmoid Th17 populations, the HIV latent reservoir, and microbial translocation in men on long-term antiretroviral therapy. AIDS 25(6):741–749

Cheung AKL, Huang Y, Kwok HY, Chen M, Chen Z (2017) Latent human cytomegalovirus enhances HIV-1 infection in CD34+ progenitor cells. Blood Adv 1(5):306–318

Chevalier MF, Weiss L (2013) The split personality of regulatory T cells in HIV infection. Blood 121(1):29–37

Chew KW et al (2014) The effect of hepatitis C virologic clearance on cardiovascular disease biomarkers in human immunodeficiency virus/hepatitis C virus coinfection. Open Forum Infect Dis 1(3):ofu104

Chomont N et al (2009) HIV reservoir size and persistence are driven by T cell survival and homeostatic proliferation. Nat Med 15(8):893–900

Choudhary SK et al (2007) Low immune activation despite high levels of pathogenic human immunodeficiency virus type 1 results in long-term asymptomatic disease. J Virol 81(16): 8838–8842

Chun TW et al (1997) Quantification of latent tissue reservoirs and total body viral load in HIV-1 infection. Nature 387(6629):183–188

Chun TW et al (1998a) Early establishment of a pool of latently infected, resting CD4(+) T cells during primary HIV-1 infection. Proc Natl Acad Sci U S A 95(15):8869–8873

Chun TW et al (1998b) Induction of HIV-1 replication in latently infected CD4+ T cells using a combination of cytokines. J Exp Med 188(1):83–91

Chun TW et al (2005) HIV-infected individuals receiving effective antiviral therapy for extended periods of time continually replenish their viral reservoir. J Clin Invest 115(11):3250–3255

Chung CY et al (2014) Progressive proximal-to-distal reduction in expression of the tight junction complex in colonic epithelium of virally-suppressed HIV+ individuals. PLoS Pathog 10(6): e1004198

Churchill MJ et al (2006) Use of laser capture microdissection to detect integrated HIV-1 DNA in macrophages and astrocytes from autopsy brain tissues. J Neurovirol 12(2):146–152

Cockerham LR et al (2014) CD4+ and CD8+ T cell activation are associated with HIV DNA in resting CD4+ T cells. PLoS ONE 9(10):e110731

Cohn LB et al (2015) HIV-1 integration landscape during latent and active infection. Cell 160(3): 420–432

Coleman CM, Wu L (2009) HIV interactions with monocytes and dendritic cells: viral latency and reservoirs. Retrovirology 6:51

Crowe S, Zhu T, Muller WA (2003) The contribution of monocyte infection and trafficking to viral persistence, and maintenance of the viral reservoir in HIV infection. J Leukoc Biol 74(5): 635–641

Dahl V et al (2014) An example of genetically distinct HIV type 1 variants in cerebrospinal fluid and plasma during suppressive therapy. J Infect Dis 209(10):1618–1622

De Martinis M et al (2006) Inflammation markers predicting frailty and mortality in the elderly. Exp Mol Pathol 80(3):219–227

Deeks SG (2011) HIV infection, inflammation, immunosenescence, and aging. Annu Rev Med 62:141–155

Deeks SG (2012) HIV: shock and kill. Nature 487(7408):439–440

Deeks SG et al (2004) Immune activation set point during early HIV infection predicts subsequent CD4+ T-cell changes independent of viral load. Blood 104(4):942–947

Descours B et al (2017) CD32a is a marker of a CD4 T-cell HIV reservoir harbouring replication-competent proviruses. Nature 543(7646):564–567

Desquilbet L et al (2007) HIV-1 infection is associated with an earlier occurrence of a phenotype related to frailty. J Gerontol A Biol Sci Med Sci 62(11):1279–1286

Dillon SM et al (2014) An altered intestinal mucosal microbiome in HIV-1 infection is associated with mucosal and systemic immune activation and endotoxemia. Mucosal Immunol 7(4): 983–994

Dion ML et al (2004) HIV infection rapidly induces and maintains a substantial suppression of thymocyte proliferation. Immunity 21(6):757–768

Doitsh G et al (2010) Abortive HIV infection mediates CD4 T cell depletion and inflammation in human lymphoid tissue. Cell 143(5):789–801

Doitsh G et al (2014) Cell death by pyroptosis drives CD4 T-cell depletion in HIV-1 infection. Nature 505(7484):509–514

Dowd JB et al (2013) Cytomegalovirus is associated with reduced telomerase activity in the Whitehall II cohort. Exp Gerontol 48(4):385–390

Eden A et al (2010) HIV-1 viral escape in cerebrospinal fluid of subjects on suppressive antiretroviral treatment. J Infect Dis 202(12):1819–1825

Effros RB et al (1996) Shortened telomeres in the expanded CD28-CD8+ cell subset in HIV disease implicate replicative senescence in HIV pathogenesis. AIDS 10(8):F17–F22

Eisele E, Siliciano RF (2012) Redefining the viral reservoirs that prevent HIV-1 eradication. Immunity 37(3):377–388

Ericsen AJ et al (2016) Microbial translocation and inflammation occur in hyperacute immunodeficiency virus infection and compromise host control of virus replication. PLoS Pathog 12(12):e1006048

Estes JD (2013) Pathobiology of HIV/SIV-associated changes in secondary lymphoid tissues. Immunol Rev 254(1):65–77

Estes JD, Haase AT, Schacker TW (2008) The role of collagen deposition in depleting CD4+ T cells and limiting reconstitution in HIV-1 and SIV infections through damage to the secondary lymphoid organ niche. Semin Immunol 20(3):181–186

Falk S, Stutte HJ (1990) The spleen in HIV infection–morphological evidence of HIV-associated macrophage dysfunction. Res Virol 141(2):161–169

Folkvord JM, Armon C, Connick E (2005) Lymphoid follicles are sites of heightened human immunodeficiency virus type 1 (HIV-1) replication and reduced antiretroviral effector mechanisms. AIDS Res Hum Retroviruses 21(5):363–370

Freeman ML et al (2016) CD8 T-Cell expansion and inflammation linked to CMV coinfection in ART-treated HIV infection. Clin Infect Dis 62(3):392–396

Fuchs D et al (1984) Urinary neopterin in the diagnosis of acquired immune deficiency syndrome. Eur J Clin Microbiol 3(1):70–71

Fulop T, Larbi A, Pawelec G (2013) Human T cell aging and the impact of persistent viral infections. Front Immunol 4:271

George VK et al (2015) HIV infection Worsens Age-Associated Defects in Antibody Responses to Influenza Vaccine. J Infect Dis 211(12):1959–1968

Gianella S et al (2013) Cytomegalovirus DNA in semen and blood is associated with higher levels of proviral HIV DNA. J Infect Dis 207(6):898–902

Gianella S et al (2014) Cytomegalovirus replication in semen is associated with higher levels of proviral HIV DNA and CD4+ T cell activation during antiretroviral treatment. J Virol 88(14): 7818–7827

Gianella S et al (2016) Replication of human herpesviruses is associated with higher HIV DNA levels during antiretroviral therapy started at early phases of HIV infection. J Virol 90(8): 3944–3952

Gioannini TL, Weiss JP (2007) Regulation of interactions of gram-negative bacterial endotoxins with mammalian cells. Immunol Res 39(1–3):249–260

Giorgi JV et al (1993) Elevated levels of CD38+CD8+ T cells in HIV infection add to the prognostic value of low CD4+ T cell levels: results of 6 years of follow-up. The Los Angeles Center, Multicenter AIDS Cohort Study. J Acquir Immune Defic Syndr 6(8):904–912

Giorgi JV et al (1999) Shorter survival in advanced human immunodeficiency virus type 1 infection is more closely associated with T lymphocyte activation than with plasma virus burden or virus chemokine coreceptor usage. J Infect Dis 179(4):859–870

Giron-Gonzalez JA et al (2007) Natural history of compensated and decompensated HCV-related cirrhosis in HIV-infected patients: a prospective multicentre study. Antivir Ther 12(6):899–907

Gonzalez VD et al (2009) High levels of chronic immune activation in the T-cell compartments of patients coinfected with hepatitis C virus and human immunodeficiency virus type 1 and on highly active antiretroviral therapy are reverted by alpha interferon and ribavirin treatment. J Virol 83(21):11407–11411

Gonzalez-Hernandez LA et al (2012) Synbiotic therapy decreases microbial translocation and inflammation and improves immunological status in HIV-infected patients: a double-blind randomized controlled pilot trial. Nutr J 11:90

Grieco MH et al (1984) Elevated beta 2-microglobulin and lysozyme levels in patients with acquired immune deficiency syndrome. Clin Immunol Immunopathol 32(2):174–184

Guadalupe M et al (2003) Severe CD4+ T-cell depletion in gut lymphoid tissue during primary human immunodeficiency virus type 1 infection and substantial delay in restoration following highly active antiretroviral therapy. J Virol 77(21):11708–11717

Haase AT et al (1996) Quantitative image analysis of HIV-1 infection in lymphoid tissue. Science 274(5289):985–989

Hadrup SR et al (2006) Longitudinal studies of clonally expanded CD8 T cells reveal a repertoire shrinkage predicting mortality and an increased number of dysfunctional cytomegalovirus-specific T cells in the very elderly. J Immunol 176(4):2645–2653

Han Y et al (2008) Orientation-dependent regulation of integrated HIV-1 expression by host gene transcriptional readthrough. Cell Host Microbe 4(2):134–146

Haraguchi S et al (2006) LMP-420, a small-molecule inhibitor of TNF-alpha, reduces replication of HIV-1 and Mycobacterium tuberculosis in human cells. AIDS Res Ther 3:8

Hardy GA et al (2013) Interferon-alpha is the primary plasma type-I IFN in HIV-1 infection and correlates with immune activation and disease markers. PLoS ONE 8(2):e56527

Hatano H et al (2013a) Comparison of HIV DNA and RNA in gut-associated lymphoid tissue of HIV-infected controllers and noncontrollers. AIDS 27(14):2255–2260

Hatano H et al (2013b) Cell-based measures of viral persistence are associated with immune activation and programmed cell death protein 1 (PD-1)-expressing CD4+ T cells. J Infect Dis 208(1):50–56

Hatano H et al (2013c) Increase in 2-long terminal repeat circles and decrease in D-dimer after raltegravir intensification in patients with treated HIV infection: a randomized, placebo-controlled trial. J Infect Dis 208(9):1436–1442

Hazenberg MD et al (2003) Persistent immune activation in HIV-1 infection is associated with progression to AIDS. AIDS 17(13):1881–1888

Ho DD et al (1995) Rapid turnover of plasma virions and CD4 lymphocytes in HIV-1 infection. Nature 373(6510):123–126

Ho YC et al (2013) Replication-competent noninduced proviruses in the latent reservoir increase barrier to HIV-1 cure. Cell 155(3):540–551

Hunt PW et al (2008) Relationship between T cell activation and CD4+ T cell count in HIV-seropositive individuals with undetectable plasma HIV RNA levels in the absence of therapy. J Infect Dis 197(1):126–133

Hunt PW et al (2011a) Impact of CD8+ T-cell activation on CD4+ T-cell recovery and mortality in HIV-infected Ugandans initiating antiretroviral therapy. AIDS 25(17):2123–2131

Hunt PW et al (2011b) Valganciclovir reduces T cell activation in HIV-infected individuals with incomplete CD4+ T cell recovery on antiretroviral therapy. J Infect Dis 203(10):1474–1483

Jacquelin B et al (2009) Nonpathogenic SIV infection of African green monkeys induces a strong but rapidly controlled type I IFN response. J Clin Invest 119(12):3544–3555

Jarvis MA, Nelson JA (2007) Human cytomegalovirus tropism for endothelial cells: not all endothelial cells are created equal. J Virol 81(5):2095–2101

Jiang W et al (2009) Plasma levels of bacterial DNA correlate with immune activation and the magnitude of immune restoration in persons with antiretroviral-treated HIV infection. J Infect Dis 199(8):1177–1185

Josefsson L et al (2012) Hematopoietic precursor cells isolated from patients on long-term suppressive HIV therapy did not contain HIV-1 DNA. J Infect Dis 206(1):28–34

Josefsson L et al (2013) The HIV-1 reservoir in eight patients on long-term suppressive antiretroviral therapy is stable with few genetic changes over time. Proc Natl Acad Sci U S A 110(51):E4987–E4996

Kearney MF et al (2014) Lack of detectable HIV-1 molecular evolution during suppressive antiretroviral therapy. PLoS Pathog 10(3):e1004010

Kelley CF et al (2009) Incomplete peripheral CD4+ cell count restoration in HIV-infected patients receiving long-term antiretroviral treatment. Clin Infect Dis 48(6):787–794

Kim SK et al (1997) Activation and migration of CD8 T cells in the intestinal mucosa. J Immunol 159(9):4295–4306

Kinter AL et al (2008) The common gamma-chain cytokines IL-2, IL-7, IL-15, and IL-21 induce the expression of programmed death-1 and its ligands. J Immunol 181(10):6738–6746

Kirk GD et al (2007) HIV infection is associated with an increased risk for lung cancer, independent of smoking. Clin Infect Dis 45(1):103–110

Klatt NR, Funderburg NT, Brenchley JM (2006) Microbial translocation, immune activation, and HIV disease. Trends Microbiol 21(1):6–13

Klatt NR et al (2011) SIV infection of rhesus macaques results in dysfunctional T- and B-cell responses to neo and recall Leishmania major vaccination. Blood 118(22):5803–5812

Klatt NR et al (2013) Probiotic/prebiotic supplementation of antiretrovirals improves gastrointestinal immunity in SIV-infected macaques. J Clin Invest 123(2):903–907

Kolte L et al (2002) Association between larger thymic size and higher thymic output in human immunodeficiency virus-infected patients receiving highly active antiretroviral therapy. J Infect Dis 185(11):1578–1585

Kovacs A et al (1999) Cytomegalovirus infection and HIV-1 disease progression in infants born to HIV-1-infected women. Pediatric pulmonary and cardiovascular complications of vertically transmitted HIV Infection Study Group. N Engl J Med 341(2):77–84

Kuller LH et al (2008) Inflammatory and coagulation biomarkers and mortality in patients with HIV infection. PLoS Med 5(10):e203

Kulpa DA et al (2013) The immunological synapse: the gateway to the HIV reservoir. Immunol Rev 254(1):305–325

Kushner LE et al (2013) Immune biomarker differences and changes comparing HCV mono-infected, HIV/HCV co-infected, and HCV spontaneously cleared patients. PLoS ONE 8(4):e60387

Lafeuillade A et al (2001) Persistence of HIV-1 resistance in lymph node mononuclear cell RNA despite effective HAART. AIDS 15(15):1965–1969

Lassen K et al (2004) The multifactorial nature of HIV-1 latency. Trends Mol Med 10(11):525–531

Lee SA et al (2014) Low proportions of CD28- CD8+ T cells expressing CD57 can be reversed by early ART initiation and predict mortality in treated HIV infection. J Infect Dis 210(3):374–382

Lelievre JD et al (2012) Initiation of c-ART in HIV-1 infected patients is associated with a decrease of the metabolic activity of the thymus evaluated using FDG-PET/computed tomography. J Acquir Immune Defic Syndr 61(1):56–63

Levy Y et al (2012) Effects of recombinant human interleukin 7 on T-cell recovery and thymic output in HIV-infected patients receiving antiretroviral therapy: results of a phase I/IIa randomized, placebo-controlled, multicenter study. Clin Infect Dis 55(2):291–300

Li JZ et al (2015) Differential levels of soluble inflammatory markers by human immunodeficiency virus controller status and demographics. Open Forum Infect Dis 2(1):ofu117

Lichtner M et al (2015) Cytomegalovirus coinfection is associated with an increased risk of severe non-AIDS-defining events in a large cohort of HIV-infected patients. J Infect Dis 211(2): 178–186

Lifson AR et al (2008) Determination of the underlying cause of death in three multicenter international HIV clinical trials. HIV Clin Trials 9(3):177–185

Lindqvist M et al (2012) Expansion of HIV-specific T follicular helper cells in chronic HIV infection. J Clin Invest 122(9):3271–3280

Long AT et al (2016) Contact system revisited: an interface between inflammation, coagulation, and innate immunity. J Thromb Haemost 14(3):427–437

Longenecker CT et al (2014) Soluble CD14 is independently associated with coronary calcification and extent of subclinical vascular disease in treated HIV infection. AIDS 28(7): 969–977

Lorenzo-Redondo R et al (2016) Persistent HIV-1 replication maintains the tissue reservoir during therapy. Nature 530(7588):51–56

Lu W et al (2015) CD4:CD8 ratio as a frontier marker for clinical outcome, immune dysfunction and viral reservoir size in virologically suppressed HIV-positive patients. J Int AIDS Soc 18:20052

Luo R et al (2013) Modelling HIV-1 2-LTR dynamics following raltegravir intensification. J R Soc Interface 10(84):20130186

Maidji E et al (2017) Replication of CMV in the gut of HIV-infected individuals and epithelial barrier dysfunction. PLoS Pathog 13(2):e1006202

Maldarelli F et al (2014) HIV latency. Specific HIV integration sites are linked to clonal expansion and persistence of infected cells. Science 345(6193):179–183

Marchetti G, Tincati C, Silvestri G (2013) Microbial translocation in the pathogenesis of HIV infection and AIDS. Clin Microbiol Rev 26(1):2–18

Marcos V et al (2010) Expression, regulation and clinical significance of soluble and membrane CD14 receptors in pediatric inflammatory lung diseases. Respir Res 11:32

Marriott JB et al (1997) A double-blind placebo-controlled phase II trial of thalidomide in asymptomatic HIV-positive patients: clinical tolerance and effect on activation markers and cytokines. AIDS Res Hum Retroviruses 13(18):1625–1631

Mehandru S et al (2004) Primary HIV-1 infection is associated with preferential depletion of CD4 + T lymphocytes from effector sites in the gastrointestinal tract. J Exp Med 200(6):761–770

Melnick JL, Adam E, Debakey ME (1993) Cytomegalovirus and atherosclerosis. Eur Heart J 14 Suppl K:30–38

Micci L et al (2015) Interleukin-21 combined with ART reduces inflammation and viral reservoir in SIV-infected macaques. J Clin Invest 125(12):4497–4513

Miles B, Connick E (2016) TFH in HIV latency and as sources of replication-competent virus. Trends Microbiol 24(5):338–344

Moir S et al (2008) Evidence for HIV-associated B cell exhaustion in a dysfunctional memory B cell compartment in HIV-infected viremic individuals. J Exp Med 205(8):1797–1805

Mousseau G, Valente ST (2016) Didehydro-cortistatin a: a new player in HIV-therapy? Expert Rev Anti Infect Ther 14(2):145–148

Munch J et al (2007) Nef-mediated enhancement of virion infectivity and stimulation of viral replication are fundamental properties of primate lentiviruses. J Virol 81(24):13852–13864

Mutlu EA et al (2014) A compositional look at the human gastrointestinal microbiome and immune activation parameters in HIV infected subjects. PLoS Pathog 10(2):e1003829

Nazli A et al (2010) Exposure to HIV-1 directly impairs mucosal epithelial barrier integrity allowing microbial translocation. PLoS Pathog 6(4):e1000852

Neuhaus J et al (2010) Markers of inflammation, coagulation, and renal function are elevated in adults with HIV infection. J Infect Dis 201(12):1788–1795

Nixon CC et al (2013) HIV-1 infection of hematopoietic progenitor cells in vivo in humanized mice. Blood 122(13):2195–2204

O'Brien M et al (2011) Spatiotemporal trafficking of HIV in human plasmacytoid dendritic cells defines a persistently IFN-alpha-producing and partially matured phenotype. J Clin Invest 121 (3):1088–1101

Olesen R et al (2015) Innate Immune Activity Correlates with CD4 T Cell-Associated HIV-1 DNA decline during latency-reversing treatment with panobinostat. J Virol 89(20):10176–10189

Palmer S et al (2003) New real-time reverse transcriptase-initiated PCR assay with single-copy sensitivity for human immunodeficiency virus type 1 RNA in plasma. J Clin Microbiol 41(10):4531–4536

Pandrea I et al (2016) Antibiotic and antiinflammatory therapy transiently reduces inflammation and hypercoagulation in acutely SIV-infected pigtailed Macaques. PLoS Pathog 12(1):e1005384

Pantaleo G, Graziosi C, Fauci AS (1993) The role of lymphoid organs in the pathogenesis of HIV infection. Semin Immunol 5(3):157–163

Pillai SK, Deeks SG (2017) Signature of the sleeper cell: a biomarker of HIV latency revealed. Trends Immunol

Ploquin MJ, Silvestri G, Muller-Trutwin M (2016) Immune activation in HIV infection: what can the natural hosts of simian immunodeficiency virus teach us? Curr Opin HIV AIDS 11(2):201–208

Reuse S et al (2009) Synergistic activation of HIV-1 expression by deacetylase inhibitors and prostratin: implications for treatment of latent infection. PLoS ONE 4(6):e6093

Revello MG, Gerna G (2010) Human cytomegalovirus tropism for endothelial/epithelial cells: scientific background and clinical implications. Rev Med Virol 20(3):136–155

Robbins GK et al (2009) Incomplete reconstitution of T cell subsets on combination antiretroviral therapy in the AIDS Clinical Trials Group protocol 384. Clin Infect Dis 48(3):350–361

Saez-Cirion A et al (2013) Post-treatment HIV-1 controllers with a long-term virological remission after the interruption of early initiated antiretroviral therapy ANRS VISCONTI Study. PLoS Pathog 9(3):e1003211

Sajadi MM et al (2012) Chronic immune activation and decreased CD4 cell counts associated with hepatitis C infection in HIV-1 natural viral suppressors. AIDS 26(15):1879–1884

Sandberg JK, Falconer K, Gonzalez VD (2010) Chronic immune activation in the T cell compartment of HCV/HIV-1 co-infected patients. Virulence 1(3):177–179

Sandler NG et al (2011) Host response to translocated microbial products predicts outcomes of patients with HBV or HCV infection. Gastroenterology 141(4):1220–1230, 1230 e1-3

Sandler NG et al (2011) Plasma levels of soluble CD14 independently predict mortality in HIV infection. J Infect Dis 203(6):780–790

Sauce D et al (2011) HIV disease progression despite suppression of viral replication is associated with exhaustion of lymphopoiesis. Blood 117(19):5142–5151

Sauter R et al (2016) CD4/CD8 ratio and CD8 counts predict CD4 response in HIV-1-infected drug naive and in patients on cART. Medicine (Baltimore) 95(42):e5094

Scaggiante R et al (2016) Epstein-Barr and cytomegalovirus DNA salivary shedding correlate with long-term plasma HIV RNA detection in HIV-infected men who have sex with men. J Med Virol 88(7):1211–1221

Schacker TW et al (2005) Amount of lymphatic tissue fibrosis in HIV infection predicts magnitude of HAART-associated change in peripheral CD4 cell count. AIDS 19(18):2169–2171

Schacker TW et al (2006) Lymphatic tissue fibrosis is associated with reduced numbers of naive CD4+ T cells in human immunodeficiency virus type 1 infection. Clin Vaccine Immunol 13(5):556–560

Schuetz A et al (2014) Initiation of ART during early acute HIV infection preserves mucosal Th17 function and reverses HIV-related immune activation. PLoS Pathog 10(12):e1004543

Sereti I, Rodger AJ, French MA (2010) Biomarkers in immune reconstitution inflammatory syndrome: signals from pathogenesis. Curr Opin HIV AIDS 5(6):504–510

Sereti I et al (2017) Persistent, albeit reduced, chronic inflammation in persons starting antiretroviral therapy in acute HIV infection. Clin Infect Dis 64(2):124–131

Serrano-Villar S et al (2014) Increased risk of serious non-AIDS-related events in HIV-infected subjects on antiretroviral therapy associated with a low CD4/CD8 ratio. PLoS ONE 9(1): e85798

Shan L et al (2011) Influence of host gene transcription level and orientation on HIV-1 latency in a primary-cell model. J Virol 85(11):5384–5393

Shan L et al (2012) Stimulation of HIV-1-specific cytolytic T lymphocytes facilitates elimination of latent viral reservoir after virus reactivation. Immunity 36(3):491–501

Siliciano RF, Greene WC (2011) HIV latency. Cold Spring Harb Perspect Med 1(1):a007096

Siliciano JD et al (2003) Long-term follow-up studies confirm the stability of the latent reservoir for HIV-1 in resting CD4+ T cells. Nat Med 9(6):727–728

Simonetti FR et al (2016) Clonally expanded CD4+ T cells can produce infectious HIV-1 in vivo. Proc Natl Acad Sci U S A 113(7):1883–1888

Sousa AE et al (2002) CD4 T cell depletion is linked directly to immune activation in the pathogenesis of HIV-1 and HIV-2 but only indirectly to the viral load. J Immunol 169 (6):3400–3406

Stacey AR et al (2009) Induction of a striking systemic cytokine cascade prior to peak viremia in acute human immunodeficiency virus type 1 infection, in contrast to more modest and delayed responses in acute hepatitis B and C virus infections. J Virol 83(8):3719–3733

Steininger C, Puchhammer-Stockl E, Popow-Kraupp T (2006) Cytomegalovirus disease in the era of highly active antiretroviral therapy (HAART). J Clin Virol 37(1):1–9

Stock PG et al (2010) Outcomes of kidney transplantation in HIV-infected recipients. N Engl J Med 363(21):2004–2014

Strategic Timing of Antiretroviral Therapy (START) (2006) CD4+ count-guided interruption of antiretroviral treatment. N Engl J Med 355(22):2283–2296

Sullivan ZA et al (2015) Latent and active tuberculosis infection increase immune activation in individuals co-infected with HIV. EBioMedicine 2(4):334–340

Sundaravaradan V et al (2013) Multifunctional double-negative T cells in sooty mangabeys mediate T-helper functions irrespective of SIV infection. PLoS Pathog 9(6):e1003441

Tabb B et al (2013) Reduced inflammation and lymphoid tissue immunopathology in rhesus macaques receiving anti-tumor necrosis factor treatment during primary simian immunodeficiency virus infection. J Infect Dis 207(6):880–892

Tenorio AR et al (2014) Soluble markers of inflammation and coagulation but not T-cell activation predict non-AIDS-defining morbid events during suppressive antiretroviral treatment. J Infect Dis 210(8):1248–1259

Thierry S et al (2015) Integrase inhibitor reversal dynamics indicate unintegrated HIV-1 dna initiate de novo integration. Retrovirology 12:24

Thompson KA et al (2011) Brain cell reservoirs of latent virus in presymptomatic HIV-infected individuals. Am J Pathol 179(4):1623–1629

Tran TA et al (2008) Resting regulatory CD4 T cells: a site of HIV persistence in patients on long-term effective antiretroviral therapy. PLoS ONE 3(10):e3305

Valantine HA et al (1999) Impact of prophylactic immediate posttransplant ganciclovir on development of transplant atherosclerosis: a post hoc analysis of a randomized, placebo-controlled study. Circulation 100(1):61–66

Vallejo A et al (2012) The effect of intensification with raltegravir on the HIV-1 reservoir of latently infected memory CD4 T cells in suppressed patients. AIDS 26(15):1885–1894

Vandergeeten C, Fromentin R, Chomont N (2012) The role of cytokines in the establishment, persistence and eradication of the HIV reservoir. Cytokine Growth Factor Rev 23(4–5): 143–149

Vandergeeten C et al (2013) Interleukin-7 promotes HIV persistence during antiretroviral therapy. Blood 121(21):4321–4329

Veazey RS et al (2000) Dynamics of CCR5 expression by CD4(+) T cells in lymphoid tissues during simian immunodeficiency virus infection. J Virol 74(23):11001–11007

Vinton C et al (2011) CD4-like immunological function by CD4-T cells in multiple natural hosts of simian immunodeficiency virus. J Virol 85(17):8702–8708

von Sydow M et al (1991) Interferon-alpha and tumor necrosis factor-alpha in serum of patients in various stages of HIV-1 infection. AIDS Res Hum Retroviruses 7(4):375–380

Wagner TA et al (2013) An increasing proportion of monotypic HIV-1 DNA sequences during antiretroviral treatment suggests proliferation of HIV-infected cells. J Virol 87(3):1770–1778

Wallet MA et al (2015) Increased inflammation but similar physical composition and function in older-aged, HIV-1 infected subjects. BMC Immunol 16:43

Whitney JB et al (2014) Rapid seeding of the viral reservoir prior to SIV viraemia in rhesus monkeys. Nature 512(7512):74–77

Yu HJ, Reuter MA, McDonald D (2008) HIV traffics through a specialized, surface-accessible intracellular compartment during trans-infection of T cells by mature dendritic cells. PLoS Pathog 4(8):e1000134

Yukl SA et al (2010) Effect of raltegravir-containing intensification on HIV burden and T-cell activation in multiple gut sites of HIV-positive adults on suppressive antiretroviral therapy. AIDS 24(16):2451–2460

Zhang J, Perelson AS (2013) Contribution of follicular dendritic cells to persistent HIV viremia. J Virol 87(14):7893–7901

# Immune Interventions to Eliminate the HIV Reservoir

Denise C. Hsu and Jintanat Ananworanich

**Abstract** Inducing HIV remission is a monumental challenge. A potential strategy is the "kick and kill" approach where latently infected cells are first activated to express viral proteins and then eliminated through cytopathic effects of HIV or immune-mediated killing. However, pre-existing immune responses to HIV cannot eradicate HIV infection due to the presence of escape variants, inadequate magnitude, and breadth of responses as well as immune exhaustion. The two major approaches to boost immune-mediated elimination of infected cells include enhancing cytotoxic T lymphocyte mediated killing and harnessing antibodies to eliminate HIV. Specific strategies include increasing the magnitude and breadth of T cell responses through therapeutic vaccinations, reversing the effects of T cell exhaustion using immune checkpoint inhibition, employing bispecific T cell targeting immunomodulatory proteins or dual-affinity re-targeting molecules to direct cytotoxic T lymphocytes to virus-expressing cells and broadly neutralizing antibody infusions. Methods to steer immune responses to tissue sites where latently infected cells are located need to be further explored. Ultimately, strategies to induce HIV remission must be tolerable, safe, and scalable in order to make a global impact.

D. C. Hsu · J. Ananworanich
U.S. Military HIV Research Program, Walter Reed Army Institute of Research,
Silver Spring, MD, USA

D. C. Hsu · J. Ananworanich
Henry M. Jackson Foundation for the Advancement of Military Medicine, Bethesda
MD, USA

D.C. Hsu
Armed Forces Research Institute of Medical Sciences, Bangkok, Thailand

J. Ananworanich (✉)
US Military HIV Research Program (MHRP), 6720-A Rockledge Drive, Suite 400,
Bethesda, MD 20817, USA
e-mail: jananworanich@hivresearch.org

Current Topics in Microbiology and Immunology (2018) 417:181–210
DOI 10.1007/82_2017_70
© Springer International Publishing AG 2017
Published Online: 26 October 2017

**Contents**

# 1   Introduction

HIV affects 36.7 million people worldwide, about 46% were on antiretroviral therapy (ART) at the end of 2015 (UNAIDS 2016). Thus, millions more are yet to access ART. The need for ART is lifelong due to the persistence of the latent viral reservoir, mainly made up by long-lived resting memory $CD4^+$ T cells that harbor stably integrated, transcriptionally silent proviruses, that are capable of producing infectious virus upon activation (Chun et al. 1995, 1997; Siliciano et al. 2003).

The report of the "Berlin patient" was a pivotal moment, demonstrating that ablation of the HIV reservoir and prevention of its reestablishment after allogeneic hematopoietic stem-cell transplant (HSCT) from a CCR5 Δ32 homozygous donor can lead to long-term control of HIV (Hutter et al. 2009). The possibility of curing HIV was greeted with much enthusiasm as it would eliminate the burden of taking medications daily, reduce ART-associated long-term toxicity, and diminish the substantial cost of lifelong ART that may be difficult to sustain economically by governments and funding agencies (Granich et al. 2016; UNAIDS 2014; Deeks et al. 2016).

However, the use of allogeneic HSCT for HIV cure is not practical given the considerable risks and expense associated with HSCT (Gooley et al. 2010; Copelan 2006), and the rarity of human leukocyte antigen (HLA)-matched CCR5 Δ32 homozygous donors. The complete elimination of the HIV reservoir, "sterilizing/eradication cure" maybe insurmountable. A functional cure or HIV remission (virologic control in the absence of ART) maybe more achievable (Cillo and Mellors 2016; Chomont and Perreau 2016). Mathematical modeling has shown that a >4 log reduction of viral reservoir may be required to prevent viral rebound following ART discontinuation (Hill et al. 2014).

Achieving HIV remission is a monumental challenge (Chomont et al. 2009; Dahabieh et al. 2015). The absence of viral protein expression in latently infected cells and the stochastic reactivation of these rare cells pose an obstacle to their elimination by host immune responses (Hill et al. 2014). Even when transcription in these latently infected cells is activated using latency reversal agents (LRA), no significant reduction in the HIV reservoir was measured (Rasmussen and Lewin 2016). This maybe secondary to suboptimal activation of transcription, insensitivity of assays that quantitate the viral reservoir, and insufficient immune responses to eliminate HIV-expressing cells (Rasmussen and Lewin 2016).

In this chapter, we will review the immune interventions that are currently being investigated for inducing HIV remission. Major strategies include enhancing cytotoxic T lymphocyte (CTL)-mediated killing and intensifying antibody-mediated antiviral effects (Fig. 1). Ultimately, strategies to induce HIV remission must be safe and scalable in order to make a global impact.

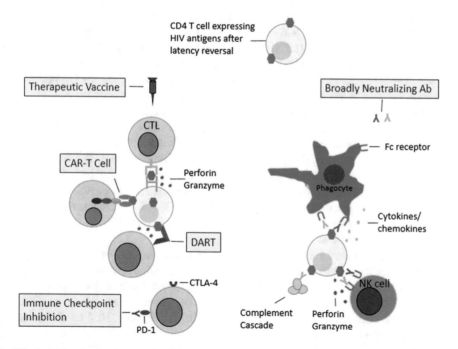

**Fig. 1** Major immune interventions to eliminate the HIV reservoir. Strategies to enhance cytotoxic T lymphocyte (CTL)-mediated killing include therapeutic vaccines, chimeric antigen receptor (CAR) T cells, bispecific T cell targeting immunomodulatory proteins/dual-affinity re-targeting (DART) molecules, and immune checkpoint inhibition. Broadly neutralizing antibodies (bnAb) can target and eliminate infected cells via antibody-dependent, cell-mediated viral inhibition (ADCVI) including antibody-dependent cell-mediated cytotoxicity (ADCC), Ab-dependent cellular phagocytosis (ADCP), Ab-mediated release of cytokines or chemokines, and complement-mediated killing

## 2  Existent Immune Responses Cannot Eliminate the HIV Reservoir

A major obstacle in HIV remission is that immune responses elicited by HIV infection cannot eradicate HIV infection. Neutralizing antibodies (nAbs) against HIV are generated in the first few months after HIV infection. However, it is unable to control HIV replication due to the rapid emergence of escape variants (Wei et al. 2003; Richman et al. 2003).

HIV-specific (sp) CD8$^+$ CTLs are also induced soon after HIV infection and are critical in restraining viral replication over the course of HIV infection. During acute infection, the magnitude and the rapidity of emergence of HIV-sp CTLs correlate with viral load set point (Borrow et al. 1994; Ndhlovu et al. 2015). HIV-sp CTLs exert substantial immune pressure on the virus, evidenced by the rapid and sustained selection of escape variants (Goonetilleke et al. 2009; Liu et al. 2013). HIV-sp CTLs continue to play a key role in bridling plasma viremia in chronic infection (Klein et al. 1995): increased capacity in HIV-sp CTL proliferation, upregulation of perforin and serine proteases, and cytotoxicity differentiate long-term non-progressors (LTNP) from progressors (Migueles and Connors 2015). The association between delay in HIV progression and expressions of HLA-B*57 and HLA-B*27 further support the role of CTLs in controlling viral replication (Migueles et al. 2000; Carrington and O'Brien 2003). The importance of CTLs in controlling viremia is reinforced by data from Rhesus Macaques (RMs) infected with Simian Immunodeficiency Virus (SIV), where CD8$^+$ T cell depletion was associated with rapid and marked increase in viremia that decreases on CD8$^+$ T cell reemergence (Schmitz et al. 1999; Jin et al. 1999). Furthermore, CTLs continue to be essential in viral control even during ART. Depletion of CD8$^+$ T cells in SIV-infected RMs, 8–32 weeks after ART initiation, resulted in increase in plasma viremia in all RMs despite ART continuation. CD8$^+$ T cell reconstitution was associated with reduction in plasma viremia to the levels prior to CD8$^+$ T cell depletion (Cartwright et al. 2016).

Despite the critical role of HIV-sp CTLs in restraining viral replication, autologous CTLs from individuals on ART are unable to eliminate the viral reservoir (Shan et al. 2012). First, unless ART is started early, the vast majority (>98%) of latent viruses carry escape mutations, rendering them unsusceptible to CTLs directed at common epitopes (Deng et al. 2015). Second, HIV-sp CTLs are exhausted, expressing inhibitory receptors including programmed cell death-1 (PD-1), CTLA-4, T cell Ig domain, and mucin domain 3 (Tim-3), 2B4, CD160 (Migueles and Connors 2015; Day et al. 2006; Hoffmann et al. 2016; Trautmann et al. 2006; Kaufmann et al. 2007; Kassu et al. 2010; Yamamoto et al. 2011; Peretz et al. 2012), and their proliferative and cytotoxic capacity is impaired despite ART (Migueles et al. 2009). Third, viral suppression with ART and resultant reduction in antigen stimulation is associated with a reduction in magnitude and breadth of

HIV-sp CTLs (Ogg et al. 1999; Casazza et al. 2001). Fourth, latently infected cells, even when induced to express viral proteins, are rare and may be overlooked by the small number of HIV-sp CTLs during surveillance. Finally, there exist sanctuary sites where infected cells are relatively out of the reach of HIV-sp CTLs (Fukazawa et al. 2015). Therefore, HIV-sp CTL responses need to be activated, expanded, and delivered to the sites of the HIV reservoir in order to contribute to the elimination of infected cells (Jones and Walker 2016).

# 3 Enhancing CTL-Mediated Killing of Infected Cells

An in vitro proof of concept study by Shan et al. demonstrated that HIV antigen stimulation or boosting of HIV-sp CTLs is necessary for elimination of infected cells upon latency reversal (Shan et al. 2012). Furthermore, ex vivo expanded, autologous, HIV-sp CTLs from individuals on ART with viral suppression were able to reduce viral recovery from infected autologous resting CD4$^+$ T cells, following exposure to vorinostat (an LRA) in vitro (Sung et al. 2015a).

Potential strategies to enhance elimination of the HIV reservoir by CTLs include therapeutic vaccines, immune checkpoint inhibition, bispecific T cell targeting immunomodulatory proteins/dual-affinity re-targeting molecules (DART), and chimeric antigen receptor (CAR)-T cell therapy (Fig. 1). We will only focus on the first three strategies in this review as CAR-T cell therapy will be discussed by Kiem et al., in the "Gene Editing For HIV Cure" chapter.

## 3.1 Therapeutic Vaccine

The goal of a therapeutic vaccine is to induce immune responses to HIV antigens that will lead to the elimination of HIV-infected cells. Over 40 therapeutic vaccine trials have been performed to date (Pantaleo and Levy 2016). These include DNA-based, RNA-based, viral vector-based, peptide-based, protein-based, and dendritic cell (DC)-based vaccines. All are phase I or II trials, investigating safety and/or immunogenicity. Most of these vaccines are relatively safe, tolerable, and do elicit or improve CD4$^+$ and CD8$^+$ T cell cytokine production, cytotoxicity, and proliferation (Pantaleo and Levy 2016; Mylvaganam et al. 2015; Garcia et al. 2012). The effects on virologic control after ART pause have also been evaluated in a number of studies. Most did not show significant effects in terms of the proportion of participants experiencing viral rebound or delaying the time to viral rebound (Kinloch-de Loes et al. 2005; Jacobson et al. 2006; Thompson et al. 2016; Pollard et al. 2014; Goldstein et al. 2012; Loret et al. 2016; Tung et al. 2016; Tubiana et al. 2005; Angel et al. 2011; Garcia et al. 2005; Ide et al. 2006; Gandhi et al. 2009;

Jacobson et al. 2016). Only several vaccines showed modest effects with lower levels of viral load when compared to pre-ART setpoint (Tung et al. 2016; Macatangay et al. 2016; Garcia et al. 2013) or to controls during viral rebound (Pollard et al. 2014; Schooley et al. 2010; Levy et al. 2014).

DC-based vaccine is one of the therapeutic vaccine strategies that showed promise. DCs are potent antigen-presenting cells that link innate and adaptive immune responses. They are sentinels of the immune system, surveying the tissues. After an immature DC encounters a pathogen, it undergoes a process of maturation, antigen processing, and migration to proximal lymphoid tissues, where it activates NK, T, and B cells (Merad et al. 2013; Mildner and Jung 2014). In addition to inducing immunity, DCs also mediate peripheral tolerance and prevent autoimmunity through T cell anergy, deletion, and the induction of regulatory T cells (Treg) (Mellor and Munn 2004; Cools et al. 2007). DC impairment is evident in HIV infection, with a reduction in frequency, diminished responsiveness, reduced cytokine production (particularly IL-2), and suboptimal activation of other immune cells (Miller and Bhardwaj 2013).

A number of DC vaccine trials have been performed, yielding varying results on the efficacy of virologic control (Garcia et al. 2013). In the study by Garcia et al., participants on ART with undetectable plasma HIV RNA and $CD4^+$ T cell count >450 cells/uL were randomized to receive three immunizations with monocyte-derived DCs pulsed with autologous, heat-inactivated, whole HIV-1; or with non-pulsed DCs. After ART pause, a decrease in plasma HIV RNA (when compared to pre-ART set point) of $\geq 1 \log_{10}$ copies/mL was observed in 12 of 22 (55%) versus 1 of 11 (9%) participants at week 12 and in 7 of 20 (35%) versus 0 of 10 (0%) participants at week 24 in the HIV-1 pulsed DC arm versus non-pulsed DC control arm, respectively. This significant decrease in plasma HIV RNA observed in recipients of HIV-1 pulsed DC was associated with a consistent increase in HIV-sp T cell responses. These data suggest that HIV-sp immune responses elicited by DC immunotherapy could significantly change plasma viral load set point after ART pause in chronic HIV-infected individuals (Garcia et al. 2013).

Insufficient potency and breadth of vaccine elicited CTLs to target escape variants may explain the only modest effects of therapeutic HIV vaccines investigated to date. In a study by Deng et al., most (>98%) of the latent viruses in individuals who started ART in chronic HIV infection carried CTL escapes that render infected cells unsusceptible to CTLs directed at common immunodominant epitopes. However, prestimulated CTL clones targeting unmutated viral epitopes can still eliminate CTL escape variants. Therefore, directing CTL responses to unmutated viral epitopes is essential to clear the HIV reservoir. However, due to bias in antigen presentation or recognition, common vaccination strategies will probably mostly restimulate immunodominant CTL clones that do not kill infected cells. Thus, boosting CTL breadth and/or modulating immunodominance will be necessary (Deng et al. 2015).

In a recent study by Borducchi et al., Ad26 prime/MVA boost vaccine (expressing $SIV_{smE543}$ *gag-pol-env*) was combined with TLR7 agonist GS-986 (that stimulates innate immune activation) in SIV-infected RMs on suppressive ART. This combination strategy was able to expand the magnitude (by >100-fold) and breadth (by >9-fold) of Gag-, Pol-, and Env-sp T cell immune responses. SIV-DNA in both lymph nodes and PBMCs were reduced to undetectable levels in the majority of RMs at the completion of the vaccination schedule at week 70. Though all RMs experienced viral rebound upon ART pause, RMs that received both the Ad26/MVA vaccine and GS-986 demonstrated a 1.74 log reduction of median plasma SIV RNA set point ($P < 0.0001$) and a delay in viral rebound from a median of 10 days, to 25 days, when compared with controls ($P = 0.003$). Furthermore, three of the nine RMs in the combination arm eventually achieved virologic control to undetectable levels, months after ART discontinuation. Importantly, this study demonstrated that breadth of cellular immune responses was associated with virologic control (Borducchi et al. 2016).

Increased breadth may also be achieved using mosaic antigens that are generated from natural sequences via computational optimization. Mosaic antigens resemble natural proteins but are engineered to include common potential epitopes, providing broad coverage (Fischer et al. 2007). In RMs, vaccination with Mosaic HIV-1 antigens resulted in 3.8-fold higher number of peptides recognized by RMs than consensus or natural sequence antigens (Barouch et al. 2010). Mosaic HIV-1 vaccines also conferred per exposure risk reduction of 90% against acquisition of SHIV infection following repetitive, intrarectal challenges in RM (Barouch et al. 2013). A phase I/IIa study of Ad26 mosaic prime and MVA mosaic boost combination is currently underway in acute HIV-1 infected individuals on ART with suppressed viral load to assess its potential to delay viral rebound following ART pause (NCT02919306).

Another strategy to address the issue of escape variants is to focus CTL responses to the conserved regions of HIV-1 proteins, common to many variants including escape variants as mutations in these regions are associated with substantial fitness costs (Hanke 2014; Rolland et al. 2007). These conserved epitopes are typically subdominant in natural infection and immunodominance hierarchy often undermines their protective potential and/or preclude their detection (Ahmed et al. 2016; Hancock et al. 2015; Hertz et al. 2013).

HIVconsv is a chimeric protein assembled from the 14 most conserved regions of the HIV-1 clades A, B, C, and D proteomes (Letourneau et al. 2007). In the study by Mothe et al., the genes encoding for the HIVconsv protein were inserted into attenuated chimpanzee adenovirus serotype 63 (ChAdV63) and modified vaccinia virus Ankara (MVA) to construct the ChAdV63.HIVconsv and MVA.HIVconsv vaccine. In this study, 24 participants who initiated ART within 6 months of HIV infection received ChAd.HIVconsv and MVA.HIVconsv prime/boost vaccinations (BCN 01). After 3 years on ART with viral suppression, 15 of the participants were immunized again with MVA.HIVconsv, followed by three weekly doses of romidepsin, and then a second MVA.HIVconsv vaccination prior to ART pause (BCN 02). Vaccinations in BCN02 boosted HIVconsv IFN-$\gamma^+$ T cell responses. Of the 13 participants who have undergone ART pause, 8 resumed ART within the first 4 weeks, but 5 continued to remain off ART, for 6, 14, 19, 21, and 28 weeks. The vaccine was relatively well tolerated with mainly grade 1 and 2 adverse events (AEs). There were two grade 4 AEs (sepsis from Shigella and CK elevation) (Mothe et al. 2017).

Therefore, therapeutic vaccines have the potential to elicit CTLs that are capable of eliminating infected cells. However, development of immunogens that can target escape variants is required.

## 3.2   Immune Checkpoint Inhibition

Immune exhaustion is evident in HIV infection characterized by the upregulation of inhibitory receptors including PD-1, CTLA4, Tim3, 2B4, and CD160 (Day et al. 2006; Trautmann et al. 2006; Kaufmann et al. 2007; Kassu et al. 2010; Yamamoto et al. 2011; Peretz et al. 2012). PD-1 and CTLA-4 blockade and resultant improvement in antitumor immune responses have led to major breakthroughs in the treatment of various malignancies (Couzin-Frankel 2013).

In RMs with untreated SIV infection, PD-1 blockade with antibody to PD-1 was associated with a rapid expansion of SIV-sp $CD8^+$ T cells with improved functional quality, proliferation of memory B cells, increases in Env-sp Abs, reductions in plasma viral load, and prolonged survival (Velu et al. 2009). PD-1 ligand blockade with a recombinant macaque PD-1 fused to a macaque Ig-Fc (rPD-1-Fc) also enhanced SIV-sp $CD4^+$ and $CD8^+$ T cell responses and delayed viral rebound during ART pause (Amancha et al. 2013). Similarly, in RM with treated SIV infection, CTLA-4 blockade was associated with increases in SIV-sp T cells. The reduction of SIV RNA in lymph node from pre-ART levels was also greater in RM treated with CTLA-4 and ART than those on ART alone. However, no effect on plasma viral rebound or plasma set point viremia was seen during ART pause (Hryniewicz et al. 2006).

A phase I trial of anti-PD-L1 monoclonal antibody (mAb, BMS-936559) in eight HIV-1 infected individuals on ART with viral suppression (NCT02028403) showed that anti-PD-L1 mAb was relatively well tolerated. Gag-sp $CD8^+$ T cell responses also increased in two individuals over 28 days post-infusion (Eron et al. 2016).

Immune checkpoint inhibition may have utility in combination with therapeutic vaccines to further boost anti-HIV immune responses. However, the use of these agents in cancer treatment has been associated with aberrant activation of autoreactive T cells and severe autoimmune-related adverse events, even resulting in death (Johnson et al. 2016a, b; Menzies et al. 2016). Thus, the risks associated with immune checkpoint inhibitions may possibly outweigh potential benefits in treated HIV infection.

## 3.3 Bispecific T Cell Targeting Immunomodulatory Proteins and Dual-Affinity Re-Targeting Molecules

Bispecific T cell targeting immunomodulatory proteins and DART are bispecific, antibody-based molecules that can be used to target HIV Env and CD3 on T cells simultaneously, facilitating the engagement of T cells with Env expressing target cells in an MHC-independent manner. This obviates the need for CTLs to be HIV-specific to mediate killing of HIV-infected cells and also bypasses the issue of CTL escape variants (Sung et al. 2015b), (Fig. 2). VRC07-αCD3 bispecific immunomodulatory proteins can target latently infected CD4$^+$ T cells and reduce the number of proviral DNA-expressing CD4$^+$ T cells in vitro (Pegu et al. 2015). Administration of VRC07-α-rhesusCD3 bispecific immunomodulatory proteins to ART-treated, SHIV infected RMs was associated with no evidence of adverse events. Plasma TNF, MIP-1b, and IL-10 levels increased 1 h post-dosing, but returned to baseline within 24 h. SHIV viral load remained suppressed on ART during the study period (Pegu et al. 2015). DARTs consisting of an HIV Env-targeting arm including broadly neutralizing Abs (bnAbs, PGT121, and PGT145) as well as non-neutralizing Abs that mediate ADCC (A32 and 7B2) with a CD3 binding arm are under investigation. In vitro data showed that these DARTs were able to induce CTL-mediated killing of HIV-1 infected CD4$^+$ T cells. Furthermore, DARTs were able to reduce viral recovery from resting CD4$^+$ T cells from HIV-infected individuals on suppressive ART following the induction of latent virus expression (Sung et al. 2015b; Sloan et al. 2015).

In summary, therapeutic vaccines have been shown to stimulate and increase the number and breadth of HIV-sp CTLs. Immune check point inhibitors have the potential to reinvigorate HIV-sp CTLs and enhance their killing. The development of bispecific T cell targeting immunomodulatory proteins and DART may possibly obviate the need for CTLs to be HIV-specific in order to target and eliminate HIV-infected cells.

## 4 Harnessing Antibodies to Induce HIV Remission

Anti-HIV Abs can eliminate HIV through direct neutralization. In addition, Abs can also target and eliminate infected cells via Fc effector functions (antibody-dependent cell-mediated viral inhibition, ADCVI), including antibody-dependent cell-mediated cytotoxicity (ADCC), Ab-dependent cellular phagocytosis (ADCP), Ab-mediated release of cytokines or chemokines, and complement-mediated killing (Forthal et al. 2013; Euler and Alter 2015) (Fig. 1).

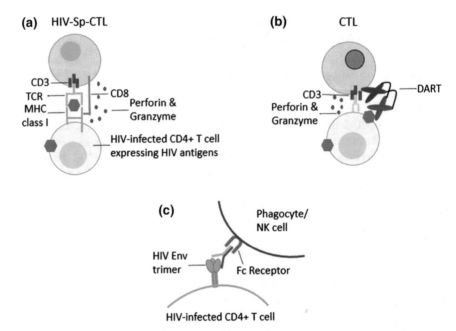

**Fig. 2** Bispecific molecules. HIV-specific (sp)-cytotoxic T lymphocytes (CTLs)-mediate killing of HIV-infected cells through T cell receptor (TCR) recognition of viral antigens presented on major histocompatibility complex (MHC), **a** (Neefjes et al. 2011). Dual-affinity re-targeting molecules (DART) are bispecific, antibody-based molecules that can be used to target HIV Env and CD3 on T cells simultaneously. This leads to the engagement of T cells with Env expressing target cells in an MHC-independent manner **b** (Sung et al. 2015; Sloan et al.2015). Bispecific antibodies targeting different regions of the HIV envelop (Env) combine the breadth and potency of two broadly neutralizing antibodies **c** The Fc regions of these antibodies can also mediate Fc effector functions (outlined in Fig. 1), (Bournazos et al. 2016)

## 4.1 Using Broadly Neutralizing Antibodies to Eliminate HIV

BnAbs are antibodies that are capable of neutralizing diverse circulating HIV-1 strains from multiple clade groups. BnAbs can be found in 20–30% of individuals with HIV-1 infection (Simek et al. 2009; Doria-Rose et al. 2009; Sather et al. 2009; Hraber et al. 2014; Landais et al. 2016), developing 2–4 years after HIV-1 infection, in the presence of continual antigen stimulation from viral replication (Sather et al. 2009; Landais et al. 2016; Gray et al. 2011). The HIV Env trimer, composed of gp120 and gp41 subunits, is the main target for bnAbs (Sadanand et al. 2016; Kwong et al. 2013). BnAbs can be directed to the CD4-binding site on gp120, the V1/V2 region, the glycans on V3 region, the membrane proximal external region (MPER) on gp41, and the gp120–gp41 interface (Sadanand et al. 2016; Kwong et al. 2013).

The impact of bnAbs in preventing viral load rebound upon ART pause has recently been investigated in three studies (Bar et al. 2016; Scheid et al. 2016). In the A5340 trial, 14 participants who were on ART with HIV RNA <50 copies/mL received VRC01 (a bnAb that targets the CD4-binding site on HIV gp120) 40 mg/kg (at week-1), prior to ART pause (week 0). Up to two further infusions were administered during ART pause (week 2 and 5). ART resumption criteria included CD4$^+$ T cell count <350 cells/mm$^3$ and HIV RNA >200 copies/mL followed by repeat HIV RNA of >1000 copies/mL. In the NIH trial, 10 participants on ART with HIV RNA <50 copies/mL received VRC01 40 mg/kg at day -3, prior to ART pause (week 0). Further infusions were administered at weeks 2, 4, 8, 12, 16, 20, and 24, as long as ART was not resumed. ART resumption criteria included a 30% decline in CD4$^+$ T cell count from baseline, CD4$^+$ T cell count <350 cells/mm$^3$, sustained ($\geq 2$ weeks) HIV RNA >1000 copies/mL, any HIV-related symptoms, or pregnancy. In both trials, VRC01 administration was associated with higher likelihood of viral suppression at week 4 of ART pause, 38% ($P = 0.04$) of A5340, and 80% ($P < 0.001$) of NIH trial participants vs 13% of historical controls. No significant differences were found at week 8 of ART pause. The median time to rebound was 4 weeks in the A5340 trial and 5.6 weeks in the NIH trial. A number of participants had baseline resistance to VRC01 and resistance to VRC01 increased at rebound in most participants (Bar et al. 2016).

The effect of another bnAb, 3BNC117, that also targets the CD4-binding site on HIV gp120 was investigated by Scheid et al. In this study, participants on ART with suppressed viremia were given 3BNC117 at 30 mg/kg, 3 weeks apart for two doses (group A) or every 2 weeks for up to four doses (group B). All participants had confirmed 3BNC117-sensitive virus in their latent reservoir prior to 3BNC117 infusions. ART pause began 2 days after the first 3BNC117 infusion. ART was reinitiated and infusions were stopped after two consecutive HIV RNA >200 copies/mL. 3BNC177 infusions were safe, well tolerated, and were associated with an average delay in viral rebound of 6.7 weeks (group A) and 9.9 weeks (group B) when compared to 2.6 weeks in historical controls ($P < 0.00001$). Viral rebound was associated with the emergence of viral escape variants in 8/13 participants where rebound occurred despite high 3BNC117 serum concentration.

BnAbs' functions are not limited to the clearance of free virus and blocking of new infection (Chun et al. 2014). Mathematical analysis of viral dynamics from HIV-infected viremic individuals given a single dose of 3BNC117 and in vivo data from humanized mice suggest that 3BNC117 can accelerate the elimination of infected cells, through Fcγ receptor-dependent mechanisms (Lu et al. 2016). The effect of bnAbs on the elimination of infected cells in chronic and treated HIV infection is less clear. In the study by Riddler et al., individuals with chronic, treated HIV infection, and suppressed viremia were administered two infusions of VRC01 (40 mg/kg). No change in the levels of residual plasma viremia, cell-associated HIV RNA/DNA ratio, or total stimulated virus production from CD4$^+$ T cells was seen. Postulated mechanisms to explain the lack of response include viral resistance to VRC01, poor penetration of VRC01 to sites of virus expression, or inherent inability of VRC01 to clear virus particles or virus-expressing cells (Riddler et al. 2017).

BnAbs may also enhance host immune response against HIV (Schoofs et al. 2016). Immunoglobulin G (IgG) from viremic HIV-infected individuals who received 3BNC117 in the study described above showed increased activity against autologous viruses as well as improvement in breadth and potency of neutralization to tier 2 HIV-1 viruses at week 24 when compared to week 0. This was also seen in HIV-infected individuals on ART who received 3BNC117, but the improvement is of a lower magnitude. Contrarily, neutralization abilities in IgG from control individuals with similar plasma HIV viral load did not change over a 6-month period. Possible explanations for this phenomenon could be that 3BNC117 infusion selected for viral variants with altered antigenic properties, that in turn stimulated new B cell lineages, or that immune complexes formed by 3BNC117 and viruses acted as immunogens thereby stimulating immune responses (Schoofs et al. 2016).

The efficacy of bnAbs is limited at this stage by the presence of baseline resistance, the rapid emergence of resistant viruses, and the need for repeated infusions and cost. Though it is unlikely that a single bnAb can maintain viral suppression, co-administration of bnAbs has been shown to improve potency and breadth in vitro. At a 50% inhibitory concentration (IC50) cutoff of 1 µg/ml per antibody, two bnAb combinations neutralized 89–98%, and three bnAb combinations neutralized 98–100% of viruses (Kong et al. 2015). In a recent study by Nishimura et al., RMs were inoculated with SHIV intrarectally at day 0, and two bnAbs (10-1074 and 3BNC117) were administered at days 3, 10, and 17. All six RMs experienced sustained viral suppression lasting 56–177 days, at which point viral rebound occurred in 5/6 RM. The time to viral rebound was directly related to the decline in plasma concentrations of bnAbs. Moreover, in 3/6 RMs, plasma viral load subsequently declined to undetectable levels. The viral suppression, however, was CTL dependent as $CD8^+$ T cell depletion was associated with an immediate increase in viremia. The authors postulated that the presence of extremely low levels of HIV replication during passive immunotherapy with a combination of two bnAbs in acute SHIV may lead to the formation of bnAb–virion immune complexes, which further stimulate CTL responses, culminating in durable viral control, even in the absence of ART (Nishimura et al. 2017).

Bispecific anti-Env bnAbs with IgG3C hinge domain variant (to increase Fab domain flexibility, thereby favoring hetero-bivalent interactions with the Env trimer) has also been engineered (Fig. 2). 3BNC117/PGT135 bispecific bnAb displays neutralization breadth and potency that is better than that of the parental bnAbs (3BNC117 and PGT135), neutralizing >93% of the tested viruses, with an average IC50 of 0.036 µg/ml. When administered to humanized mice with HIV infection, 3BNC117/PGT135 bispecific bnAb reduced viremia by an average of 1.5 $\log_{10}$ copies/ml, in comparison to a reduction of only 0.15 $\log_{10}$ copies/ml in a 1:1 mix of 3BNC117 and PGT135 (Bournazos et al. 2016).

BnAbs may also be improved by modifying the Fc region to modulate effector functions (Euler and Alter 2015). Potential engineering techniques that have been demonstrated in in vitro models include S239D/I332E/A330L mutations that can improve ADCC (Lazar et al. 2006), S239D/I332E/G236A mutations that can enhance macrophage-mediated phagocytosis (Richards et al. 2008), and H268F/S324T mutations that can increase complement-dependent cytotoxicity (Moore et al. 2010). Modification of VRC01 by M428L/N434S mutations to VRC01-LS increased its binding to the neonatal Fc receptor (FcRn) and resulted in a threefold longer serum half-life (Ko et al. 2014). In a study measuring the protective effects of bnAbs against repeated low-dose SHIV challenges in RMs, the median number of challenges required for all RMs to become infected was 14.5 for VRC01-LS vs 8 for VRC01 (Gautam et al. 2016).

A potential strategy to eliminate the need for repeated infusions of bnAb is using vector-mediated antibody gene transfer to express bnAbs (also termed vectored immunoprophylaxis, VIP). Maintenance of Ab production and protection from HIV and SIV have been seen in murine and NHP models (Johnson et al. 2009; Balazs et al. 2012, 2014). The first human trial of recombinant adeno-associated virus (rAAV) vector coding for PG9 Ab in 24 healthy men is close to completion (clinicaltrials.gov NCT01937455). A potential pitfall of this strategy is the inability to switch off Ab expression on the occurrence of adverse effects (Schnepp and Johnson 2014).

The use of bnAbs in HIV remission is promising. The safety and efficacy of using bnAbs singly or in combination and bnAbs engineered to extend half-life are currently being evaluated in a number of clinical trials. These include a phase 1 study exploring the safety and antiviral activity of PGT121 in HIV-uninfected and HIV-infected individuals on or not on ART (NCT02960581); a phase 1 study on the safety and virologic effect of VRC01 in combination with ART during acute HIV infection (NCT02591420); a study evaluating the effect of early viral reactivation with LRA (romidepsin) and/or 3BNC117 on the latent reservoir in HIV-infected individuals initiating ART (NCT03041012); a phase 1 study on the safety and efficacy of VRC01 in maintaining viral suppression during ART pause in individuals who initiated ART during acute HIV infection (NCT03036709) (Crowell et al. 2017); a phase 2 study evaluating the effect of romidepsin and/or 3BNC117 in maintaining viral suppression during ART pause (NCT02850016); a phase 1b study exploring the use of the combination of 3BNC117 and 10-1074 in reducing HIV viral load and delaying viral rebound during ART pause (NCT02825797); and studies investigating the safety and pharmacokinetics of VRC01-LS (NCT02797171, NCT02599896) and VRC07-523LS (NCT03015181) when administered to healthy individuals and the efficacy of VRC01-LS in reducing viremia in HIV-1-infected adults (NCT02840474).

## 4.2 Anti-α4β7 Integrin Ab

α4β7 integrin is expressed on immune cells, and at high levels in a subset of memory T cells (Farstad et al. 1997). It enables cell migration into the gut through interaction with mucosal addressin cell adhesion molecule-1 (MAdCAM-1) on gut endothelial cells (Erle et al. 1994). α4β7 integrin can also bind to HIV gp120 (Arthos et al. 2008), and thus α4β7 integrin$^+$ CD4$^+$ T cells are highly susceptible to productive HIV infection (Cicala et al. 2009) and are preferentially infected and depleted (Kader et al. 2009).

In the study by Byrareddy et al., RMs were initiated on ART 5 weeks post-SIV infection. At weeks 9–18, primatized mAb against α4β7 integrin was administered every 3 weeks in combination with ART. At weeks 18–32, ART was ceased while anti-α4β7 integrin mAb infusions continued every 3 weeks. At weeks 32–50, all treatments were stopped. Two out of eight anti-α4β7 integrin mAb-treated RMs never rebounded, and the remaining six out of eight rebounded but then regained control of viremia. Virologic control in all eight anti-α4β7 integrin mAb-treated RMs persisted to week 81. Proviral DNA also became undetectable in all eight anti-α4β7 integrin mAb-treated RMs. The mechanism for persistent virologic control remains to be defined. The recovery of Th17 and Th22 cells in the gut and plasma retinoic acid levels, increases in peripheral blood cytokine$^+$ NK cells and gut NKp44$^+$ innate lymphoid cells (ILC) as well as plasma V2 ab responses may have contributed to the immune control (Byrareddy et al. 2016). These data suggest that ART and anti-α4β7 integrin mAb administrations during acute SIV infection led to sustained control of plasma viremia, even months after the discontinuation of both ART and anti-α4β7 integrin mAb.

## 5   Additional Considerations for Immune Interventions

### 5.1   Sanctuary Sites

Sanctuary sites represent locations where persistent HIV replication can occur due to reduced penetration by ART or immune privilege. It is important for immune interventions that aim to induce HIV remission to reach latently infected cells in these sites. A detailed review of potential sanctuary sites is beyond the scope of this review but have been published (Wong and Yukl 2016) and will also be discussed by Clements et al., in the chapter "Latency in Non-T cells and Non-Lymphoid Tissues". We will highlight two sites, the central nervous system and lymph nodes, that have important implications for the immune interventions discussed above.

Despite viral suppression on ART, HIV RNA can still be detected in the CSF (Spudich et al. 2006; Canestri et al. 2010) and in the brain (Kumar et al. 2007; Langford et al. 2006) in a subset of HIV-infected individuals. Furthermore, there is also evolution of drug-resistant mutations in the central nervous system (CNS), independent from the peripheral blood (Canestri et al. 2010; Smit et al. 2004; Peluso et al. 2012), suggesting compartmentalization of HIV in the CNS. This is thought to be secondary to suboptimal ART levels due to impedance of penetration by the blood–brain barrier (BBB) (Letendre et al. 2008; Calcagno et al. 2015). An intact BBB also limits the passage of immune cells and antibodies (Bell and Ehlers

2014), and may reduce the effectiveness of potential immune interventions to eliminate infected cells in the brain. Recent data have shown that bispecific antibodies can be engineered so that one arm binds to endogenous BBB receptors, enabling crossing of the BBB via receptor-mediated transport (Pardridge 2015). One such receptor is the transferrin receptor (TfR). A TfR bispecific antibody platform has been shown to safely deliver therapeutic abs across the BBB in cynomolgus monkeys brain (Yu et al. 2014). Human data, however, are not yet available.

Lymph nodes have also been postulated to be a sanctuary site. ART levels in lymph nodes are lower than in peripheral blood (Fletcher et al. 2014; Lorenzo-Redondo et al. 2016) and ongoing viral evolution occurs in lymph nodes despite undetectable HIV RNA in the peripheral blood (Lorenzo-Redondo et al. 2016), suggesting that ART concentration in the lymph node may not be sufficient to completely suppress viral replication. Furthermore, the follicular regions of lymph nodes are also relatively inaccessible to the majority of effector $CD8^+$ T cells (Fukazawa et al. 2015; Connick et al. 2007) as <5% of peripheral blood $CD8^+$ T cells express CXCR5 and thus cannot home to the follicular regions (Schaerli et al. 2000; Vinuesa and Cyster 2011). Thus, infected cells in the follicular regions are relatively sheltered from CTL-mediated killing (Fukazawa et al. 2015). This may pose as a hurdle to the use of CTLs to eliminate the HIV reservoir.

A small subset of $CD8^+$ T cells are $CXCR5^{high}$ and are able to migrate to the B cell follicles in response to CXCL13 (Quigley et al. 2007). These follicular CTLs have been shown to be able to eradicate lymphocytic choriomeningitis virus (LCMV)-infected cells in murine models (He et al. 2016; Leong et al. 2016). However, relatively few of these cells are HIV-specific (Petrovas et al. 2017; Connick et al. 2014). The use of VRC07-$\alpha$CD3 bispecific immunomodulatory protein obviates the need for HIV-sp CTLs and was able to direct follicular CTLs to eliminate HIV-infected cells in vitro (Petrovas et al. 2017).

## 5.2 Individuals Treated Early in the Course of HIV Infection as Potentially Ideal Candidates for Immune Interventions

Individuals treated during acute HIV infection possess characteristics that may enhance their responses to immune interventions aimed at inducing HIV remission. First, these individuals have significantly fewer latently infected cells in the blood and tissues (Archin et al. 2012; Buzon et al. 2014; Ananworanich et al. 2016a, b). Smaller HIV reservoir size secondary to early ART initiation has been associated with longer delays in viral rebound and instances of HIV remission after ART interruption (Williams et al. 2014; Saez-Cirion et al. 2013; Li et al. 2016; Fidler et al. 2017; Goujard et al. 2012). Second, early ART limits HIV-associated $CD4^+$ T cell depletion and immune dysfunction. Individuals initiated on ART during acute infection are more likely to have near normal CD4/CD8 ratio (Ananworanich et al. 2016a; Thornhill et al. 2016) and their CTLs display superior memory and proliferative capacity, potentially limiting HIV reservoir seeding (Lecuroux et al. 2009a, b; Takata et al. 2017). Finally, individuals who initiated ART during early HIV infection have lower viral diversity and immune escape variants when compared to individuals with chronic HIV infection (Altfeld et al. 2001; Keele et al. 2008). Thus, the smaller reservoir, the preserved immunity, and the lower viral diversity associated with early treatment may possibly lead to better responses to immune interventions aimed at inducing HIV remission (Fig. 3).

Most of the human studies to date involved the administration of immune interventions aiming to induce HIV remission after ART initiation and virologic

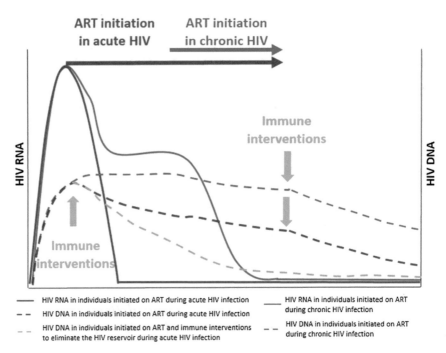

**Fig. 3** Schematic of the HIV reservoir during the course of HIV infection, treatment with antiretroviral therapy (ART), and reservoir eliminating immune interventions. After HIV infection, HIV RNA levels (*solid lines*) peak rapidly and the HIV reservoir is seeded within days. HIV DNA (*dashed lines*) can be used as a surrogate marker for the HIV reservoir. The *red lines* (HIV RNA, *solid*; HIV DNA, *dashed*) depict the scenario where HIV-infected individuals are initiated on ART during acute infection. The *blue lines* (HIV RNA, *solid*; HIV DNA, *dashed*) depict the scenario where untreated individuals experience a decline in plasma viremia to the set point that remains relatively stable for years. After ART initiation during both acute (*red lines*) and chronic HIV infections (*blue lines*), there is suppression of plasma viremia (*solid lines*) and reduction in the HIV reservoir (*dashed lines*). Concomitant implementation of immune interventions with the initiation of ART in acute HIV infection (*gray-dashed line*) or after a period of viral suppression on ART (*red- and blue-dashed lines*) to boost elimination of infected cells may potentially accelerate the decay of the HIV reservoir

suppression. In the study by Bolton et al., a single infusion of a combination of two bnAbs (VRC07-523 and PGT121), day 10 post-SHIV infection was associated with similar viremia decay and a lower level of cell-associated viral DNA in naive CD4$^+$ T cells in lymph node when compared to ART (Bolton et al. 2015). Furthermore, administration of a combination of two bnAbs (VRC07-523 and PGT121) on days 1, 4, 7, and 10 post-SHIV infection was associated with eradication of SHIV by day 14 and prevention of viral rebound even after the decay of the bnAbs (Hessell et al. 2016). Therefore, the administration of immune interventions in conjunction with ART during acute HIV infection may potentially be synergistic, boosting HIV-specific immune responses, increasing the elimination of HIV-expressing infected cells, and further limiting reservoir seeding (Fig. 3).

# 6 Conclusion

In summary, a number of immune-mediated strategies are in development to enhance the elimination of the HIV reservoir. These include the boosting of CTL-mediated killing with therapeutic vaccines, check point inhibitors, bispecific immunomodulatory proteins/DART and CAR-T cells, and harnessing antibodies to eliminate infected cells. The mutability of HIV and the emergence of escape variants may threaten the success of these strategies. Tools to overcome this challenge, including the use of mosaic or conserved immunogens for therapeutic vaccines and the use of combinations of bnAbs or bispecific bnAbs, are in clinical development. A major gap in knowledge exists on ways to direct immune effectors to sites where latently infected cells are located. Ultimately, strategies to induce HIV remission must be tolerable, safe, and scalable in order to make a global impact.

**Acknowledgements** This work was supported by a cooperative agreement (W81XWH-11-2-0174) between the Henry M. Jackson Foundation for the Advancement of Military Medicine Inc. and the U.S. Department of the Army. The views expressed herein are those of the authors and should not be construed to represent the positions of the Departments of the Army or Defense. Trade names are used for identification purposes only and do not imply endorsement.

# References

Ahmed T, Borthwick NJ, Gilmour J, Hayes P, Dorrell L, Hanke T (2016) Control of HIV-1 replication in vitro by vaccine-induced human CD8(+) T cells through conserved subdominant Pol epitopes. Vaccine 34(9):1215–1224. doi:10.1016/j.vaccine.2015.12.021

Altfeld M, Rosenberg ES, Shankarappa R, Mukherjee JS, Hecht FM, Eldridge RL, Addo MM, Poon SH, Phillips MN, Robbins GK, Sax PE, Boswell S, Kahn JO, Brander C, Goulder PJ, Levy JA, Mullins JI, Walker BD (2001) Cellular immune responses and viral diversity in individuals treated during acute and early HIV-1 infection. J Exp Med 193(2):169–180

Amancha PK, Hong JJ, Rogers K, Ansari AA, Villinger F (2013) In vivo blockade of the programmed cell death-1 pathway using soluble recombinant PD-1-Fc enhances CD4+ and CD8+ T cell responses but has limited clinical benefit. J Immunol 191(12):6060–6070. doi:10.4049/jimmunol.1302044

Ananworanich J, Sacdalan CP, Pinyakorn S, Chomont N, de Souza M, Luekasemsuk T, Schuetz A, Krebs SJ, Dewar R, Jagodzinski L, Ubolyam S, Trichavaroj R, Tovanabutra S, Spudich S, Valcour V, Sereti I, Michael N, Robb M, Phanuphak P, Kim JH, Phanuphak N (2016a) Virological and immunological characteristics of HIV-infected individuals at the earliest stage of infection. J Virus Erad 2:43–48

Ananworanich J, Chomont N, Eller LA, Kroon E, Tovanabutra S, Bose M, Nau M, Fletcher JL, Tipsuk S, Vandergeeten C, O'Connell RJ, Pinyakorn S, Michael N, Phanuphak N, Robb ML, Rv, groups RSs (2016) HIV DNA set point is rapidly established in acute HIV infection and dramatically reduced by early ART. EBioMedicine 11:68–72. doi:10.1016/j.ebiom.2016.07.024

Angel JB, Routy JP, Tremblay C, Ayers D, Woods R, Singer J, Bernard N, Kovacs C, Smaill F, Gurunathan S, Sekaly RP (2011) A randomized controlled trial of HIV therapeutic vaccination using ALVAC with or without Remune. Aids 25(6):731–739. doi:10.1097/QAD.0b013e328344cea5

Archin NM, Vaidya NK, Kuruc JD, Liberty AL, Wiegand A, Kearney MF, Cohen MS, Coffin JM, Bosch RJ, Gay CL, Eron JJ, Margolis DM, Perelson AS (2012) Immediate antiviral therapy appears to restrict resting CD4+ cell HIV-1 infection without accelerating the decay of latent infection. Proc Natl Acad Sci U S A 109(24):9523–9528. doi:10.1073/pnas.1120248109

Arthos J, Cicala C, Martinelli E, Macleod K, Van Ryk D, Wei D, Xiao Z, Veenstra TD, Conrad TP, Lempicki RA, McLaughlin S, Pascuccio M, Gopaul R, McNally J, Cruz CC, Censoplano N, Chung E, Reitano KN, Kottilil S, Goode DJ, Fauci AS (2008) HIV-1 envelope protein binds to and signals through integrin alpha4beta7, the gut mucosal homing receptor for peripheral T cells. Nat Immunol 9(3):301–309. doi:10.1038/ni1566

Balazs AB, Chen J, Hong CM, Rao DS, Yang L, Baltimore D (2012) Antibody-based protection against HIV infection by vectored immunoprophylaxis. Nature 481(7379):81–84. doi:10.1038/nature10660

Balazs AB, Ouyang Y, Hong CM, Chen J, Nguyen SM, Rao DS, An DS, Baltimore D (2014) Vectored immunoprophylaxis protects humanized mice from mucosal HIV transmission. Nat Med 20(3):296–300. doi:10.1038/nm.3471

Bar KJ, Sneller MC, Harrison LJ, Justement JS, Overton ET, Petrone ME, Salantes DB, Seamon CA, Scheinfeld B, Kwan RW, Learn GH, Proschan MA, Kreider EF, Blazkova J, Bardsley M, Refsland EW, Messer M, Clarridge KE, Tustin NB, Madden PJ, Oden K, O'Dell SJ, Jarocki B, Shiakolas AR, Tressler RL, Doria-Rose NA, Bailer RT, Ledgerwood JE, Capparelli EV, Lynch RM, Graham BS, Moir S, Koup RA, Mascola JR, Hoxie JA, Fauci AS, Tebas P, Chun TW (2016) Effect of HIV antibody VRC01 on viral rebound after treatment interruption. N Engl J Med 375(21):2037–2050. doi:10.1056/NEJMoa1608243

Barouch DH, O'Brien KL, Simmons NL, King SL, Abbink P, Maxfield LF, Sun YH, La Porte A, Riggs AM, Lynch DM, Clark SL, Backus K, Perry JR, Seaman MS, Carville A, Mansfield KG, Szinger JJ, Fischer W, Muldoon M, Korber B (2010) Mosaic HIV-1 vaccines expand the breadth and depth of cellular immune responses in rhesus monkeys. Nat Med 16(3):319–323. doi:10.1038/nm.2089

Barouch DH, Stephenson KE, Borducchi EN, Smith K, Stanley K, McNally AG, Liu J, Abbink P, Maxfield LF, Seaman MS, Dugast AS, Alter G, Ferguson M, Li W, Earl PL, Moss B, Giorgi EE, Szinger JJ, Eller LA, Billings EA, Rao M, Tovanabutra S, Sanders-Buell E, Weijtens M, Pau MG, Schuitemaker H, Robb ML, Kim JH, Korber BT, Michael NL (2013) Protective efficacy of a global HIV-1 mosaic vaccine against heterologous SHIV challenges in rhesus monkeys. Cell 155(3):531–539. doi:10.1016/j.cell.2013.09.061

Bell RD, Ehlers MD (2014) Breaching the blood-brain barrier for drug delivery. Neuron 81(1):1–3. doi:10.1016/j.neuron.2013.12.023

Bolton DL, Pegu A, Wang K, McGinnis K, Nason M, Foulds K, Letukas V, Schmidt SD, Chen X, Todd JP, Lifson JD, Rao S, Michael NL, Robb ML, Mascola JR, Koup RA (2015) Human immunodeficiency virus type 1 monoclonal antibodies suppress acute simian-human immunodeficiency virus viremia and limit seeding of cell-associated viral reservoirs. J Virol 90 (3):1321–1332. doi:10.1128/JVI.02454-15

Borducchi EN, Cabral C, Stephenson KE, Liu J, Abbink P, Ng'ang'a D, Nkolola JP, Brinkman AL, Peter L, Lee BC, Jimenez J, Jetton D, Mondesir J, Mojta S, Chandrashekar A, Molloy K, Alter G, Gerold JM, Hill AL, Lewis MG, Pau MG, Schuitemaker H, Hesselgesser J, Geleziunas R, Kim JH, Robb ML, Michael NL, Barouch DH (2016) Ad26/MVA therapeutic vaccination with TLR7 stimulation in SIV-infected rhesus monkeys. Nature 540(7632): 284–287. doi:10.1038/nature20583

Borrow P, Lewicki H, Hahn BH, Shaw GM, Oldstone MB (1994) Virus-specific CD8+ cytotoxic T-lymphocyte activity associated with control of viremia in primary human immunodeficiency virus type 1 infection. J Virol 68(9):6103–6110

Bournazos S, Gazumyan A, Seaman MS, Nussenzweig MC, Ravetch JV (2016) Bispecific anti-HIV-1 antibodies with enhanced breadth and potency. Cell 165(7):1609–1620. doi:10. 1016/j.cell.2016.04.050

Buzon MJ, Martin-Gayo E, Pereyra F, Ouyang Z, Sun H, Li JZ, Piovoso M, Shaw A, Dalmau J, Zangger N, Martinez-Picado J, Zurakowski R, Yu XG, Telenti A, Walker BD, Rosenberg ES, Lichterfeld M (2014) Long-term antiretroviral treatment initiated at primary HIV-1 infection affects the size, composition, and decay kinetics of the reservoir of HIV-1-infected CD4 T cells. J Virol 88(17):10056–10065. doi:10.1128/JVI.01046-14

Byrareddy SN, Arthos J, Cicala C, Villinger F, Ortiz KT, Little D, Sidell N, Kane MA, Yu J, Jones JW, Santangelo PJ, Zurla C, McKinnon LR, Arnold KB, Woody CE, Walter L, Roos C, Noll A, Van Ryk D, Jelicic K, Cimbro R, Gumber S, Reid MD, Adsay V, Amancha PK, Mayne AE, Parslow TG, Fauci AS, Ansari AA (2016) Sustained virologic control in SIV + macaques after antiretroviral and alpha4beta7 antibody therapy. Science 354(6309): 197–202. doi:10.1126/science.aag1276

Calcagno A, Simiele M, Alberione MC, Bracchi M, Marinaro L, Ecclesia S, Di Perri G, D'Avolio A, Bonora S (2015) Cerebrospinal fluid inhibitory quotients of antiretroviral drugs in HIV-infected patients are associated with compartmental viral control. Clin Infect Dis 60 (2):311–317. doi:10.1093/cid/ciu773

Canestri A, Lescure FX, Jaureguiberry S, Moulignier A, Amiel C, Marcelin AG, Peytavin G, Tubiana R, Pialoux G, Katlama C (2010) Discordance between cerebral spinal fluid and plasma HIV replication in patients with neurological symptoms who are receiving suppressive antiretroviral therapy. Clin Infect Dis 50(5):773–778. doi:10.1086/650538

Carrington M, O'Brien SJ (2003) The influence of HLA genotype on AIDS. Annu Rev Med 54:535–551. doi:10.1146/annurev.med.54.101601.152346

Cartwright EK, Spicer L, Smith SA, Lee D, Fast R, Paganini S, Lawson BO, Nega M, Easley K, Schmitz JE, Bosinger SE, Paiardini M, Chahroudi A, Vanderford TH, Estes JD, Lifson JD, Derdeyn CA, Silvestri G (2016) CD8(+) lymphocytes are required for maintaining viral suppression in siv-infected macaques treated with short-term antiretroviral therapy. Immunity 45(3):656–668. doi:10.1016/j.immuni.2016.08.018

Casazza JP, Betts MR, Picker LJ, Koup RA (2001) Decay kinetics of human immunodeficiency virus-specific CD8+ T cells in peripheral blood after initiation of highly active antiretroviral therapy. J Virol 75(14):6508–6516. doi:10.1128/JVI.75.14.6508-6516.2001

Chomont N, Perreau M (2016) Strategies for targeting residual HIV infection. Curr Opin HIV AIDS 11(4):359–361. doi:10.1097/COH.0000000000000291

Chomont N, El-Far M, Ancuta P, Trautmann L, Procopio FA, Yassine-Diab B, Boucher G, Boulassel MR, Ghattas G, Brenchley JM, Schacker TW, Hill BJ, Douek DC, Routy JP, Haddad EK, Sekaly RP (2009) HIV reservoir size and persistence are driven by T cell survival and homeostatic proliferation. Nat Med 15(8):893–900. doi:10.1038/nm.1972

Chun TW, Finzi D, Margolick J, Chadwick K, Schwartz D, Siliciano RF (1995) In vivo fate of HIV-1-infected T cells: quantitative analysis of the transition to stable latency. Nat Med 1 (12):1284–1290

Chun TW, Carruth L, Finzi D, Shen X, DiGiuseppe JA, Taylor H, Hermankova M, Chadwick K, Margolick J, Quinn TC, Kuo YH, Brookmeyer R, Zeiger MA, Barditch-Crovo P, Siliciano RF (1997) Quantification of latent tissue reservoirs and total body viral load in HIV-1 infection. Nature 387(6629):183–188. doi:10.1038/387183a0

Chun TW, Murray D, Justement JS, Blazkova J, Hallahan CW, Fankuchen O, Gittens K, Benko E, Kovacs C, Moir S, Fauci AS (2014) Broadly neutralizing antibodies suppress HIV in the persistent viral reservoir. Proc Natl Acad Sci U S A 111(36):13151–13156. doi:10.1073/pnas. 1414148111

Cicala C, Martinelli E, McNally JP, Goode DJ, Gopaul R, Hiatt J, Jelicic K, Kottilil S, Macleod K, O'Shea A, Patel N, Van Ryk D, Wei D, Pascuccio M, Yi L, McKinnon L, Izulla P, Kimani J, Kaul R, Fauci AS, Arthos J (2009) The integrin alpha4beta7 forms a complex with cell-surface CD4 and defines a T-cell subset that is highly susceptible to infection by HIV-1. Proc Natl Acad Sci U S A 106(49):20877–20882. doi:10.1073/pnas.0911796106

Cillo AR, Mellors JW (2016) Which therapeutic strategy will achieve a cure for HIV-1? Curr Opin Virol 18:14–19. doi:10.1016/j.coviro.2016.02.001

Crowell TA, Colby DJ, Pinyakorn S, Intasan J, Benjapornpong K, Tanjnareel K et al (2017) HIV-specific broadly-neutralizing monoclonal antibody, VRC01, minimally impacts time to viral rebound following treatment interruption in virologically-suppressed, HIV-infected participants who initiated antiretroviral therapy during acute HIV infection. IAS 2017, Paris

Connick E, Mattila T, Folkvord JM, Schlichtemeier R, Meditz AL, Ray MG, McCarter MD, Mawhinney S, Hage A, White C, Skinner PJ (2007) CTL fail to accumulate at sites of HIV-1 replication in lymphoid tissue. J Immunol 178(11):6975–6983

Connick E, Folkvord JM, Lind KT, Rakasz EG, Miles B, Wilson NA, Santiago ML, Schmitt K, Stephens EB, Kim HO, Wagstaff R, Li S, Abdelaal HM, Kemp N, Watkins DI, MaWhinney S, Skinner PJ (2014) Compartmentalization of simian immunodeficiency virus replication within secondary lymphoid tissues of rhesus macaques is linked to disease stage and inversely related to localization of virus-specific CTL. J Immunol 193(11):5613–5625. doi:10.4049/jimmunol. 1401161

Cools N, Ponsaerts P, Van Tendeloo VF, Berneman ZN (2007) Balancing between immunity and tolerance: an interplay between dendritic cells, regulatory T cells, and effector T cells. J Leukoc Biol 82(6):1365–1374. doi:10.1189/jlb.0307166

Copelan EA (2006) Hematopoietic stem-cell transplantation. N Engl J Med 354(17):1813–1826. doi:10.1056/NEJMra052638

Couzin-Frankel J (2013) Breakthrough of the year 2013. Cancer Immunother Sci 342(6165):1432–1433. doi:10.1126/science.342.6165.1432

Dahabieh MS, Battivelli E, Verdin E (2015) Understanding HIV latency: the road to an HIV cure. Annu Rev Med 66:407–421. doi:10.1146/annurev-med-092112-152941

Day CL, Kaufmann DE, Kiepiela P, Brown JA, Moodley ES, Reddy S, Mackey EW, Miller JD, Leslie AJ, DePierres C, Mncube Z, Duraiswamy J, Zhu B, Eichbaum Q, Altfeld M, Wherry EJ, Coovadia HM, Goulder PJ, Klenerman P, Ahmed R, Freeman GJ, Walker BD (2006) PD-1 expression on HIV-specific T cells is associated with T-cell exhaustion and disease progression. Nature 443(7109):350–354. doi:10.1038/nature05115

Deeks SG, Lewin SR, Ross AL, Ananworanich J, Benkirane M, Cannon P, Chomont N, Douek D, Lifson JD, Lo YR, Kuritzkes D, Margolis D, Mellors J, Persaud D, Tucker JD, Barre-Sinoussi F, International ASTaCWG, Alter G, Auerbach J, Autran B, Barouch DH, Behrens G, Cavazzana M, Chen Z, Cohen EA, Corbelli GM, Eholie S, Eyal N, Fidler S, Garcia L, Grossman C, Henderson G, Henrich TJ, Jefferys R, Kiem HP, McCune J, Moodley K, Newman PA, Nijhuis M, Nsubuga MS, Ott M, Palmer S, Richman D, Saez-Cirion A, Sharp M, Siliciano J, Silvestri G, Singh J, Spire B, Taylor J, Tolstrup M, Valente S, van Lunzen J, Walensky R, Wilson I, Zack J (2016) International AIDS Society global scientific strategy: towards an HIV cure 2016. Nat Med. doi:10.1038/nm.4108

Deng K, Pertea M, Rongvaux A, Wang L, Durand CM, Ghiaur G, Lai J, McHugh HL, Hao H, Zhang H, Margolick JB, Gurer C, Murphy AJ, Valenzuela DM, Yancopoulos GD, Deeks SG, Strowig T, Kumar P, Siliciano JD, Salzberg SL, Flavell RA, Shan L, Siliciano RF (2015) Broad CTL response is required to clear latent HIV-1 due to dominance of escape mutations. Nature 517(7534):381–385. doi:10.1038/nature14053

Doria-Rose NA, Klein RM, Manion MM, O'Dell S, Phogat A, Chakrabarti B, Hallahan CW, Migueles SA, Wrammert J, Ahmed R, Nason M, Wyatt RT, Mascola JR, Connors M (2009) Frequency and phenotype of human immunodeficiency virus envelope-specific B cells from patients with broadly cross-neutralizing antibodies. J Virol 83(1):188–199. doi:10.1128/JVI. 01583-08

Erle DJ, Briskin MJ, Butcher EC, Garcia-Pardo A, Lazarovits AI, Tidswell M (1994) Expression and function of the MAdCAM-1 receptor, integrin alpha4beta7, on human leukocytes. J Immunol 153(2):517–528

Eron JJ, Gay C, Bosch R, Ritz J, Hataye JM, Hwang C, Tressler RL, Mason SW, Koup RA, Mellors JW (2016) Safety, immunologic and virologic activity of anti-PD-L1 in HIV-1 participants on ART. Abstract 25. In: Conference on Retroviruses and Opportunistic Infections CROI, pp 22–25

Euler Z, Alter G (2015) Exploring the potential of monoclonal antibody therapeutics for HIV-1 eradication. AIDS Res Hum Retroviruses 31(1):13–24. doi:10.1089/AID.2014.0235

Farstad IN, Halstensen TS, Kvale D, Fausa O, Brandtzaeg P (1997) Topographic distribution of homing receptors on B and T cells in human gut-associated lymphoid tissue: relation of L-selectin and integrin alpha4beta7 to naive and memory phenotypes. Am J Pathol 150 (1):187–199

Fidler S, Olson AD, Bucher HC, Fox J, Thornhill J, Morrison C, Muga R, Phillips A, Frater J, Porter K (2017) Virological blips and predictors of post treatment viral control after stopping ART started in primary HIV infection. J Acquir Immune Defic Syndr 74(2):126–133. doi:10. 1097/QAI.0000000000001220

Fischer W, Perkins S, Theiler J, Bhattacharya T, Yusim K, Funkhouser R, Kuiken C, Haynes B, Letvin NL, Walker BD, Hahn BH, Korber BT (2007) Polyvalent vaccines for optimal coverage of potential T-cell epitopes in global HIV-1 variants. Nat Med 13(1):100–106. doi:10.1038/ nm1461

Fletcher CV, Staskus K, Wietgrefe SW, Rothenberger M, Reilly C, Chipman JG, Beilman GJ, Khoruts A, Thorkelson A, Schmidt TE, Anderson J, Perkey K, Stevenson M, Perelson AS, Douek DC, Haase AT, Schacker TW (2014) Persistent HIV-1 replication is associated with lower antiretroviral drug concentrations in lymphatic tissues. Proc Natl Acad Sci U S A 111 (6):2307–2312. doi:10.1073/pnas.1318249111

Forthal D, Hope TJ, Alter G (2013) New paradigms for functional HIV-specific nonneutralizing antibodies. Curr Opin HIV AIDS 8(5):393–401. doi:10.1097/COH.0b013e328363d486

Fukazawa Y, Lum R, Okoye AA, Park H, Matsuda K, Bae JY, Hagen SI, Shoemaker R, Deleage C, Lucero C, Morcock D, Swanson T, Legasse AW, Axthelm MK, Hesselgesser J, Geleziunas R, Hirsch VM, Edlefsen PT, Piatak M Jr, Estes JD, Lifson JD, Picker LJ (2015) B cell follicle sanctuary permits persistent productive simian immunodeficiency virus infection in elite controllers. Nat Med 21(2):132–139. doi:10.1038/nm.3781

Gandhi RT, O'Neill D, Bosch RJ, Chan ES, Bucy RP, Shopis J, Baglyos L, Adams E, Fox L, Purdue L, Marshak A, Flynn T, Masih R, Schock B, Mildvan D, Schlesinger SJ, Marovich MA, Bhardwaj N, Jacobson JM, team ACTGA (2009) A randomized therapeutic vaccine trial of canarypox-HIV-pulsed dendritic cells vs. canarypox-HIV alone in HIV-1-infected patients on antiretroviral therapy. Vaccine 27(43):6088–6094. doi:10.1016/j. vaccine.2009.05.016

Garcia F, Lejeune M, Climent N, Gil C, Alcami J, Morente V, Alos L, Ruiz A, Setoain J, Fumero E, Castro P, Lopez A, Cruceta A, Piera C, Florence E, Pereira A, Libois A, Gonzalez N, Guila M, Caballero M, Lomena F, Joseph J, Miro JM, Pumarola T, Plana M, Gatell JM, Gallart T (2005) Therapeutic immunization with dendritic cells loaded with heat-inactivated autologous HIV-1 in patients with chronic HIV-1 infection. J Infect Dis 191 (10):1680–1685. doi:10.1086/429340

Garcia F, Leon A, Gatell JM, Plana M, Gallart T (2012) Therapeutic vaccines against HIV infection. Hum Vaccin Immunother 8(5):569–581. doi:10.4161/hv.19555

Garcia F, Climent N, Guardo AC, Gil C, Leon A, Autran B, Lifson JD, Martinez-Picado J, Dalmau J, Clotet B, Gatell JM, Plana M, Gallart T, Group DMOS (2013) A dendritic cell-based vaccine elicits T cell responses associated with control of HIV-1 replication. Sci Transl Med 5 (166):166ra162. doi:10.1126/scitranslmed.3004682

Garcia F, Plana M, Climent N, Leon A, Gatell JM, Gallart T (2013b) Dendritic cell based vaccines for HIV infection: the way ahead. Hum Vaccin Immunother 9(11):2445–2452

Gautam R, Nishimura Y, Pegu A, Nason MC, Klein F, Gazumyan A, Golijanin J, Buckler-White A, Sadjadpour R, Wang K, Mankoff Z, Schmidt SD, Lifson JD, Mascola JR, Nussenzweig MC, Martin MA (2016) A single injection of anti-HIV-1 antibodies protects against repeated SHIV challenges. Nature 533(7601):105–109. doi:10.1038/nature17677

Goldstein G, Damiano E, Donikyan M, Pasha M, Beckwith E, Chicca J (2012) HIV-1 Tat B-cell epitope vaccination was ineffectual in preventing viral rebound after ART cessation: HIV rebound with current ART appears to be due to infection with new endogenous founder virus and not to resurgence of pre-existing Tat-dependent viremia. Hum Vaccine Immunother 8 (10):1425–1430. doi:10.4161/hv.21616

Gooley TA, Chien JW, Pergam SA, Hingorani S, Sorror ML, Boeckh M, Martin PJ, Sandmaier BM, Marr KA, Appelbaum FR, Storb R, McDonald GB (2010) Reduced mortality after allogeneic hematopoietic-cell transplantation. N Engl J Med 363(22):2091–2101. doi:10. 1056/NEJMoa1004383

Goonetilleke N, Liu MK, Salazar-Gonzalez JF, Ferrari G, Giorgi E, Ganusov VV, Keele BF, Learn GH, Turnbull EL, Salazar MG, Weinhold KJ, Moore S, CCC B, Letvin N, Haynes BF, Cohen MS, Hraber P, Bhattacharya T, Borrow P, Perelson AS, Hahn BH, Shaw GM, Korber BT, McMichael AJ (2009) The first T cell response to transmitted/founder virus contributes to the control of acute viremia in HIV-1 infection. J Exp Med 206(6):1253–1272. doi:10.1084/jem.20090365

Goujard C, Girault I, Rouzioux C, Lecuroux C, Deveau C, Chaix ML, Jacomet C, Talamali A, Delfraissy JF, Venet A, Meyer L, Sinet M, Group ACPS (2012) HIV-1 control after transient antiretroviral treatment initiated in primary infection: role of patient characteristics and effect of therapy. Antivir Ther 17(6):1001–1009. doi:10.3851/IMP2273

Granich R, Gupta S, Montaner J, Williams B, Zuniga JM (2016) Pattern, determinants, and impact of HIV spending on care and treatment in 38 high-burden low- and middle-income countries. J Int Assoc Provid AIDS Care 15(2):91–100. doi:10.1177/2325957415623261

Gray ES, Madiga MC, Hermanus T, Moore PL, Wibmer CK, Tumba NL, Werner L, Mlisana K, Sibeko S, Williamson C, Abdool Karim SS, Morris L, Team CS (2011) The neutralization breadth of HIV-1 develops incrementally over four years and is associated with CD4+ T cell decline and high viral load during acute infection. J Virol 85(10):4828–4840. doi:10.1128/JVI. 00198-11

Hancock G, Yang H, Yorke E, Wainwright E, Bourne V, Frisbee A, Payne TL, Berrong M, Ferrari G, Chopera D, Hanke T, Mothe B, Brander C, McElrath MJ, McMichael A, Goonetilleke N, Tomaras GD, Frahm N, Dorrell L (2015) Identification of effective subdominant anti-HIV-1 CD8+ T cells within entire post-infection and post-vaccination immune responses. PLoS Pathog 11(2):e1004658. doi:10.1371/journal.ppat.1004658

Hanke T (2014) Conserved immunogens in prime-boost strategies for the next-generation HIV-1 vaccines. Expert Opin Biol Ther 14(5):601–616. doi:10.1517/14712598.2014.885946

He R, Hou S, Liu C, Zhang A, Bai Q, Han M, Yang Y, Wei G, Shen T, Yang X, Xu L, Chen X, Hao Y, Wang P, Zhu C, Ou J, Liang H, Ni T, Zhang X, Zhou X, Deng K, Chen Y, Luo Y, Xu J, Qi H, Wu Y, Ye L (2016) Follicular CXCR5-expressing CD8+ T cells curtail chronic viral infection. Nature 537(7620):412–428. doi:10.1038/nature19317

Hertz T, Ahmed H, Friedrich DP, Casimiro DR, Self SG, Corey L, McElrath MJ, Buchbinder S, Horton H, Frahm N, Robertson MN, Graham BS, Gilbert P (2013) HIV-1 vaccine-induced

T-cell responses cluster in epitope hotspots that differ from those induced in natural infection with HIV-1. PLoS Pathog 9(6):e1003404. doi:10.1371/journal.ppat.1003404

Hessell AJ, Jaworski JP, Epson E, Matsuda K, Pandey S, Kahl C, Reed J, Sutton WF, Hammond KB, Cheever TA, Barnette PT, Legasse AW, Planer S, Stanton JJ, Pegu A, Chen X, Wang K, Siess D, Burke D, Park BS, Axthelm MK, Lewis A, Hirsch VM, Graham BS, Mascola JR, Sacha JB, Haigwood NL (2016) Early short-term treatment with neutralizing human monoclonal antibodies halts SHIV infection in infant macaques. Nat Med 22(4):362–368. doi:10.1038/nm.4063

Hill AL, Rosenbloom DI, Fu F, Nowak MA, Siliciano RF (2014) Predicting the outcomes of treatment to eradicate the latent reservoir for HIV-1. Proc Natl Acad Sci U S A 111(37):13475–13480. doi:10.1073/pnas.1406663111

Hoffmann M, Pantazis N, Martin GE, Hickling S, Hurst J, Meyerowitz J, Willberg CB, Robinson N, Brown H, Fisher M, Kinloch S, Babiker A, Weber J, Nwokolo N, Fox J, Fidler S, Phillips R, Frater J, Spartac Investigators C (2016) Exhaustion of activated CD8 T cells predicts disease progression in primary HIV-1 infection. PLoS Pathog 12(7):e1005661. doi:10.1371/journal.ppat.1005661

Hraber P, Seaman MS, Bailer RT, Mascola JR, Montefiori DC, Korber BT (2014) Prevalence of broadly neutralizing antibody responses during chronic HIV-1 infection. AIDS 28(2):163–169. doi:10.1097/QAD.0000000000000106

Hryniewicz A, Boasso A, Edghill-Smith Y, Vaccari M, Fuchs D, Venzon D, Nacsa J, Betts MR, Tsai WP, Heraud JM, Beer B, Blanset D, Chougnet C, Lowy I, Shearer GM, Franchini G (2006) CTLA-4 blockade decreases TGF-beta, IDO, and viral RNA expression in tissues of SIVmac251-infected macaques. Blood 108(12):3834–3842. doi:10.1182/blood-2006-04-010637

Hutter G, Nowak D, Mossner M, Ganepola S, Mussig A, Allers K, Schneider T, Hofmann J, Kucherer C, Blau O, Blau IW, Hofmann WK, Thiel E (2009) Long-term control of HIV by CCR5 Delta32/Delta32 stem-cell transplantation. N Engl J Med 360(7):692–698. doi:10.1056/NEJMoa0802905

Ide F, Nakamura T, Tomizawa M, Kawana-Tachikawa A, Odawara T, Hosoya N, Iwamoto A (2006) Peptide-loaded dendritic-cell vaccination followed by treatment interruption for chronic HIV-1 infection: a phase 1 trial. J Med Virol 78(6):711–718. doi:10.1002/jmv.20612

Jacobson JM, Pat Bucy R, Spritzler J, Saag MS, Eron JJ, Jr., Coombs RW, Wang R, Fox L, Johnson VA, Cu-Uvin S, Cohn SE, Mildvan D, O'Neill D, Janik J, Purdue L, O'Connor DK, Vita CD, Frank I, National Institute of A, Infectious Diseases ACTGPT (2006) Evidence that intermittent structured treatment interruption, but not immunization with ALVAC-HIV vCP1452, promotes host control of HIV replication: the results of AIDS Clinical Trials Group 5068. J Infect Dis 194(5):623–632. doi:10.1086/506364

Jacobson JM, Routy JP, Welles S, DeBenedette M, Tcherepanova I, Angel JB, Asmuth DM, Stein DK, Baril JG, McKellar M, Margolis DM, Trottier B, Wood K, Nicolette C (2016) Dendritic cell immunotherapy for HIV-1 infection using autologous HIV-1 RNA: a randomized, double-blind, placebo-controlled clinical trial. J Acquir Immune Defic Syndr 72(1):31–38. doi:10.1097/QAI.0000000000000926

Jin X, Bauer DE, Tuttleton SE, Lewin S, Gettie A, Blanchard J, Irwin CE, Safrit JT, Mittler J, Weinberger L, Kostrikis LG, Zhang L, Perelson AS, Ho DD (1999) Dramatic rise in plasma viremia after CD8(+) T cell depletion in simian immunodeficiency virus-infected macaques. J Exp Med 189(6):991–998

Johnson PR, Schnepp BC, Zhang J, Connell MJ, Greene SM, Yuste E, Desrosiers RC, Clark KR (2009) Vector-mediated gene transfer engenders long-lived neutralizing activity and protection against SIV infection in monkeys. Nat Med 15(8):901–906. doi:10.1038/nm.1967

Johnson DB, Balko JM, Compton ML, Chalkias S, Gorham J, Xu Y, Hicks M, Puzanov I, Alexander MR, Bloomer TL, Becker JR, Slosky DA, Phillips EJ, Pilkinton MA, Craig-Owens L, Kola N, Plautz G, Reshef DS, Deutsch JS, Deering RP, Olenchock BA, Lichtman AH, Roden DM, Seidman CE, Koralnik IJ, Seidman JG, Hoffman RD, Taube JM, Diaz LA Jr,

Anders RA, Sosman JA, Moslehi JJ (2016a) Fulminant myocarditis with combination immune checkpoint blockade. N Engl J Med 375(18):1749–1755. doi:10.1056/NEJMoa1609214

Johnson DB, Sullivan RJ, Ott PA, Carlino MS, Khushalani NI, Ye F, Guminski A, Puzanov I, Lawrence DP, Buchbinder EI, Mudigonda T, Spencer K, Bender C, Lee J, Kaufman HL, Menzies AM, Hassel JC, Mehnert JM, Sosman JA, Long GV, Clark JI (2016b) Ipilimumab therapy in patients with advanced melanoma and preexisting autoimmune disorders. JAMA Oncol 2(2):234–240. doi:10.1001/jamaoncol.2015.4368

Jones RB, Walker BD (2016) HIV-specific CD8(+) T cells and HIV eradication. J Clin Invest 126 (2):455–463. doi:10.1172/JCI80566

Kader M, Wang X, Piatak M, Lifson J, Roederer M, Veazey R, Mattapallil JJ (2009) Alpha4(+) beta7(hi) CD4(+) memory T cells harbor most Th-17 cells and are preferentially infected during acute SIV infection. Mucosal Immunol 2(5):439–449. doi:10.1038/mi.2009.90

Kassu A, Marcus RA, D'Souza MB, Kelly-McKnight EA, Golden-Mason L, Akkina R, Fontenot AP, Wilson CC, Palmer BE (2010) Regulation of virus-specific CD4+ T cell function by multiple costimulatory receptors during chronic HIV infection. J Immunol 185(5):3007–3018. doi:10.4049/jimmunol.1000156

Kaufmann DE, Kavanagh DG, Pereyra F, Zaunders JJ, Mackey EW, Miura T, Palmer S, Brockman M, Rathod A, Piechocka-Trocha A, Baker B, Zhu B, Le Gall S, Waring MT, Ahern R, Moss K, Kelleher AD, Coffin JM, Freeman GJ, Rosenberg ES, Walker BD (2007) Upregulation of CTLA-4 by HIV-specific CD4+ T cells correlates with disease progression and defines a reversible immune dysfunction. Nat Immunol 8(11):1246–1254. doi:10.1038/ni1515

Keele BF, Giorgi EE, Salazar-Gonzalez JF, Decker JM, Pham KT, Salazar MG, Sun C, Grayson T, Wang S, Li H, Wei X, Jiang C, Kirchherr JL, Gao F, Anderson JA, Ping LH, Swanstrom R, Tomaras GD, Blattner WA, Goepfert PA, Kilby JM, Saag MS, Delwart EL, Busch MP, Cohen MS, Montefiori DC, Haynes BF, Gaschen B, Athreya GS, Lee HY, Wood N, Seoighe C, Perelson AS, Bhattacharya T, Korber BT, Hahn BH, Shaw GM (2008) Identification and characterization of transmitted and early founder virus envelopes in primary HIV-1 infection. Proc Natl Acad Sci U S A 105(21):7552–7557. doi:10.1073/pnas. 0802203105

Kinloch-de Loes S, Hoen B, Smith DE, Autran B, Lampe FC, Phillips AN, Goh LE, Andersson J, Tsoukas C, Sonnerborg A, Tambussi G, Girard PM, Bloch M, Battegay M, Carter N, El Habib R, Theofan G, Cooper DA, Perrin L, Group QS (2005) Impact of therapeutic immunization on HIV-1 viremia after discontinuation of antiretroviral therapy initiated during acute infection. J Infect Dis 192(4):607–617. doi:10.1086/432002

Klein MR, van Baalen CA, Holwerda AM, Kerkhof Garde SR, Bende RJ, Keet IP, Eeftinck-Schattenkerk JK, Osterhaus AD, Schuitemaker H, Miedema F (1995) Kinetics of Gag-specific cytotoxic T lymphocyte responses during the clinical course of HIV-1 infection: a longitudinal analysis of rapid progressors and long-term asymptomatics. J Exp Med 181 (4):1365–1372

Ko SY, Pegu A, Rudicell RS, Yang ZY, Joyce MG, Chen X, Wang K, Bao S, Kraemer TD, Rath T, Zeng M, Schmidt SD, Todd JP, Penzak SR, Saunders KO, Nason MC, Haase AT, Rao SS, Blumberg RS, Mascola JR, Nabel GJ (2014) Enhanced neonatal Fc receptor function improves protection against primate SHIV infection. Nature 514(7524):642–645. doi:10.1038/nature13612

Kong R, Louder MK, Wagh K, Bailer RT, deCamp A, Greene K, Gao H, Taft JD, Gazumyan A, Liu C, Nussenzweig MC, Korber B, Montefiori DC, Mascola JR (2015) Improving neutralization potency and breadth by combining broadly reactive HIV-1 antibodies targeting major neutralization epitopes. J Virol 89(5):2659–2671. doi:10.1128/JVI.03136-14

Kumar AM, Borodowsky I, Fernandez B, Gonzalez L, Kumar M (2007) Human immunodeficiency virus type 1 RNA Levels in different regions of human brain: quantification using real-time reverse transcriptase-polymerase chain reaction. J Neurovirol 13(3):210–224. doi:10.1080/13550280701327038

Kwong PD, Mascola JR, Nabel GJ (2013) Broadly neutralizing antibodies and the search for an HIV-1 vaccine: the end of the beginning. Nat Rev Immunol 13(9):693–701. doi:10.1038/nri3516

Landais E, Huang X, Havenar-Daughton C, Murrell B, Price MA, Wickramasinghe L, Ramos A, Bian CB, Simek M, Allen S, Karita E, Kilembe W, Lakhi S, Inambao M, Kamali A, Sanders EJ, Anzala O, Edward V, Bekker LG, Tang J, Gilmour J, Kosakovsky-Pond SL, Phung P, Wrin T, Crotty S, Godzik A, Poignard P (2016) Broadly neutralizing antibody responses in a large longitudinal sub-saharan HIV primary infection cohort. PLoS Pathog 12 (1):e1005369. doi:10.1371/journal.ppat.1005369

Langford D, Marquie-Beck J, de Almeida S, Lazzaretto D, Letendre S, Grant I, McCutchan JA, Masliah E, Ellis RJ (2006) Relationship of antiretroviral treatment to postmortem brain tissue viral load in human immunodeficiency virus-infected patients. J Neurovirol 12(2):100–107. doi:10.1080/13550280600713932

Lazar GA, Dang W, Karki S, Vafa O, Peng JS, Hyun L, Chan C, Chung HS, Eivazi A, Yoder SC, Vielmetter J, Carmichael DF, Hayes RJ, Dahiyat BI (2006) Engineered antibody Fc variants with enhanced effector function. Proc Natl Acad Sci U S A 103(11):4005–4010. doi:10.1073/pnas.0508123103

Lecuroux C, Girault I, Boutboul F, Urrutia A, Goujard C, Meyer L, Lambotte O, Chaix ML, Martinez V, Autran B, Sinet M, Venet A, Anrs Primo Cohort AHICSG, Cohort AA, Group AHS (2009a) Antiretroviral therapy initiation during primary HIV infection enhances both CD127 expression and the proliferative capacity of HIV-specific CD8+ T cells. Aids 23 (13):1649–1658. doi:10.1097/QAD.0b013e32832e6634

Lecuroux C, Girault I, Urrutia A, Doisne JM, Deveau C, Goujard C, Meyer L, Sinet M, Venet A (2009b) Identification of a particular HIV-specific CD8+ T-cell subset with a CD27 + CD45RO-/RA+ phenotype and memory characteristics after initiation of HAART during acute primary HIV infection. Blood 113(14):3209–3217. doi:10.1182/blood-2008-07-167601

Leong YA, Chen Y, Ong HS, Wu D, Man K, Deleage C, Minnich M, Meckiff BJ, Wei Y, Hou Z, Zotos D, Fenix KA, Atnerkar A, Preston S, Chipman JG, Beilman GJ, Allison CC, Sun L, Wang P, Xu J, Toe JG, Lu HK, Tao Y, Palendira U, Dent AL, Landay AL, Pellegrini M, Comerford I, McColl SR, Schacker TW, Long HM, Estes JD, Busslinger M, Belz GT, Lewin SR, Kallies A, Yu D (2016) CXCR5(+) follicular cytotoxic T cells control viral infection in B cell follicles. Nat Immunol 17(10):1187–1196. doi:10.1038/ni.3543

Letendre S, Marquie-Beck J, Capparelli E, Best B, Clifford D, Collier AC, Gelman BB, McArthur JC, McCutchan JA, Morgello S, Simpson D, Grant I, Ellis RJ, Group C (2008) Validation of the CNS penetration-effectiveness rank for quantifying antiretroviral penetration into the central nervous system. Arch Neurol 65(1):65–70. doi:10.1001/archneurol.2007.31

Letourneau S, Im EJ, Mashishi T, Brereton C, Bridgeman A, Yang H, Dorrell L, Dong T, Korber B, McMichael AJ, Hanke T (2007) Design and pre-clinical evaluation of a universal HIV-1 vaccine. PLoS ONE 2(10):e984. doi:10.1371/journal.pone.0000984

Levy Y, Thiebaut R, Montes M, Lacabaratz C, Sloan L, King B, Perusat S, Harrod C, Cobb A, Roberts LK, Surenaud M, Boucherie C, Zurawski S, Delaugerre C, Richert L, Chene G, Banchereau J, Palucka K (2014) Dendritic cell-based therapeutic vaccine elicits polyfunctional HIV-specific T-cell immunity associated with control of viral load. Eur J Immunol 44(9):2802–2810. doi:10.1002/eji.201344433

Li JZ, Etemad B, Ahmed H, Aga E, Bosch RJ, Mellors JW, Kuritzkes DR, Lederman MM, Para M, Gandhi RT (2016) The size of the expressed HIV reservoir predicts timing of viral rebound after treatment interruption. Aids 30(3):343–353. doi:10.1097/QAD.0000000000000953

Liu MK, Hawkins N, Ritchie AJ, Ganusov VV, Whale V, Brackenridge S, Li H, Pavlicek JW, Cai F, Rose-Abrahams M, Treurnicht F, Hraber P, Riou C, Gray C, Ferrari G, Tanner R, Ping LH, Anderson JA, Swanstrom R, B CC, Cohen M, Karim SS, Haynes B, Borrow P, Perelson AS, Shaw GM, Hahn BH, Williamson C, Korber BT, Gao F, Self S, McMichael A, Goonetilleke N (2013) Vertical T cell immunodominance and epitope entropy determine HIV-1 escape. J Clin Invest 123(1):380–393. doi:10.1172/JCI65330

Lorenzo-Redondo R, Fryer HR, Bedford T, Kim EY, Archer J, Kosakovsky Pond SL, Chung YS, Penugonda S, Chipman JG, Fletcher CV, Schacker TW, Malim MH, Rambaut A, Haase AT, McLean AR, Wolinsky SM (2016) Persistent HIV-1 replication maintains the tissue reservoir during therapy. Nature 530(7588):51–56. doi:10.1038/nature16933

Loret EP, Darque A, Jouve E, Loret EA, Nicolino-Brunet C, Morange S, Castanier E, Casanova J, Caloustian C, Bornet C, Coussirou J, Boussetta J, Couallier V, Blin O, Dussol B, Ravaux I (2016) Intradermal injection of a Tat Oyi-based therapeutic HIV vaccine reduces of 1.5 log copies/mL the HIV RNA rebound median and no HIV DNA rebound following cART interruption in a phase I/II randomized controlled clinical trial. Retrovirology 13:21. doi:10.1186/s12977-016-0251-3

Lu CL, Murakowski DK, Bournazos S, Schoofs T, Sarkar D, Halper-Stromberg A, Horwitz JA, Nogueira L, Golijanin J, Gazumyan A, Ravetch JV, Caskey M, Chakraborty AK, Nussenzweig MC (2016) Enhanced clearance of HIV-1-infected cells by broadly neutralizing antibodies against HIV-1 in vivo. Science 352(6288):1001–1004. doi:10.1126/science.aaf1279

Macatangay BJ, Riddler SA, Wheeler ND, Spindler J, Lawani M, Hong F, Buffo MJ, Whiteside TL, Kearney MF, Mellors JW, Rinaldo CR (2016) Therapeutic vaccination with dendritic cells loaded with autologous HIV type 1-infected apoptotic cells. J Infect Dis 213(9):1400–1409. doi:10.1093/infdis/jiv582

Mellor AL, Munn DH (2004) IDO expression by dendritic cells: tolerance and tryptophan catabolism. Nat Rev Immunol 4(10):762–774. doi:10.1038/nri1457

Menzies AM, Johnson DB, Ramanujam S, Atkinson VG, Wong AN, Park JJ, McQuade JL, Shoushtari AN, Tsai KK, Eroglu Z, Klein O, Hassel JC, Sosman JA, Guminski A, Sullivan RJ, Ribas A, Carlino MS, Davies MA, Sandhu SK, Long GV (2016) Anti-PD-1 therapy in patients with advanced melanoma and preexisting autoimmune disorders or major toxicity with ipilimumab. Ann Oncol. doi:10.1093/annonc/mdw443

Merad M, Sathe P, Helft J, Miller J, Mortha A (2013) The dendritic cell lineage: ontogeny and function of dendritic cells and their subsets in the steady state and the inflamed setting. Annu Rev Immunol 31:563–604. doi:10.1146/annurev-immunol-020711-074950

Migueles SA, Connors M (2015) Success and failure of the cellular immune response against HIV-1. Nat Immunol 16(6):563–570. doi:10.1038/ni.3161

Migueles SA, Sabbaghian MS, Shupert WL, Bettinotti MP, Marincola FM, Martino L, Hallahan CW, Selig SM, Schwartz D, Sullivan J, Connors M (2000) HLA B*5701 is highly associated with restriction of virus replication in a subgroup of HIV-infected long term nonprogressors. Proc Natl Acad Sci U S A 97(6):2709–2714. doi:10.1073/pnas.050567397

Migueles SA, Weeks KA, Nou E, Berkley AM, Rood JE, Osborne CM, Hallahan CW, Cogliano-Shutta NA, Metcalf JA, McLaughlin M, Kwan R, Mican JM, Davey RT Jr, Connors M (2009) Defective human immunodeficiency virus-specific CD8+ T-cell polyfunctionality, proliferation, and cytotoxicity are not restored by antiretroviral therapy. J Virol 83(22):11876–11889. doi:10.1128/JVI.01153-09

Mildner A, Jung S (2014) Development and function of dendritic cell subsets. Immunity 40(5):642–656. doi:10.1016/j.immuni.2014.04.016

Miller E, Bhardwaj N (2013) Dendritic cell dysregulation during HIV-1 infection. Immunol Rev 254(1):170–189. doi:10.1111/imr.12082

Moore GL, Chen H, Karki S, Lazar GA (2010) Engineered Fc variant antibodies with enhanced ability to recruit complement and mediate effector functions. MAbs 2(2):181–189

Mothe B, Moltó J, Manzardo C, Coll J, Puertas MC, Martinez-Picado J, Hanke T, Clotet B, Brander C (2017) Viral control induced by HIVconsv vaccines & romidepsin in early treated individuals. Paper presented at the CROI, Seattle

Mylvaganam GH, Silvestri G, Amara RR (2015) HIV therapeutic vaccines: moving towards a functional cure. Curr Opin Immunol 35:1–8. doi:10.1016/j.coi.2015.05.001

Ndhlovu ZM, Kamya P, Mewalal N, Kloverpris HN, Nkosi T, Pretorius K, Laher F, Ogunshola F, Chopera D, Shekhar K, Ghebremichael M, Ismail N, Moodley A, Malik A, Leslie A, Goulder PJ, Buus S, Chakraborty A, Dong K, Ndung'u T, Walker BD (2015) Magnitude and kinetics of CD8+ T cell activation during hyperacute HIV infection impact viral set point. Immunity 43(3):591–604. doi:10.1016/j.immuni.2015.08.012

Neefjes J, Jongsma ML, Paul P, Bakke O (2011) Towards a systems understanding of MHC class I and MHC class II antigen presentation. Nat Rev Immunol 11(12):823–836. doi:10.1038/nri3084

Nishimura Y, Gautam R, Chun TW, Sadjadpour R, Foulds KE, Shingai M, Klein F, Gazumyan A, Golijanin J, Donaldson M, Donau OK, Plishka RJ, Buckler-White A, Seaman MS, Lifson JD, Koup RA, Fauci AS, Nussenzweig MC, Martin MA (2017) Early antibody therapy can induce long-lasting immunity to SHIV. Nature 543(7646):559–563. doi:10.1038/nature21435

Ogg GS, Jin X, Bonhoeffer S, Moss P, Nowak MA, Monard S, Segal JP, Cao Y, Rowland-Jones SL, Hurley A, Markowitz M, Ho DD, McMichael AJ, Nixon DF (1999) Decay kinetics of human immunodeficiency virus-specific effector cytotoxic T lymphocytes after combination antiretroviral therapy. J Virol 73(1):797–800

Pantaleo G, Levy Y (2016) Therapeutic vaccines and immunological intervention in HIV infection: a paradigm change. Curr Opin HIV AIDS 11(6):576–584. doi:10.1097/COH.0000000000000324

Pardridge WM (2015) Blood-brain barrier drug delivery of IgG fusion proteins with a transferrin receptor monoclonal antibody. Expert Opin Drug Deliv 12(2):207–222. doi:10.1517/17425247.2014.952627

Pegu A, Asokan M, Wu L, Wang K, Hataye J, Casazza JP, Guo X, Shi W, Georgiev I, Zhou T, Chen X, O'Dell S, Todd JP, Kwong PD, Rao SS, Yang ZY, Koup RA, Mascola JR, Nabel GJ (2015) Activation and lysis of human CD4 cells latently infected with HIV-1. Nat Commun 6:8447. doi:10.1038/ncomms9447

Peluso MJ, Ferretti F, Peterson J, Lee E, Fuchs D, Boschini A, Gisslen M, Angoff N, Price RW, Cinque P, Spudich S (2012) Cerebrospinal fluid HIV escape associated with progressive neurologic dysfunction in patients on antiretroviral therapy with well controlled plasma viral load. Aids 26(14):1765–1774. doi:10.1097/QAD.0b013e328355e6b2

Peretz Y, He Z, Shi Y, Yassine-Diab B, Goulet JP, Bordi R, Filali-Mouhim A, Loubert JB, El-Far M, Dupuy FP, Boulassel MR, Tremblay C, Routy JP, Bernard N, Balderas R, Haddad EK, Sekaly RP (2012) CD160 and PD-1 co-expression on HIV-specific CD8 T cells defines a subset with advanced dysfunction. PLoS Pathog 8(8):e1002840. doi:10.1371/journal.ppat.1002840

Petrovas C, Ferrando-Martinez S, Gerner MY, Casazza JP, Pegu A, Deleage C, Cooper A, Hataye J, Andrews S, Ambrozak D, Del Rio Estrada PM, Boritz E, Paris R, Moysi E, Boswell KL, Ruiz-Mateos E, Vagios I, Leal M, Ablanedo-Terrazas Y, Rivero A, Gonzalez-Hernandez LA, McDermott AB, Moir S, Reyes-Teran G, Docobo F, Pantaleo G, Douek DC, Betts MR, Estes JD, Germain RN, Mascola JR, Koup RA (2017) Follicular CD8 T cells accumulate in HIV infection and can kill infected cells in vitro via bispecific antibodies. Sci Transl Med 9(373). doi:10.1126/scitranslmed.aag2285

Pollard RB, Rockstroh JK, Pantaleo G, Asmuth DM, Peters B, Lazzarin A, Garcia F, Ellefsen K, Podzamczer D, van Lunzen J, Arasteh K, Schurmann D, Clotet B, Hardy WD, Mitsuyasu R, Moyle G, Plettenberg A, Fisher M, Fatkenheuer G, Fischl M, Taiwo B, Baksaas I, Jolliffe D, Persson S, Jelmert O, Hovden AO, Sommerfelt MA, Wendel-Hansen V, Sorensen B (2014) Safety and efficacy of the peptide-based therapeutic vaccine for HIV-1, Vacc-4x: a phase 2 randomised, double-blind, placebo-controlled trial. Lancet Infect Dis 14(4):291–300. doi:10.1016/S1473-3099(13)70343-8

Quigley MF, Gonzalez VD, Granath A, Andersson J, Sandberg JK (2007) CXCR5+ CCR7-CD8 T cells are early effector memory cells that infiltrate tonsil B cell follicles. Eur J Immunol 37(12):3352–3362. doi:10.1002/eji.200636746

Rasmussen TA, Lewin SR (2016) Shocking HIV out of hiding: where are we with clinical trials of latency reversing agents? Curr Opin HIV AIDS 11(4):394–401. doi:10.1097/COH.0000000000000279

Richards JO, Karki S, Lazar GA, Chen H, Dang W, Desjarlais JR (2008) Optimization of antibody binding to FcgammaRIIa enhances macrophage phagocytosis of tumor cells. Mol Cancer Ther 7(8):2517–2527. doi:10.1158/1535-7163.MCT-08-0201

Richman DD, Wrin T, Little SJ, Petropoulos CJ (2003) Rapid evolution of the neutralizing antibody response to HIV type 1 infection. Proc Natl Acad Sci U S A 100(7):4144–4149. doi:10.1073/pnas.0630530100

Riddler S, Durand C, Zheng L, Ritz J, Koup RA, Ledgerwood J, Macatangay B, Cyktor JC, Mellors JW (2017) VRC01 Infusion has no effect on HIV-1 persistence in ART-suppressed chronic infection. In: Conference on Retroviruses and Opportunistic Infections

Rolland M, Nickle DC, Mullins JI (2007) HIV-1 group M conserved elements vaccine. PLoS Pathog 3(11):e157. doi:10.1371/journal.ppat.0030157

Sadanand S, Suscovich TJ, Alter G (2016) Broadly neutralizing antibodies against HIV: new insights to inform vaccine design. Annu Rev Med 67:185–200. doi:10.1146/annurev-med-091014-090749

Saez-Cirion A, Bacchus C, Hocqueloux L, Avettand-Fenoel V, Girault I, Lecuroux C, Potard V, Versmisse P, Melard A, Prazuck T, Descours B, Guergnon J, Viard JP, Boufassa F, Lambotte O, Goujard C, Meyer L, Costagliola D, Venet A, Pancino G, Autran B, Rouzioux C, Group AVS (2013) Post-treatment HIV-1 controllers with a long-term virological remission after the interruption of early initiated antiretroviral therapy ANRS VISCONTI Study. PLoS Pathog 9(3):e1003211. doi:10.1371/journal.ppat.1003211

Sather DN, Armann J, Ching LK, Mavrantoni A, Sellhorn G, Caldwell Z, Yu X, Wood B, Self S, Kalams S, Stamatatos L (2009) Factors associated with the development of cross-reactive neutralizing antibodies during human immunodeficiency virus type 1 infection. J Virol 83(2):757–769. doi:10.1128/JVI.02036-08

Schaerli P, Willimann K, Lang AB, Lipp M, Loetscher P, Moser B (2000) CXC chemokine receptor 5 expression defines follicular homing T cells with B cell helper function. J Exp Med 192(11):1553–1562

Scheid JF, Horwitz JA, Bar-On Y, Kreider EF, Lu CL, Lorenzi JC, Feldmann A, Braunschweig M, Nogueira L, Oliveira T, Shimeliovich I, Patel R, Burke L, Cohen YZ, Hadrigan S, Settler A, Witmer-Pack M, West AP Jr, Juelg B, Keler T, Hawthorne T, Zingman B, Gulick RM, Pfeifer N, Learn GH, Seaman MS, Bjorkman PJ, Klein F, Schlesinger SJ, Walker BD, Hahn BH, Nussenzweig MC (2016) HIV-1 antibody 3BNC117 suppresses viral rebound in humans during treatment interruption. Nature 535(7613):556–560. doi:10.1038/nature18929

Schmitz JE, Kuroda MJ, Santra S, Sasseville VG, Simon MA, Lifton MA, Racz P, Tenner-Racz K, Dalesandro M, Scallon BJ, Ghrayeb J, Forman MA, Montefiori DC, Rieber EP, Letvin NL, Reimann KA (1999) Control of viremia in simian immunodeficiency virus infection by CD8 + lymphocytes. Science 283(5403):857–860

Schnepp BC, Johnson PR (2014) Adeno-associated virus delivery of broadly neutralizing antibodies. Curr Opin HIV AIDS 9(3):250–256. doi:10.1097/COH.0000000000000056

Schoofs T, Klein F, Braunschweig M, Kreider EF, Feldmann A, Nogueira L, Oliveira T, Lorenzi JC, Parrish EH, Learn GH, West AP Jr, Bjorkman PJ, Schlesinger SJ, Seaman MS, Czartoski J, McElrath MJ, Pfeifer N, Hahn BH, Caskey M, Nussenzweig MC (2016) HIV-1 therapy with monoclonal antibody 3BNC117 elicits host immune responses against HIV-1. Science 352(6288):997–1001. doi:10.1126/science.aaf0972

Schooley RT, Spritzler J, Wang H, Lederman MM, Havlir D, Kuritzkes DR, Pollard R, Battaglia C, Robertson M, Mehrotra D, Casimiro D, Cox K, Schock B, Team ACTGS (2010) AIDS clinical trials group 5197: a placebo-controlled trial of immunization of HIV-1-infected persons with a replication-deficient adenovirus type 5 vaccine expressing the HIV-1 core protein. J Infect Dis 202(5):705–716. doi:10.1086/655468

Shan L, Deng K, Shroff NS, Durand CM, Rabi SA, Yang HC, Zhang H, Margolick JB, Blankson JN, Siliciano RF (2012) Stimulation of HIV-1-specific cytolytic T lymphocytes facilitates elimination of latent viral reservoir after virus reactivation. Immunity 36(3):491–501. doi:10.1016/j.immuni.2012.01.014

Siliciano JD, Kajdas J, Finzi D, Quinn TC, Chadwick K, Margolick JB, Kovacs C, Gange SJ, Siliciano RF (2003) Long-term follow-up studies confirm the stability of the latent reservoir for HIV-1 in resting CD4+ T cells. Nat Med 9(6):727–728. doi:10.1038/nm880

Simek MD, Rida W, Priddy FH, Pung P, Carrow E, Laufer DS, Lehrman JK, Boaz M, Tarragona-Fiol T, Miiro G, Birungi J, Pozniak A, McPhee DA, Manigart O, Karita E, Inwoley A, Jaoko W, Dehovitz J, Bekker LG, Pitisuttithum P, Paris R, Walker LM, Poignard P, Wrin T, Fast PE, Burton DR, Koff WC (2009) Human immunodeficiency virus type 1 elite neutralizers: individuals with broad and potent neutralizing activity identified by using a high-throughput neutralization assay together with an analytical selection algorithm. J Virol 83(14):7337–7348. doi:10.1128/JVI.00110-09

Sloan DD, Lam CY, Irrinki A, Liu L, Tsai A, Pace CS, Kaur J, Murry JP, Balakrishnan M, Moore PA, Johnson S, Nordstrom JL, Cihlar T, Koenig S (2015) Targeting HIV reservoir in infected CD4 T cells by dual-affinity re-targeting molecules (DARTs) that bind HIV envelope and recruit cytotoxic T cells. PLoS Pathog 11(11):e1005233. doi:10.1371/journal.ppat.1005233

Smit TK, Brew BJ, Tourtellotte W, Morgello S, Gelman BB, Saksena NK (2004) Independent evolution of human immunodeficiency virus (HIV) drug resistance mutations in diverse areas of the brain in HIV-infected patients, with and without dementia, on antiretroviral treatment. J Virol 78(18):10133–10148. doi:10.1128/JVI.78.18.10133-10148.2004

Spudich S, Lollo N, Liegler T, Deeks SG, Price RW (2006) Treatment benefit on cerebrospinal fluid HIV-1 levels in the setting of systemic virological suppression and failure. J Infect Dis 194(12):1686–1696. doi:10.1086/508750

Sung JA, Lam S, Garrido C, Archin N, Rooney CM, Bollard CM, Margolis DM (2015a) Expanded cytotoxic T-cell lymphocytes target the latent HIV reservoir. J Infect Dis 212(2):258–263. doi:10.1093/infdis/jiv022

Sung JA, Pickeral J, Liu L, Stanfield-Oakley SA, Lam CY, Garrido C, Pollara J, LaBranche C, Bonsignori M, Moody MA, Yang Y, Parks R, Archin N, Allard B, Kirchherr J, Kuruc JD, Gay CL, Cohen MS, Ochsenbauer C, Soderberg K, Liao HX, Montefiori D, Moore P, Johnson S, Koenig S, Haynes BF, Nordstrom JL, Margolis DM, Ferrari G (2015b) Dual-Affinity Re-Targeting proteins direct T cell-mediated cytolysis of latently HIV-infected cells. J Clin Invest 125(11):4077–4090. doi:10.1172/JCI82314

Takata H, Buranapraditkun S, Kessing C, Fletcher JL, Muir R, Tardif V, Cartwright P, Vandergeeten C, Bakeman W, Nichols CN, Pinyakorn S, Hansasuta P, Kroon E, Chalermchai T, O'Connell R, Kim J, Phanuphak N, Robb ML, Michael NL, Chomont N, Haddad EK, Ananworanich J, Trautmann L, Rv254/Search, the RVSSG (2017) Delayed differentiation of potent effector CD8+ T cells reducing viremia and reservoir seeding in acute HIV infection. Sci Transl Med 9 (377). doi:10.1126/scitranslmed.aag1809

Thompson M, Heath SL, Sweeton B, Williams K, Cunningham P, Keele BF, Sen S, Palmer BE, Chomont N, Xu Y, Basu R, Hellerstein MS, Kwa S, Robinson HL (2016) DNA/MVA vaccination of HIV-1 infected participants with viral suppression on antiretroviral therapy, followed by treatment interruption: elicitation of immune responses without control of re-emergent virus. PLoS ONE 11(10):e0163164. doi:10.1371/journal.pone.0163164

Thornhill J, Inshaw J, Kaleebu P, Cooper D, Ramjee G, Schechter M, Tambussi G, Fox J, Samuel M, Miro JM, Weber J, Porter K, Fidler S (2016) Brief report: enhanced normalization of CD4/CD8 ratio with earlier antiretroviral therapy at primary HIV infection. J Acquir Immune Defic Syndr 73(1):69–73. doi:10.1097/QAI.0000000000001013

Trautmann L, Janbazian L, Chomont N, Said EA, Gimmig S, Bessette B, Boulassel MR, Delwart E, Sepulveda H, Balderas RS, Routy JP, Haddad EK, Sekaly RP (2006) Upregulation of PD-1 expression on HIV-specific CD8+ T cells leads to reversible immune dysfunction. Nat Med 12(10):1198–1202. doi:10.1038/nm1482

Tubiana R, Carcelain G, Vray M, Gourlain K, Dalban C, Chermak A, Rabian C, Vittecoq D, Simon A, Bouvet E, El Habib R, Costagliola D, Calvez V, Autran B, Katlama C, Vacciter Study g (2005) Therapeutic immunization with a human immunodeficiency virus (HIV) type

1-recombinant canarypox vaccine in chronically HIV-infected patients: the vacciter study (ANRS 094). Vaccine 23(34):4292–4301. doi:10.1016/j.vaccine.2005.04.013

Tung FY, Tung JK, Pallikkuth S, Pahwa S, Fischl MA (2016) A therapeutic HIV-1 vaccine enhances anti-HIV-1 immune responses in patients under highly active antiretroviral therapy. Vaccine 34(19):2225–2232. doi:10.1016/j.vaccine.2016.03.021

UNAIDS (2014) The gap report. http://files.unaids.org/en/media/unaids/contentassets/documents/unaidspublication/2014/UNAIDS_Gap_report_en.pdf. 2016

UNAIDS (2016) Gobal AIDS update 2016. http://www.who.int/hiv/pub/arv/global-aids-update-2016-pub/en/. 2016

Velu V, Titanji K, Zhu B, Husain S, Pladevega A, Lai L, Vanderford TH, Chennareddi L, Silvestri G, Freeman GJ, Ahmed R, Amara RR (2009) Enhancing SIV-specific immunity in vivo by PD-1 blockade. Nature 458(7235):206–210. doi:10.1038/nature07662

Vinuesa CG, Cyster JG (2011) How T cells earn the follicular rite of passage. Immunity 35(5):671–680. doi:10.1016/j.immuni.2011.11.001

Wei X, Decker JM, Wang S, Hui H, Kappes JC, Wu X, Salazar-Gonzalez JF, Salazar MG, Kilby JM, Saag MS, Komarova NL, Nowak MA, Hahn BH, Kwong PD, Shaw GM (2003) Antibody neutralization and escape by HIV-1. Nature 422(6929):307–312. doi:10.1038/nature01470

Williams JP, Hurst J, Stohr W, Robinson N, Brown H, Fisher M, Kinloch S, Cooper D, Schechter M, Tambussi G, Fidler S, Carrington M, Babiker A, Weber J, Koelsch KK, Kelleher AD, Phillips RE, Frater J, Investigators SP (2014) HIV-1 DNA predicts disease progression and post-treatment virological control. Elife 3:e03821. doi:10.7554/eLife.03821

Wong JK, Yukl SA (2016) Tissue reservoirs of HIV. Curr Opin HIV AIDS 11(4):362–370. doi:10.1097/COH.0000000000000293

Yamamoto T, Price DA, Casazza JP, Ferrari G, Nason M, Chattopadhyay PK, Roederer M, Gostick E, Katsikis PD, Douek DC, Haubrich R, Petrovas C, Koup RA (2011) Surface expression patterns of negative regulatory molecules identify determinants of virus-specific CD8+ T-cell exhaustion in HIV infection. Blood 117(18):4805–4815. doi:10.1182/blood-2010-11-317297

Yu YJ, Atwal JK, Zhang Y, Tong RK, Wildsmith KR, Tan C, Bien-Ly N, Hersom M, Maloney JA, Meilandt WJ, Bumbaca D, Gadkar K, Hoyte K, Luk W, Lu Y, Ernst JA, Scearce-Levie K, Couch JA, Dennis MS, Watts RJ (2014) Therapeutic bispecific antibodies cross the blood-brain barrier in nonhuman primates. Sci Transl Med 6(261):261ra154. doi:10.1126/scitranslmed.3009835

# Cell and Gene Therapy for HIV Cure

**Christopher W. Peterson and Hans-Peter Kiem**

**Abstract** As the HIV pandemic rapidly spread worldwide in the 1980s and 1990s, a new approach to treat cancer, genetic diseases, and infectious diseases was also emerging. Cell and gene therapy strategies are connected with human pathologies at a fundamental level, by delivering DNA and RNA molecules that could correct and/or ameliorate the underlying genetic factors of any illness. The history of HIV gene therapy is especially intriguing, in that the virus that was targeted was soon co-opted to become part of the targeting strategy. Today, HIV-based lentiviral vectors, along with many other gene delivery strategies, have been used to evaluate HIV cure approaches in cell culture, small and large animal models, and in patients. Here, we trace HIV cell and gene therapy from the earliest clinical trials, using genetically unmodified cell products from the patient or from matched donors, through current state-of-the-art strategies. These include engineering HIV-specific immunity in T-cells, gene editing approaches to render all blood cells in the body HIV-resistant, and most importantly, combination therapies that draw from both of these respective "offensive" and "defensive" approaches. It is widely agreed upon that combinatorial approaches are the most promising route to functional cure/remission of HIV infection. This chapter outlines cell and gene therapy strategies that are poised to play an essential role in eradicating HIV-infected cells in vivo.

## Contents

C. W. Peterson · H. -P. Kiem
Fred Hutchinson Cancer Research Center, University of Washington School of Medicine, Seattle, WA 98109-1024, USA

H. -P. Kiem (✉)
Fred Hutchinson Cancer Research Center, 1100 Fairview Avenue N, D1-100, Seattle, WA 98109-1024, USA
e-mail: hkiem@fredhutch.org

Current Topics in Microbiology and Immunology (2018) 417:211–248
DOI 10.1007/82_2017_71
© Springer International Publishing AG 2017
Published Online: 19 December 2017

# 1   Introduction

The advent of cell and gene therapy approaches to treat various genetic and
infectious diseases emerged in parallel with the outbreak of the global HIV pandemic. Similarly, the increased efficacy of cell and gene therapies in clinical trials
has mirrored the vast amounts of information gained regarding the requirements for
an HIV cure. Broadly defined, "Cell Therapy" approaches for human disease
involve the administration of a genetically unmodified cell product into a patient,
whereas "Gene Therapy" approaches involve any delivery of foreign genetic
material to diseased or effector cells. Some of the earliest clinical attempts at gene
therapy were in the early 1970s. In two patients with hyperargininemia, the
administration of wild-type Shope papillomavirus did not ameliorate these patients'
nitrogen metabolism disorders, despite crude predictions regarding the enzymatic
activities of the virus (Friedmann 1992). Subsequent trials for other diseases lacked
efficacy, and also raised bioethical questions regarding the genetic modification of
human cells (Sheridan 2011). As the HIV-1 pandemic reached horrific levels in the
late 1980s and early 1990s, the field of gene therapy made striking advances,
chiefly in the treatment of severe combined immunodeficiency (SCID) syndromes
(Blaese et al. 1995; Bordignon et al. 1995). During this period, many pioneering
laboratories set out to apply gene therapy techniques to HIV-infected patients as
well. Most early strategies involved the use of an engineered viral vector carrying a
transgene of interest, which transduces a targeted cell type, then expresses the
genetic cargo either under the control of a viral or eukaryotic promoter. In parallel,
transplantation biologists designed clinical studies designed to assess the impact of
transplanted cells from $HIV^+$ patients themselves (autologous transplantation), from
identical twins (syngeneic transplantation), or from genetically matched donors
(allogeneic transplantation). In this chapter, we will review the history of cell and

gene therapies for HIV, including therapeutic approaches, mechanisms of various anti-HIV transgenes, and strategies for delivering these molecules to cells. We emphasize treatments that present the greatest promise for future HIV Cure studies, including those that have progressively developed over time, and those that may be worthy of revisiting in what is an explosive and exciting era of human cell and gene therapy research.

# 2 Hematopoietic Cell Transplantation in HIV⁺ Patients

## 2.1 Allogeneic and Syngeneic Cell Therapies in the Pre-cART Era

Recent reports make clear that the impact of cell and gene therapy approaches in HIV⁺ patients should be divided into those attempted prior to the era of suppressive combination antiretroviral therapy (cART), and those conducted in stably cART-suppressed patients, including the functional cure/remission of HIV infection in the Berlin Patient (Hutter et al. 2009). As with many HIV cure approaches, early cell and gene therapy approaches for HIV cure were often complicated by patients' high viral loads and onset of AIDS. As such, the outcomes of these early studies were often ineffective and disappointing. Nevertheless, the findings in these early trials, and their potential reapplication in stably suppressed patients, offer many promising paths forward.

In the pre-cART era, the majority of trials in HIV⁺ patients involved the transplantation of autologous, syngeneic, or allogeneic hematopoietic stem and progenitor cells (HSPCs) or T-cells. The first bone marrow transplantation trial for what would later be described as HIV/AIDS occurred in two patients in 1983, who presented with Kaposi's Sarcoma and a history of opportunistic infections (Hassett et al. 1983). Allogeneic transplantation had no impact in these patients, although notably, a conditioning regimen, which includes cytotoxic chemotherapy or radiation designed to broadly kill leukocytes primarily to "make room" for the transplanted cells, was excluded based on the patients' health status. One year later, a third patient received syngeneic bone marrow and lymphocyte products from his identical twin; although CD4 counts improved, the patient died of CMV infection one year later (Lane et al. 1984). Again, a conditioning regimen was excluded from this patient's treatment, due to the risks of administering a cytotoxic, immuno-suppressive therapy to an immunocompromised, viremic patient. In 1989, the case study of a 41-year-old male infected with HIV-1 and non-Hodgkin lymphoma was reported (Holland et al. 1989). This was the first transplantation trial in an HIV⁺ patient to include effective antiretroviral therapy: an intensive Zidovudine regimen was administered, consisting of an intravenous dose of 5 mg/kg every 4 h for 2 weeks prior to allogeneic bone marrow infusion, then 1.33 mg/kg beginning on the day of bone marrow infusion. This trial also included an initial round of chemotherapy prior to transplantation, and a conditioning regimen including total body irradiation and cyclophosphamide that was administered 1–9 days prior to infusion of bone marrow cells. Donor chimerism measurements (quantifying the extent to which donor cells engrafted in the recipient) showed at least 99.9% donor chimerism by 17 days post-transplant. This patient died 47 days after transplant due to recurrence of the malignancy. HIV was detected in some post-transplant samples, but was absent in many others, leading the authors to posit that eradication may have been achieved at 6 weeks post-transplant. Although HIV$^+$ cells may have persisted below the level of detection available at that time, like many transplantation studies in AIDS Lymphoma patients, recurrence of the malignancy prevented an extended test for rebound HIV viremia. A larger follow-up study in 15 symptomatic and 1 asymptomatic HIV$^+$ patients without associated malignancies was published in 1990 (Lane et al. 1990). Here, donor products were syngeneic, and peripheral blood lymphocyte infusions were combined with bone marrow transplantation. Transplantation-dependent improvements in CD4$^+$ T-cell counts were transient, and, no change in replication-competent virus was observed either in patients ± transplantation, or ± Zidovudine. Although these and other studies (Aboulafia et al. 1991; Torlontano et al. 1992; Turner et al. 1992) suggested that administration of Zidovudine to syngeneic and allogeneic transplant patients was safe and not associated with impaired rates of donor cell engraftment, combinations of viral suppression and cell transplantation protocols did not induce HIV Cure, and in the case of AIDS lymphoma patients, were further complicated by recurrent malignant disease.

## 2.2 Autologous Transplantation in the Pre-cART Era

Early cellular therapies for HIV-infected patients focused on products from matched donors or identical twins. The potential benefits of a "graft versus HIV reservoir" effect from allogeneic donor products were noted as early as 1993 (Contu et al. 1993), although as with these products' use in the treatment of hematological malignancies, it quickly became clear that balancing the benefits and risks of graft versus host disease (GVHD) would be difficult. Although using a patient's own cells offered a safer alternative, trials designed to test the impact of autologous transplantation on HIV viremia were not immediately undertaken. This was mainly due to the fact that without a full understanding of the viral replication cycle and susceptible cellular subsets, readministration of a patient's own cells could also reintroduce infected cells and/or virus. In order for the curative potential of autologous transplantation to be tested, two barriers needed to be overcome.

First, the inherent risks of bone marrow and organ transplantation for HIV acquisition needed to be clarified. Case studies from the early 1990s suggested that patients may have acquired HIV-tainted blood products as early as the 1960s (Jacobson et al. 1991). Bone marrow transplant patients, especially those with hematologic diseases, were at particularly high risk. This was due to increased exposure to potentially tainted blood and cell products, and the unenviable choice of proceeding with transplantation procedures that were essential for the treatment of the genetic disease, versus side effects that often exacerbated HIV viremia (Angelucci et al. 1990; Rubin et al. 1987). Between 1980 and 1993, HIV acquisition had been reported in 50 cases of kidney transplantation, 13 for liver, 6 for heart, 1 for pancreas, 4 for bone, and 1 for skin; the advent of more standardized and rigorous screening protocols for donor blood and tissue products significantly decreased the exposure risk from these products (Simonds 1993).

The other barrier to autologous transplantation for HIV$^+$ patients concerned the ability of bone marrow cells, specifically CD34$^+$ HSPCs, to become infected and carry replication-competent HIV proviruses through the transplantation procedure; this was, and continues to be controversial. Early in vitro experiments suggested that CD34$^+$ cells were HIV targets, but alluded to the rarity of this cell population in vitro and in vivo (Kaushal et al. 1996; Kearns et al. 1997). Later experiments argued that even CD34$^+$ HSPCs expressing CD4, CCR5, and CXCR4 were HIV-resistant (Shen et al. 1999). Most recently, detailed phylogenetic tracking suggests that although rare CD34 infection events may occur in vivo (Onafuwa-Nuga et al. 2010; McNamara et al. 2013), only CD4$^+$ T-cells should be considered a primary target for HIV cure studies (Eisele and Siliciano 2012; Durand et al. 2012).

In 1996, autologous transplantation was performed in an HIV$^+$ patient with non-Hodgkin's lymphoma as a rescue therapy; the advanced stage of HIV disease prior to transplantation was consistent with a poor outcome following transplant (Gabarre et al. 1996). In subsequent HIV/malignancy patients who were in asymptomatic phases of infection, the transplant procedure was determined to be safe, but not efficacious in terms of HIV cure (Schneider et al. 1997; Trickett et al. 1998).

In short, the results of cellular therapies for HIV$^+$ patients in the pre-cART era were mixed. Although patients responded surprisingly well to these interventions, they were limited by the inability to robustly suppress plasma viremia, and by a lack of assays able to detect persistent infected cells with high sensitivity. As described below, a clearer interpretation of these trials emerged as suppressive cART regimens and sophisticated viral reservoir assays were developed in the 2000s and 2010s. Subsequent studies suggested that autologous transplantation alone is unlikely to be curative, and although allogeneic transplantation could lead to large decreases in infected cells, this procedure was incapable of HIV eradication. As such, the implementation of gene therapy approaches on top of cell-based therapies became a primary focus of the HIV gene therapy field.

## 2.3  HSPC Transplantation in Stably cART-Suppressed Patients

The advent of highly active cART in the late 1990s opened doors to more creative and aggressive therapeutic approaches to eradicate HIV. Although gene therapy interventions are frequently employed, the HIV gene and cell therapy field continues to be defined by the clinical cases of three cART-suppressed HIV⁺ patients who did not receive gene-modified cell products. In this section, we review advances in gene and cell therapy-based HIV cure approaches in stably suppressed patients. The specific mechanisms of the various anti-HIV gene therapy strategies will be discussed in depth in Sect. 3.

The ability of HIV⁺ patients to live "otherwise healthy lives" when adhering to potent cART regimens called into question the risks associated with myeloablative conditioning-based transplantation trials. Therefore, infected and suppressed patients that presented with associated hematological malignancies, which did call for more aggressive transplantation protocols, became the modicum for HIV gene therapy trials in the early 2000s (Johnston et al. 2016). In two patients with acute myelogenous leukemia (AML) and Hodgkin disease, respectively, allogeneic HSPC products were transduced with a gammaretroviral vector expressing a transdominant negative mutant of HIV Rev (Kang et al. 2002a, b). As before, one patient was lost to recurrent malignancy; the other died of thrombotic thrombocytopenic purpura over 3 years post-transplant (Hayakawa et al. 2009). HIV eradication was not observed in these patients, and the persistence of gene-modified cells ("gene marking") was low. Low levels of gene marking in vivo were also observed when using an HIV-based lentiviral vector that transduced autologous HSPCs with (i) small hairpin RNA (shRNA) against HIV Tat/Rev, (ii) a decoy molecule against the HIV transactivation response element (TAR), and (iii) a CCR5 RNA-targeting RNA cleaving enzyme (CCR5 ribozyme) (DiGiusto et al. 2010). These and other trials in well-suppressed HIV⁺ patients suggested some correlates of efficacy, but due to the limited efficiencies of gene transfer and persistence of gene-modified cells in vivo, their true potential for HIV cure was unclear.

The seminal publication for the HIV gene therapy field was published in February 2009. This case study detailed the treatment of a 40-year-old male, who would come to be known as the "Berlin Patient." This patient had contracted HIV in the mid 1990s, initiated cART in the early 2000s, and was diagnosed with AML in June of 2006 (Hutter et al. 2009; Hutter and Ganepola 2011). During an initial course of chemotherapy for AML, liver and kidney toxicities developed, which were treated by short withdraw of cART; during this period, plasma viral load rebound was observed. Following readministration of cART, viremia returned to undetectable levels, and the patient underwent allogeneic transplantation in early 2007. Importantly, the allogeneic bone marrow product that the Berlin Patient received was not only MHC-matched, but also carried a homozygous deletion in the HIV-1 coreceptor gene CCR5, known as CCR5Δ32 (Huang et al. 1996). This mutation has an intriguing history, including potential selection in northern Europe during the Bubonic Plague of the Middle Ages (Goldrick 2003). Both CCR5Δ32 carriers and homozygotes are not limited by any major health complications, establishing CCR5 as a primary target for HIV gene therapy strategies. Among the many other remarkable aspects of this case study are (i) that the Berlin Patient was administered his last dose of cART in early 2007, concurrent with his first transplant, and (ii) that like many HIV/malignancy patients, his AML rebounded, necessitating a second transplant from the same CCR5Δ32 donor in March of 2008 (Allers et al. 2011). Following arguably the most aggressive therapy administered to an HIV⁺ patient to date, it has now been more than 10 years since the Berlin Patient's last cART dose, with no replication-competent HIV-detected. Viruses predicted to utilize the alternate HIV coreceptor CXCR4 were present in this patient prior to transplant, but did not grow out post-transplant, potentially due to the fact that these variants still required CCR5 for replication in vitro (Symons et al. 2014). In 2013, a summary of intensive sampling of this patient (peripheral blood, cerebrospinal fluid, lymph nodes, and gut tissue) revealed no strong evidence of any residual HIV RNA or DNA, although qualitative, low-level positives were sporadically detected (Yukl et al. 2013). The authors could not rule out the possibility that these results were false positives, for example, due to low-level laboratory contamination. In short, the Berlin Patient quickly emerged as the first patient to be cured of HIV, although other terms, including "functional cure/remission," have been proposed, analogous to long-term remission in cancer patients.

Despite the excitement upon publication of these results, a key limitation was immediately clear: the allogeneic transplant procedure was toxic and risky, and the ability to find MHC-matched, CCR5Δ32 donors for every HIV⁺ patient worldwide was highly unlikely. The fields of HIV gene and cell therapy and transplantation therefore quickly moved to address a key question: which aspect of the Berlin Patient's cure was most important: (i) the infection-resistant CCR5Δ32 donor cells, (ii) the allogeneic "graft versus reservoir" effect, or (iii) the myeloablative conditioning regimen? The impact of the graft versus reservoir effect was demonstrated in 2013, with the report of the Boston patients (Henrich et al. 2013). These two patients' treatment differed significantly from the Berlin patient in two ways: first, the donor cell product was CCR5 wild type, i.e. not HIV-resistant. Second, the patients received a reduced intensity conditioning regimen, in contrast to the myeloablative regimen received by the Berlin Patient. Importantly, all 3 patients had in common an MHC-matched, allogeneic cell source. Therefore, a comparison of the Boston Patients and the Berlin Patient would directly address the role of allogeneic transplantation versus infection-resistant cells in HIV Cure (Table 1). Although dramatic decreases in the size of the viral reservoir were observed in each of the Boston Patients following transplantation, including undetectable HIV in peripheral blood and gut tissue, cART treatment interruption led to HIV rebound in 12 and 32 weeks, respectively (Henrich et al. 2014). As with the Berlin Patient, viral sequence prediction algorithms identified minority viral variants capable of using alternative coreceptors in vitro, though the relevance of these minority variants was unclear (Henrich et al. 2015). Currently, both patients have resumed cART, and are healthy. The findings from these case studies make clear that a graft versus reservoir effect is likely insufficient to eradicate HIV. Furthermore, the plethora of autologous transplant trials in patients and animal models suggest that although myeloablative conditioning is safe (Alvarnas et al. 2016), it is also insufficient for a cure (Mavigner et al. 2014; Peterson et al. 2017). Collectively, these results suggest that gene-protected cells, either CCR5Δ32 in the case of the Berlin Patient, or a gene-engineered autologous cell product, will be essential to

designing necessary cohort sizes and frequencies of sampling needed in future cure studies (Hill et al. 2016). The contribution of the Berlin and Boston Patients to HIV cure efforts cannot be understated. As the next generation of HSPC transplantation trials for HIV cure develops, the experience of these patients will continue to strongly influence study design and interpretation of results.

## 2.4  Transplantation of Differentiated Hematopoietic Cells

The impact of transplanted T-cells and Dendritic Cells (DCs) has also been extensively evaluated in HIV$^+$ patients. Much like HSPC-based cell therapies, the efficacy of these approaches was initially limited by their application in unsuppressed patients or those treated only with Zidovudine. The earliest T-cell transplants built on the observation that CD8$^+$ cells could suppress infection of CD4$^+$ T-cells ex vivo. In early phase I trials, leukapheresis products were collected from HIV$^+$ patients, and CD8$^+$ cells were collected, activated, expanded, and reinfused; despite escalating cell doses and addition of IL2 stimulation, persistence was not observed (Herberman 1992; Whiteside et al. 1993; Ho et al. 1993; Lieberman et al. 1997; Tan et al. 1999; Klimas 1992). In one intriguing case study, CD8$^+$ T-cells from an uninfected individual that had received an early HIV vaccine were collected, similarly expanded, and infused into his HIV$^+$ identical twin. The cells were delivered in two doses, the first without, and the second with stimulation by the HIV envelope vaccine antigen; although CD4$^+$ counts transiently increased, so too did plasma viral load (Bex et al. 1994). These and other early attempts to infuse autologous or syngeneic T-cells into HIV$^+$ patients were ineffective, and in some cases, were associated with disease progression (Koenig et al. 1995).

Further obstacles were encountered when autologous CD8$^+$ T-cell products were gene-modified; modified cells were rapidly cleared, which was tied to the immunogenicity of the gene therapy vector and transgene (Riddell et al. 1992, 1996; Berger et al. 2006). To increase the in vivo persistence of these products, efforts were undertaken to decrease their immunogenicity (Woffendin et al. 1996; Ranga et al. 1998), and to identify novel vectors and transgenes that were less immunogenic. One such transgene, first reported in 1994, was initially referred to as "Universal Chimeric T-Cell Receptors," but would later come to be known as chimeric antigen receptors (CARs, described in detail below). CD4-CAR transgenes could be delivered to primary CD8$^+$ T-cells (Roberts et al. 1994), persisted in HIV$^+$ patients for months following infusion in an early Phase I/II clinical trial (Mitsuyasu et al. 2000), and persisted for at least 11 years, with correlates of efficacy, in a randomized phase II clinical trial (Deeks et al. 2002; Scholler et al. 2012). Most recently, CD4-CAR transduced HSPCs have also shown promise in a humanized mouse model (Zhen et al. 2015). However, despite continued development of CARs in the HIV setting (for example the addition of CD28 costimulation (Levine et al. 1996), development of HIV antibody-based CARs (Masiero et al. 2005; Ali et al. 2016; Hale et al. 2017b) and the enormous success of CAR

T-cells in patients with hematological malignancies, almost no clinical trials are currently listed for HIV-specific CARs.

More recently, Bollard and colleagues have pioneered a novel T-cell-based gene therapy approach that redirects virus-specific immunity. Their work is based on the observation that the natural progression of adaptive immunity against HIV cannot keep pace with the rate of virus evolution (Patel et al. 2016a). In short, primary DCs are pulsed with HIV peptides and co-cultured with autologous T-cells, generating a polyclonal T-cell pool containing HIV peptide-specific clones. This allows virus-specific T-cells to be generated against any portion of the virus. Intriguingly, when applied to other infectious diseases, this approach generates virus-specific cells with atypical epitope specificities (Hanley et al. 2009, 2015). Redirected T-cell products generated from HIV$^+$ patients (Lam et al. 2015) and from uninfected donors (Patel et al. 2016b) are currently under evaluation in clinical trials. Furthermore, induced pluripotent stem cell (iPSC) approaches are also being applied to "reinvigorate" virus-specific T-cells from infected patients (Nishimura et al. 2013).

Finally, it is important to note that T-cells are not the only other candidate cell type for anti-HIV gene therapy. As described above, DCs are a particularly promising subset to target, in order to redirect/improve anti-HIV immunity. Infusion of HIV peptide-pulsed DCs into HIV$^+$ patients was first attempted in 1998 (Kundu et al. 1998). Similar to early T-cell therapies, the DC infusion was generally well tolerated except for a transient increase in viral load, but showed little efficacy. More recent iterations using autologous, heat-inactivated virus to pulse DCs from HIV$^+$ donors showed modest efficacy in vivo (Garcia et al. 2013; Andres et al. 2015). Modified lentiviral vectors designed to efficiently transduce DCs have also facilitated improved expression of HIV peptides, and are currently being evaluated in vivo (Norton et al. 2015).

# 3  Anti-HIV Gene Modification Strategies

The respective functional cure and substantial but incomplete clearance of viral reservoirs in the Berlin and Boston Patients have set the stage for autologous gene therapy-based strategies to eradicate HIV. These approaches are based on the premise that genetic modification of a patient's own cells will avoid toxicities associated with allogeneic transplantation. Importantly, a complete understanding of the mechanisms underlying the "graft versus reservoir" effect in the allogeneic setting could also lead to novel approaches that maximize the ability to destroy latently infected cells, while minimizing collateral damage to uninfected cells. In this section, we review various transgenic and gene editing-based strategies that are designed to protect cells against infection, and/or actively clear infected cells. In the next section, we will outline the various delivery strategies to introduce these products into cells.

## 3.1 Decoys and Dominant Negatives

One of the first promising in vitro strategies to inhibit HIV replication was to overexpress portions of the virus in *trans*. These included RNA signals required for packaging/virus expression, and dominant negative proteins capable of inhibiting multiple steps in the viral replication cycle. The HIV-1 Tat ("Trans-Activator of Transcription") protein was identified as a key modulator of HIV transcription, via binding to the RNA stem-loop structure named TAR ("Tat activation response element"). By overexpressing this element, replication of HIV and SIV, but not of an unrelated murine retrovirus, was reduced by nearly 100% in a CEM cell line (Sullenger et al. 1990). Other groups focused on HIV-1 Rev ("Regulator of Expression of Virion proteins"), which binds a REV response element (RRE) in HIV transcripts, facilitating their export from the nucleus for protein translation. RRE decoys as short as 13 nucleotides were capable of inhibiting Rev function and viral replication (Lee et al. 1992), although interestingly, their activity differed between CEM cell line clones (Lee et al. 1994). Both TAR and RRE decoy strategies showed promise in primary CD34$^+$ HSPCs in vitro (Bahner et al. 1996; Li et al. 2003). In contrast, efforts to characterize packaging signal (psi sequence) decoys were hampered by the fact that their efficiency did not appear to correlate either with the packaging of the viral genomic RNA, or expression of the decoy (Dorman and Lever 2001). In 1999, a clinical trial in four HIV$^+$ children was conducted, using autologous CD34$^+$ HSPCs that were transduced with a Moloney Murine Leukemia Virus (MoMLV) vector expressing an RRE decoy; these cells were infused without a conditioning regimen (Kohn et al. 1999). Although gene transfer efficiencies were low and gene-marked cells were difficult to measure in the transplanted patients, the demonstration of safety in this early trial was highly valuable for the HIV gene therapy field.

Early protein-based strategies to inhibit HIV included both decoy and dominant negative designs. The best-characterized protein decoy for HIV was soluble CD4; similar to RNA decoys, soluble CD4 protein approaches were designed to saturate the extracellular space with a CD4 ligand that HIV viruses could not utilize to enter a cell. This approach showed early promise in vitro and was valuable in characterizing the mechanisms underlying HIV entry into target cells (Smith et al. 1987). However, more recent experiments suggest that the efficacy and safety of this approach may depend on several factors, including the amount of cell-free virus, and local concentrations of the soluble CD4 protein (Haim et al. 2009; Sullivan et al. 1998). A greater focus has been placed on the generation of proteins that act in a dominant negative fashion to destabilize the nascent virion. These include dominant negative versions of Gag (Trono et al. 1989), Tat (Green et al. 1989; Rustanti et al. 2017), Env (Buchschacher et al. 1992), and Rev (Bevec et al. 1992). In particular, the potency of the RevM10 mutant was extensively characterized in multiple vector systems (Bahner et al. 2007; Plavec et al. 1996). However, this was also among the first gene therapy strategy for which virus escape mutations were noted. Subsequent findings from several groups made clear that similar to combination antiretroviral therapy, combination gene therapy approaches were most likely to restrict viral replication while minimizing the possibility of viral escape (Hamm et al. 1999; Legiewicz et al. 2008).

A final protein-based strategy to inhibit HIV replication utilized proteins closely related to those encoded by HIV-1 as dominant negatives. The Vpx protein is not expressed by HIV-1, but is expressed by HIV-2 and SIV; when packaged by HIV-1, the resulting viral particles are non-infectious (Matsuda et al. 1993). This strategy was further used to recruit a mutant of HIV-2 protease to nascent HIV-2 virions, providing proof-of-principle to inactivate HIV-1 by similar methods (Wu et al. 1996). Importantly, more recent understanding of Vpx function, including antagonism of the restriction factor SAMHD1, calls into question the safety of generating chimeric, replication-competent HIV-1/Vpx viruses (Schaller et al. 2014).

## 3.2   RNA Therapies: Structural and Enzymatic Inhibitors

In addition to RNA decoys, gene therapy-based strategies are also well suited to express RNA molecules that inhibit HIV replication via novel structural and/or enzymatic properties. For example, Ribozymes are a class of RNA molecules that possess unique secondary structural patterns that facilitate energy-independent cleavage of RNA substrates. "Hammerhead" ribozymes were first characterized as RNA pathogens in plants, and were adapted to target HIV Gag RNA in 1990 (Sarver et al. 1990). Potent ribozymes directed against the 5' long terminal repeat of HIV-1 RNA soon followed (Weerasinghe et al. 1991), along with combination strategies using multicistronic retroviral vectors to express HIV ribozymes in combination with TAR decoys (Lisziewicz et al. 1993), or RRE decoys and the RevM10 dominant negative protein (Bauer et al. 1997). In 2004, a Phase I clinical trial in 10 HIV[+] patients without conditioning used a Tat/Vpr ribozyme-expressing MoMLV vector to transduce autologous CD34[+] HSPCs (Amado et al. 2004). As with contemporary HSPC trials, the gene transfer efficiency was low, although gene-marked T-cells were detected up to 36 months following transplantation, suggesting that gene-modified cells could persist. The largest cell-delivered gene therapy trial to date followed up on these results: using improved HSPC collection methods, this approach significantly decreased viral loads and increased CD4[+] T-cell counts relative to controls, and was safe in a Phase II trial (Mitsuyasu et al. 2009). Surprisingly, subsequent data on anti-HIV ribozymes in clinical and preclinical experiments has been limited over the past 8 years.

Aptamers are related to ribozymes, as both are RNA molecules that derive their function from unique secondary structures and RNA base pairing. However, instead of catalyzing RNA hydrolysis, aptamers combine the ligand specificity properties of an antibody with the delivery potential of a viral vector. For example, an aptamer designed to bind HIV-1 env gp120 and conjugated to a tat/rev small interfering RNA (siRNA), was internalized into target cells, degraded tat/rev transcripts, and suppressed viral replication in vitro and in humanized mice (Zhou et al. 2008; Neff et al. 2011). Aptamers with affinity to CD4 (Wheeler et al. 2011) and CCR5 (Zhou et al. 2015) have also been shown to be internalized into HIV target cells and decrease HIV infectivity in vitro and in vivo. However, as with many of the approaches outlined above, the efficacy of these molecules in patients is unclear. The toxicity, half-life, and potency of a systemically administered aptamer-siRNA conjugate in patients remain to be tested (Zhou and Rossi 2014).

A final class of RNA-based therapies for HIV cure, distinct from decoys and structural RNAs, are antisense and RNA interference (RNAi) molecules. Both classes silence RNA translation, although the distinction between the mechanisms for each has become less clear in recent years. Generally, antisense approaches, which can take the form of DNA, RNA, or a chemically modified intermediate, are designed to bind homologous mRNAs and sterically hinder translation. In HIV, Tat antisense molecules were among the earliest gene therapy strategies tested in vitro (Zaia et al. 1988). Potent antisense molecules against TAR (Chatterjee et al. 1992), reverse transcriptase (Meyer et al. 1993), and Gag were characterized in vitro during the same period (Veres et al. 1996). Rhesus macaques that received autologous CD4$^+$ lymphocytes transduced with a retroviral vector expressing a tat/rev antisense molecule showed lower viral loads, increased CD4$^+$ T-cell counts, and maintained lymph node architecture, relative to controls (Donahue et al. 1998). Analogous clinical studies in patients receiving env antisense-expressing T-cells demonstrated safety; although escape mutations against the antisense product were detected, efficacy was low (Tebas et al. 2013).

In contrast to the steric inhibition hypothesized for antisense molecules, RNAi strategies utilize siRNAs, as outlined above, to target a homologous mRNA for destruction via an enzymatic pathway. This raises the possibility that at least a subset of antisense approaches may, in fact, utilize RNAi pathways (Kole et al. 2012). Regardless, siRNA-based approaches rapidly expanded in the early 2000s, targeting multiple viral RNAs in vitro (Coburn and Cullen 2002; Lee et al. 2002; Novina et al. 2002; Qin et al. 2003). Combination strategies were frequently pursued using short hairpin RNAs (shRNAs) that can be expressed from retroviral vectors, in conjunction with ribozymes and TAR decoys, again demonstrating the benefits of a multi-pronged approach to target HIV replication. This is especially important in the context of RNAi strategies, where escape mutations in HIV have been identified (Das et al. 2004), and multiplex approaches have shown impressive potency (McIntyre et al. 2009; Herrera-Carrillo et al. 2014). Although RNAi-based methods were promising in immunocompromised mouse models (Banerjea et al. 2003), experiments in nonhuman primates were limited by the transduction efficiency, engraftment, and persistence of the gene-modified CD34$^+$ HSPC product (An et al. 2007; Trobridge et al. 2009).

To summarize, ribozyme, aptamer, antisense, and RNAi approaches have all demonstrated potency in vitro, and in some cases, limited efficacy in patients. The co-expression of multiple RNA-based molecules enhances these effects. However, a common limitation is the ability to deliver one or multiple RNA-based gene therapy molecules to a sufficient proportion of cells in order to mediate a curative effect in vivo.

## 3.3   HIV-Inducible Suicide Genes

Specific transcriptional elements and machinery, including the HIV-1 Tat protein, and promoter sequences including TAR, are required for the efficient expression of protein-coding and genomic HIV-1 RNAs. Early innovative gene therapy strategies for HIV cure sought to exploit these requirements to destroy infected cells, by placing cytotoxic and/or antiviral genes under the control of Tat/TAR-mediated transcription. These approaches should selectively kill cells in which viral replication is active. For example, a transgenic cassette containing the herpes simplex type 1 thymidine kinase gene (HSV1-TK) under the control of TAR was introduced into cultured HIV$^+$ cells in the presence of acyclovir, which generates a cytotoxic product when phosphorylated by HSV1-TK protein. Using therapeutically relevant concentrations of acyclovir, viral replication was completely inhibited (Caruso and Klatzmann 1992). An analogous approach was recently reported that utilized an HSV1-TK variant with increased potency, and was delivered on an integration-deficient lentiviral vector designed to avoid permanent genomic effects on uninfected cells (Garg and Joshi 2016). Another candidate HIV$^+$ cell suicide strategy employed HIV Tat/Rev-dependent expression of diphtheria toxin A chain (DT-A) (Harrison et al. 1992). More advanced iterations again employed integration-deficient lentiviral vectors, and coupled Rev-dependent expression of DT-A and tumor necrosis factor receptor-associated factor 6 (TRAF6) to drive apoptosis of actively infected cells (Wang et al. 2010). An important unanswered question for these approaches is how to deliver these suicide cargoes to every infected cell in a patient, especially those in difficult-to-target viral reservoir sites.

To circumvent these limitations, other groups focused on expression of secreted antiviral factors that could inhibit viral replication both in the HIV$^+$ cells from which they were expressed, and in neighboring infected cells. The most notable gene products in these studies belonged to the interferon family of antiviral cytokines. Interferon α2 (IFNα2) under control of the HIV-1 LTR was expressed at 6–20 fold higher levels in infected cells relative to uninfected cells, completely inhibiting viral replication (Bednarik et al. 1989). In a humanized mouse model, Tat-dependent expression of IFNα, IFNβ, and IFNγ proteins led to a drastic but incomplete reduction in proviral DNA (Sanhadji et al. 1997). Autologous lymphocytes from SIV-infected nonhuman primates were transduced with a retroviral vector expressing IFNβ, this time under the control of a constitutive murine promoter to drive continuous low-level expression of the antiviral cytokine. These cells did not persist, and efficacy was not observed (Gay et al. 2004). New hypotheses suggest that the beneficial effects of IFN signaling during early HIV infection may be countered by a role in driving immune dysfunction during chronic infection (Zhen et al. 2017; Cheng et al. 2017). As such, the use of gene therapy-based HIV cure strategies involving expression of interferon genes must be rethought, namely for infected and stably cART-suppressed patients.

## 3.4  Peptide Inhibitors, Intrabodies, and Antibodies

Our group and others have pioneered the introduction of HIV fusion inhibitor peptides into hematopoietic stem cells to block the earliest steps in virus replication in vivo. A membrane-anchored version of the HIV gp41 envelope-derived fusion inhibitor T20/Enfuvirtide was first reported in 2001 (Hildinger et al. 2001). This peptide effectively blocks the ability of gp41 sequences to interact and drive the membrane fusion step of HIV entry into target cells. Following correlates of success in a primary cell model (Perez et al. 2005), experiments were undertaken to generate an optimal, minimally immunogenic membrane-bound derivative of T20, named mC46 (Hermann et al. 2009; Brauer et al. 2013). We have shown that mC46-containing retroviral vectors protect against virus challenge ex vivo (Kiem et al. 2010; Trobridge et al. 2009) and in vivo (Younan et al. 2013; Peterson et al. 2016a). Other clever approaches have fused similar T20-related peptides to the CXCR4 coreceptor for HIV-1, thus placing the inhibitor directly at the site of virus fusion (Leslie et al. 2016). The small size of the promoter-mC46 cassette makes it an ideal "defensive" component to include in multicistronic anti-HIV vectors that express transgenes designed to actively detect and destroy infected cells (see below).

Protein-based inhibitors have been designed to target post-entry steps in the viral life cycle as well. Specifically, intracellular antibodies known as "intrabodies" were designed to target viral proteins such as gp120 envelope, and inactivate various steps of virus infection and assembly. Using Tat-inducible strategies as described above, gp120 antibodies had a partial effect on HIV replication in vitro (Chen et al. 1994). Intrabodies against CCR5 were designed to mimic the inability of CCR5Δ32 proteins to express at the cell surface (Steinberger et al. 2000), while other intrabodies blocked the function of HIV proteins Tat, Rev, and Vif (Braun et al. 2012; Vercruysse et al. 2010; Aires da Silva et al. 2004). Progress in this field has waned in recent years, in part due to the lack of efficacious single chain antibody moieties (Marschall et al. 2015). Importantly, the advent of broadly neutralizing antibodies (bNAbs) against HIV-1 may reinvigorate intrabody-dependent HIV cure approaches.

Indeed, arguably the most promising new strategy for HIV cure is bNAbs. The history, mechanism, and passive administration of these molecules are discussed in depth in other chapters. As the name suggests, bNAbs are capable of neutralizing a broad array of HIV variants with considerable potency, and can be adapted to any approach requiring protein-based recognition of viral antigen. Gene therapy approaches (vector-mediated delivery) have been evaluated in parallel to passive administration (usually an intravenous injection of bNAb protein), in order to increase the persistence of bNAbs in vivo. A popular approach is to deliver cargoes to muscle tissue via adeno-associated virus (AAV) vectors, creating "factories" that supply transgenic protein to the periphery. This strategy has been successful in nonhuman primates, protecting 6 of 9 animals against SIV challenge in one study (Johnson et al. 2009), and demonstrating dose-dependent protection in another study (Fuchs et al. 2015). Importantly, host antibody responses against the transgenic antibody product, referred to as "anti-anti" responses, were shown to be a key barrier in the latter study (Martinez-Navio et al. 2016). To address this, Farzan and colleagues (Gardner et al. 2015) used a rationally designed bNAb referred to as eCD4-Ig, which contains the D1D2 extracellular domains of CD4 for HIV envelope binding, an IgG1 Fc region, and a CCR5 mimetic peptide. This molecule exploits the requirement of viral particles to bind CD4 and CCR5, resulting in increased potency and breadth, and decreased immunogenicity relative to traditional bNAbs. These features should allow for therapeutic serum concentrations of eCD4-Ig that are substantially lower than bNAbs, a key point in facilitating efficacious gene therapy-mediated approaches.

In summary, protein-protein binding properties have been used to inhibit HIV-1 replication at multiple stages in the viral life cycle. As HIV cure approaches continue in the bNAb age, these strategies promise to take center stage, facilitating neutralization of free virions, recognition of cells expressing viral antigen, and/or recruitment of immune complexes to destroy infected cells through mechanisms such as antibody-dependent cellular cytotoxicity (ADCC).

### 3.5   Chimeric Restriction Factors and Antigen-Binding Proteins

As described above, multicistronic retroviral vectors expressing several anti-HIV transgenes have been a popular means of targeting multiple stages of the viral life cycle with a single gene therapy product. One such multicistronic vector expressed a TAR decoy and CCR5 shRNA, along with a third cassette, a chimeric version of the TRIM5$\alpha$ antiviral restriction factor. This strategy inhibited viral replication ex vivo and protected human CD4$^+$ T-cells in humanized mice (Anderson et al. 2009; Walker et al. 2012). The chimeric TRIM5$\alpha$ allele was constructed by substituting 11 amino acids of human TRIM5$\alpha$ with 13 amino acids from rhesus TRIM5$\alpha$ (Anderson and Akkina 2008). This short amino acid stretch enables old world nonhuman primate cells to destroy HIV-1 virions following virus entry (Sawyer et al. 2005), and is just one example of how rational engineering of human restriction factors can provide a significant advantage in the "arms race" between host and viral factors. In addition to a 13-amino acid graft from the rhesus sequence, TRIM5$\alpha$ has also been shown to gain anti-HIV-1 function when fused to Cyclophilin A (Neagu et al. 2009), and when amino acid substitutions were introduced at as few as two proximal residues (Pham et al. 2010). Importantly, modified TRIM5$\alpha$ monotherapy is associated with viral escape, reinforcing the need to combine this approach with other gene therapies, as described above (Setiawan and Kootstra 2015). Other restriction factors, including APOBEC3 proteins and BST2/Tetherin, have been similarly engineered to evade HIV-dependent inactivation and efficiently inhibit viral replication. APOBEC3 proteins deaminate cytidine to uridine in viral genomic

RNA, but are inactivated by HIV-1 Vif, which targets these proteins for proteasomal degradation. Negative regulation by Vif can be counteracted by a single point mutation in APOBEC3 proteins (Xu et al. 2004); its antiviral function can be further enhanced by fusion with a virus-targeting polypeptide known as R88 (Ao et al. 2011). Intriguingly, other APOBEC family members show virus-inhibitory properties that are independent of their enzymatic function, suggesting other applications for these factors (Katuwal et al. 2014). Although the ability of HIV virions to evade BST2/Tetherin have been well-characterized (Douglas et al. 2010), this third well-characterized HIV-1 restriction factor has not been widely evaluated in gene therapy-based HIV cure strategies, in part because its best-characterized function is to restrict virus budding late in the virus replication cycle. Nevertheless, understanding the basic biology of antiviral restriction factors, including cognate inhibitory proteins expressed by HIV, will be essential in order to further apply these proteins in future HIV gene therapy studies.

Other chimeric proteins deviate further from evolutionarily selected restriction factors, mixing and matching specific binding moieties and immune signaling functions to generate potent, HIV-specific immune cells. As mentioned above, the best known of these molecules are chimeric antigen receptors (CARs), which have shown immense promise in the treatment of hematological malignancies (Fig. 1) (Geyer and Brentjens 2016; Wang et al. 2017). The general principle of CAR therapy is to link an extracellular ligand-binding domain to an intracellular T-cell signaling domain; as such, any $CAR^+$ cell will redirect immune responses against the antigen specified by the extracellular binding domain. $HIV^+$ patients were among the first to receive CAR T-cells in clinical trials. Although efficacy was modest, cells expressing the above-described CD4-CAR persisted for over 10 years in vivo (Scholler et al. 2012). More recently, the same CAR construct was shown to be effective when expressed from HSPCs in a humanized mouse model (Zhen et al. 2015). Other groups have adapted the variable fragment from promising bNAbs as the extracellular domain in anti-HIV CARs, showing killing activity in vitro (Hale et al. 2017b; Liu et al. 2016), and in combined binding moieties containing CD4 D1D2 (Liu et al. 2015). In addition, T-Cell Receptors (TCRs) have been cloned from $CD8^+$ T-cells that recognize conserved HIV epitopes and potently restrict infected cells (Kitchen et al. 2012; Varela-Rohena et al. 2008). Although this approach has the advantage of using an extracellular recognition domain that may be less immunogenic, it is limited by the need for a compatible MHC haplotype for proper function. A third and final class of chimeric proteins, which can be administered as soluble intravenous agents rather than being expressed from HSPCs or immune cells, are Dual-Affinity Re-Targeting proteins (DARTs). Rather than merging protein-protein recognition modules with cell signaling molecules, DARTs link an antigen-binding arm with a CD3-binding arm, in order to place infected cells in close proximity to immune effectors (Sung et al. 2015; Sloan et al. 2015). Although DARTs are in their infancy, an important obstacle that must be overcome with these molecules, as well as with CARs, is the possibility that binding of free virions will facilitate infection of the CAR-containing, or DART-linked cell,

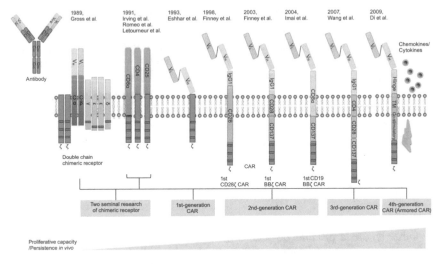

**Fig. 1** Evolution of chimeric antigen receptors for cancer and HIV patients. Early precursors of modern chimeric antigen receptors consisted of MHC-unrestricted chimeric TCR (cTCR) complexes, adapted for antigen specificity using the variable heavy and light chain ($V_H$ and $V_L$) domains from a candidate antibody. The first bonafide CARs linked an extracellular ligand-binding domain (for example a portion of the CD4 protein that interacts with HIV envelope) to an intracellular CD3$\zeta$ signaling domain, rendering the CAR$^+$ cell specific for the specified ligand. "1st Generation" CARs built on cTCR concepts by adapting antibody $V_H$ and $V_L$ domains for ligand binding. The potency of these molecules was enhanced by the addition of one or more costimulatory domains, including those from the CD28 and CD137 (4-1BB) proteins. Currently under investigation are the sequence and length of the spacer/hinge region that links the antigen-binding and transmembrane domains, and the utility of co-expressing cytokines and co-stimulatory ligands to increase CAR efficacy. Figure originally published in *Protein & Cell*, Wang et al. 2017, "Current status and perspectives of chimeric antigen receptor modified T cells for cancer treatment," [Epub ahead of print 2017 May 2]. Reproduced with permission via Copyright Clearance Center (Wang et al. 2017)

respectively. Analogous to anti-cancer "armored CARs" that co-express transgenes designed to increase functional persistence of these cells (Fig. 1) (Jaspers and Brentjens 2017), co-expression of CCR5 shRNA molecules or peptide inhibitors such as mC46 are leading strategies to protect CAR- or DART-expressing cells against infection.

In summary, a wide array of gene therapy approaches to inhibit HIV replication in vivo have been proposed over the last 3 decades. Many of these, including decoys, dominant negatives, RNA-based approaches, HIV-inducible suicide genes, and intrabodies, carry two common limitations: (i) the need to deliver to a high proportion of infected cells, and (ii) a purely defensive function which potently protects against infection, but does not actively target persistent viral reservoirs in stably suppressed patients. Merging these approaches with bNAbs, CARs, and other immunotherapies that should actively seek out and destroy cells expressing viral antigen represents an ideal combination HIV gene therapy strategy.

# 4 Delivering Curative Gene Therapy Cargoes

## 4.1 Permanent Gene Modification: Retroviral Vectors

The anti-HIV transgenes described above must be efficiently delivered into target cells, for example HSPCs and T-cells, in order to play an efficacious role in infected patients. Broadly, delivery strategies can be either transient or permanent. Among approaches to permanently gene-modify target cells, retroviral vectors have been most reliable; in the 1980s and 1990s, gammaretroviral vectors such as MoMLV were frequently used. Gammaretroviruses are straightforward to produce, utilizing a range of stable virus-producing cell lines, and a vector backbone that is distinct from HIV-1 (Maetzig et al. 2011). The major disadvantages are that gammaretroviruses can only infect mitotically active cells, and more importantly, that transduced cells may proliferate in an unrestricted manner, leading to myelodysplasia and leukemia in some patients (Howe et al. 2008; Stein et al. 2010; Hacein-Bey-Abina et al. 2008).

HIV-based gene therapy vectors were first constructed to understand HIV biology rather than as delivery vehicles per se, investigating parameters including virus tropism and the role of individual viral proteins in viral replication (Page et al. 1990; Planelles et al. 1995). From these findings, bona fide lentiviral gene therapy vectors were constructed, focusing on safety, delivery efficiency, and transgene expression. The adaptation of HIV-based vectors for gene therapy, especially in HIV$^+$ patients, required a careful examination of the ability of safety-engineered vectors to recombine with replication-competent HIV virus. By incorporating deletions in Gag (Parolin et al. 1994) and LTR sequences (Miyoshi et al. 1998; Iwakuma et al. 1999; Zufferey et al. 1998), self-inactivating ("SIN") vectors with robust safety profiles were developed (Escarpe et al. 2003). In addition to safety, efficient entry of these vectors into target cells was key. As such, a number of "pseudotyped" envelopes were evaluated; although vesicular stomatitis virus G glycoprotein (VSVG) is the most widely used lentiviral envelope, many others have been evaluated for a number of context-dependent approaches (Burns et al. 1993; Relander et al. 2005; Eleftheriadou et al. 2017; Trobridge et al. 2010). Finally, the ability to control the level of transgene expression from engineered lentiviral vectors was examined, both for maximal and titratable expression. Elements such as the central polypurine tract (cPPT) (Zennou et al. 2000; Follenzi et al. 2000) and Woodchuck hepatitis virus posttranscriptional regulatory element (WPRE) (Zufferey et al. 1999) have become common additions to HIV-based gene therapy vectors. Internal promoters that regulate expression levels and tissue specificity have been identified, and variants that utilize other lentivirus backbones (HIV-2, SIV, and Feline Immunodeficiency Virus) have also been reported (Barker and Planelles 2003). We and others have further improved the performance of lentiviral gene therapy vectors by optimizing the culture conditions for transduced cells (Santoni de Sio et al. 2006; Wang et al. 2014; Horn et al. 2004; Trobridge et al. 2008). Despite the marked improvements in lentivirus gene therapy, several limitations remain. The vector itself, as well as the pseudotyped envelope, may be immunogenic in vivo (Breckpot et al. 2010; Pichlmair et al. 2007), although the effect on engraftment of gene-modified cells is still unclear. Most pressing for HIV gene therapy is the fact that HIV-based lentiviral vectors will carry the same sensitivities as replication-competent virus, both to cART regimens and to transgenes that may inhibit the assembly of the vectors themselves. Hence, cell products from cART-treated patients will require drug washouts or alternative means to facilitate delivery of cART-loaded cells (Younan et al. 2015).

In cases where a retroviral vector is needed, without the risks and limitations associated with HIV-based vectors, a closely related retrovirus known as foamy virus has emerged as a promising alternative (Taylor et al. 2008). Foamy viruses persist in quiescent cells, and have increased packaging capacities and improved transduction efficiencies in cART-treated cells, relative to lentiviruses (Olszko and Trobridge 2013; Younan et al. 2015). Including gammaretrovirus, lentivirus, and foamy virus-based vectors, retroviral gene transfer is the best studied, safest, and most effective means of permanent gene modification in target cells.

## 4.2   Adenovirus and AAV Vectors: Transient Expression of Gene Therapy Cargoes

Retroviral vectors can be engineered for "non-permanent" gene modification by mutating the integrase gene, resulting in an integration-deficient vector (Yanez-Munoz et al. 2006). However, the best-characterized means of delivering a gene therapy cargo without permanently modifying the host genome is through the use of adenoviral and AAV vectors. Transient expression is ideal when the transgene of interest is only necessary for a defined time period, or in gene editing approaches, where the transgene itself can make a permanent modification to the genome without itself requiring permanent expression.

Adenoviral vectors, in addition to being non-integrating, possess an unrivaled packaging capacity relative to other viral vector gene therapy platforms, which has increased even further with recent helper-dependent (HD) varieties (Segura et al. 2008). Adenoviral delivery strategies were used in early HIV gene therapy experiments to deliver inhibitory transgenes in vitro, for example Tat-dependent HSV1-Tk suicide genes (Venkatesh et al. 1990), and TAR decoy/Gag dominant negative/Gag antisense multicistronic vectors (Lori et al. 1994). However, because these vectors are more difficult and time-consuming to produce than retroviral vectors, the latter became the strategy of choice for transducing target cells ex vivo. Nevertheless, the use of adenoviral vectors has emerged along with CMV-based vectors as a key tool in next-generation vaccines for HIV and other diseases (Borducchi et al. 2016; Hansen et al. 2011; Barouch and Picker 2014). These newer, engineered, chimeric vectors demonstrated reduced immunogenicity and toxicity in vivo (Wang et al. 2008; DiPaolo et al. 2006; Capasso et al. 2014) and have been further applied for ex vivo gene editing in autologous, HIV⁺ T-cells (Tebas et al. 2014), and HSPCs (Saydaminova et al. 2015). Adenoviral vectors are also leading candidates for in vivo delivery approaches, which dispense entirely with ex vivo cell manipulations in favor of direct, intravenous administration of the gene therapy vector in order to transduce cells in vivo (Richter et al. 2016). The promise of these vectors to bring portable, easy-to administer treatments to HIV⁺ patients, especially in the developing world, makes adenovirus strategies an essential area of continued study for HIV gene therapy.

Another popular viral vector for non-integrating delivery of anti-HIV transgenes is Adeno-Associated Virus (AAV). As described above, AAV vectors have shown efficacy in multiple nonhuman primate AIDS studies by delivering bNAb transgenes to muscle cells in vivo, establishing antibody "factories" that generate therapeutically relevant concentrations of anti-HIV bNAbs in the periphery, and protect against challenge with HIV-like viruses (Johnson et al. 2009; Fuchs et al. 2015; Gardner et al. 2015). AAV strategies were used in early HIV gene therapy experiments to deliver antisense RNA against HIV-1 polyadenylation signals and TAR (Chatterjee et al. 1992), and ribozymes directed against the 5′ LTR and Gag regions of viral RNA (Horster et al. 1999). More recently, AAV vectors have emerged in a new capacity, as part of multi-step strategies to substitute or correct genes at any gene edited locus (see below). As with adenoviral vectors, an important obstacle for AAV gene therapy is immune response against the vector itself. Neutralizing antibodies against 3 AAV serotypes were found in as many as 96.6% of healthy individuals in a Chinese study, although the authors found that samples from North American and European subjects had significantly lower anti-AAV antibody titers than Chinese subjects (Liu et al. 2014). Consistent with the above observations of anti-AAV responses in a nonhuman primate bNAb study (Martinez-Navio et al. 2016), future strategies designed to decrease AAV immunogenicity in vivo will be essential. As with adenoviral vectors, the ability to apply these vectors in multiple delivery schemes make them an attractive candidate for future studies.

## 4.3 Gene Editing for HIV Cure

Arguably the most important advent in the gene therapy field over the past decade has been that of gene editing, which alters host genomic DNA specifically at a locus of interest, using one of several available site-specific nuclease platforms. The advantage of this approach over integrating viral vectors is that random genome integration events that may lead to oncogenesis, as has been shown for gammaretroviral vectors, should be avoided. Although early strategies were explored in the 1990s (Singwi and Joshi 2000), it was the Zinc Finger Nuclease (ZFN) platform that first entered clinical trials. Gene editing ZFNs were first reported by Chandrasegaran and colleagues in 1996 (Kim et al. 1996). Carroll and colleagues further developed this technology in multiple in vivo models (Bibikova et al. 2003;

Morton et al. 2006), which was then optimized for the clinic through a partnership led by Sangamo Biosciences (Urnov et al. 2005). Early efforts focused on delivery via non-integrating lentiviral vectors (Lombardo et al. 2007), and adenoviral vectors, which were applied in the first-in-human clinical trial (Tebas et al. 2014). The target locus chosen for these first HIV gene editing studies was CCR5, based on the effective HIV protection afforded by the CCR5Δ32 allele (Dean et al. 1996; Jin et al. 2008), and extensive evidence related to the safety of inactivating this gene (Nazari and Joshi 2008). Results of the first ZFN trial, which transplanted CCR5-edited CD4$^+$ T-cells into suppressed HIV$^+$ patients, showed that the approach was safe, and although not curative, suppressed the magnitude of viral rebound in a CCR5Δ32 heterozygous patient. This proof-of-principle study emboldened development of numerous gene editing platforms with the ultimate goal of application not only to HIV$^+$ patients, but those with other infectious and genetic diseases. Among the other platforms developed to edit CCR5 were Transcription Activator-Like Effector Nucleases (TALENs), which are structurally similar to ZFNs, but easier to assemble (Benjamin et al. 2016; Mock et al. 2015), megaTALs, which combine the enzymatic/DNA binding domains of a meganuclease with added site specificity conferred by an additional TALEN DNA binding domain (Sather et al. 2015; Hale et al. 2017a), and CRISPR-Cas9, for which the ease of design led to targeting of many other HIV-relevant targets (Park et al. 2017). Across each of these platforms, applicability must be considered on a per-target basis. For example, although CXCR4 can be efficiently targeted by ZFNs (Yuan et al. 2012), it is required for proper hematopoietic development, meaning that it cannot be safely targeted in HSPCs.

Other groups expanded the use of gene editing to target the provirus itself. The most straightforward approach involved nuclease-dependent mutation of integrated proviral DNA in order to generate replication-incompetent virions (Kaminski et al. 2016; Lebbink et al. 2017). However, a critical finding from these experiments was that infectious, nuclease-resistant virions could still be generated (De Silva Feelixge et al. 2016), consistent with the remarkable mutability of the HIV-1 genome. Still other groups repurposed gene editing platforms to alter chromatin structure at HIV-1 integration sites. ZFN fusions with DNA methyltransferase 1 silenced HIV proviruses in vitro (Deng et al. 2017), suggesting that this approach could induce permanent latency in vivo. In contrast, other groups used CRISPR-Cas9 complexes fused to potent transcriptional activator proteins to reactivate latent proviruses as part of a "kick and kill" approach to reveal and clear latently infected cells (Saayman et al. 2016; Limsirichai et al. 2016; Ji et al. 2016).

Several in-progress clinical trials are applying the rapidly growing knowledge base for gene editing to suppressed, HIV$^+$ patients. The previous CCR5 ZFN trial using CD4$^+$ T-cells has been augmented to include cyclophosphamide conditioning, which should increase engraftment of gene edited T-cells with minimal toxicity (NCT02225665, NCT02388594). In parallel, the same CCR5 ZFN product is being delivered to autologous HSPCs ex vivo via electroporation of ZFN-encoding mRNA rather than adenoviral vectors (DiGiusto et al. 2016). This trial (NCT02500849) is particularly exciting, as it will be among the first HSPC gene therapy clinical trials conducted in otherwise healthy HIV$^+$ patients (i.e. no associated malignancies). Based on our preclinical data in nonhuman primates using a macaque CCR5-specific ZFN pair, which suggest that autologous CCR5 gene edited cells engraft and persist in vivo (Peterson et al. 2016b), findings in NCT02500849 will be hugely instructive for future HIV gene therapy studies, and for clinical gene editing approaches more generally. Finally, as mentioned above, AAV vectors promise to further accelerate the gene editing field by carrying DNA templates capable of recombining at a targeted locus (Fig. 2). This technology has shown promise in conjunction with nuclease-dependent approaches, which generate the double strand break into which the AAV-derived donor DNA template recombines (Sather et al. 2015; Hale et al. 2017a; Wang et al. 2015), and may also function in a nuclease-independent manner at some loci (Gaj et al. 2016; Riolobos et al. 2013). The goal of these experiments is to take gene editing approaches beyond nonhomologous end joining (NHEJ) pathway endpoints, which can only generate a mutation at a targeted locus that usually leads to inactivation of the gene product. Instead, by providing a donor DNA template, homologous donor recombination (HDR) pathways can be applied, facilitating "correction" of any gene in the genome, including targeted inactivation mutations in HIV proviruses, and gain

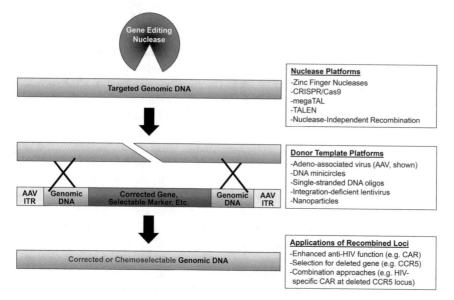

**Fig. 2** Homologous donor recombination approaches for gene correction. Following generation of a double strand break, several platforms are available to deliver a recombination template to the targeted locus. Flanking homology regions that will facilitate precise recombination of the donor vector at the targeted locus are a fundamental requirement for this template DNA. Maximizing the efficiency of recombination is key in order to skew gene editing away from nonhomologous end joining (NHEJ) pathways that randomly mutate the gene, and toward HDR pathways, capable of correcting the gene to any desired sequence

of function mutations in HIV restriction factor genes (Fig. 2). The future is bright for gene editing-based HIV cure approaches, and will include parallel work in vitro, in animal models, and in clinical trials.

## 4.4 Future Approaches: Nanoparticles and in Vivo Selection

Viral vectors, and more recently electroporation of gene editing nuclease mRNA, have been the predominant methods of delivery of anti-HIV gene therapy cargoes into cells. However, as the need to target both small molecules and gene therapy products to specific physiological sites in suppressed HIV$^+$ patients has emerged, so too have the advantages of nanoparticles, defined here as a molecular carrier with at least 1 dimension measuring <100 nm (Fig. 3) (Roy et al. 2015). A primary focus of nanoparticles for HIV$^+$ patients has been in the design of sustained release cART formulations that facilitate less frequent/more flexible dosing regimens (Edagwa et al. 2017). A huge array of nanoparticle molecules and anti-HIV gene therapy cargoes has been evaluated in vitro, but few have progressed to in vivo studies. As far back as 1990, liposomes coated with anti-CD3 antibodies were used to target antisense RNA against HIV env to infected cells in vitro (Renneisen et al. 1990). Pan-leukocyte-targeted liposomes carrying CCR5 siRNAs were efficacious in a humanized mouse model (Kim et al. 2010). Synthetic branching polymers, referred to as dendrimers, were linked to anti-HIV RNA therapeutics and were effective in vitro, although the chemistry of these complexes could be limiting (Szewczyk et al. 2012; Briz et al. 2012). RNA aptamer-siRNA conjugates (described above) also qualify as nanoparticles, offering scalability, and less environmental toxicity than other nanoparticle approaches (Zhou et al. 2009; Rossi 2011). Another biodegradable option, Poly-lactic-co-glycolic acid nanoparticles carrying triplex-forming peptide nucleic acids, were capable of mutating the CCR5 gene in vivo; although the efficiency of this process was low, the nanoparticle vector has a proven safety record in patients (Schleifman et al. 2013). Finally, vault nanoparticles have been described that are capable of carrying latency reversing agents to reactive latently HIV$^+$ cells (Buehler et al. 2014). This same approach could be used to deliver DARTs, or anti-HIV peptides such as mC46. In short, a number of nanoparticle candidates, each with distinct strengths and weaknesses, have been evaluated in vitro. Molecules that can be modified to traffic to viral reservoir sites such as lymph node B-cell follicles and gut-associated lymphoid tissue, and have the best record for safety in vivo, will constitute the most attractive candidates for evaluation in patients.

Finally, our group and others have developed techniques to increase the number of anti-HIV gene-modified cells in vivo, through a process known as "in vivo selection." We have shown that gene therapy vectors that carry the P140K mutant of the methylguanine methyltransferase (MGMT) gene can undergo positive

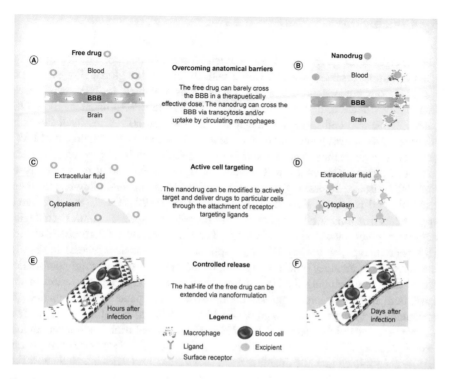

**Fig. 3** Use of nanoparticles for HIV cure. Nanoparticle-based formulations offer several advantages in the delivery of small molecule and gene therapies to targeted sites in vivo. These include (**a–b**) the ability to cross otherwise impenetrable anatomical barriers such as the blood brain barrier (BBB), (**c–d**) targeting to cell types of interest, namely CD4$^+$ T-Cells, and (**e–f**) increasing (or decreasing) the in vivo half-life of the gene therapy cargo. Figure originally published in *Nanomedicine*, Roy et al. 2015, 10(24):3597–3609, by FUTURE MEDICINE LTD. Reproduced with permission of FUTURE MEDICINE LTD in the format Book via Copyright Clearance Center (Roy et al. 2015)

selection following low-dose chemotherapy. The percentage of gene-marked cells can approach 100% in selected peripheral blood subsets following this treatment (Beard et al. 2010), resulting in populations that are highly resistant to ex vivo virus challenge (Trobridge et al. 2009). Importantly, different chemoselection approaches may be required in order to select for different transgene-expressing cellular subsets, e.g. HSPCs vs. T-cells. For example, selection for a variant of inosine monophosphate dehydrogenase 2 (IMPDH2) expanded and protected gene-marked T-cells in vitro (Yam et al. 2006), which was enhanced when combined with selection for a mutant of dihydrofolate reductase (DHFR) (Jonnalagadda et al. 2013). Although requiring delivery of anti-HIV and selection gene cargoes in tandem, use of MGMT$^{P140K}$, IMPDH2, and/or DHFR markers would augment the delivery efficiency of any of the above HIV Cure strategies, by selecting for gene-marked cells and against unmarked cells in vivo.

## 5 Conclusions

The emergence of gene therapy as a clinically translatable approach for the treatment of human diseases was concurrent with the rise of the global HIV pandemic. As such, our understanding of the mechanisms involved in HIV persistence, immune dysfunction, and treatment has paralleled the development of more efficacious gene and cell therapy strategies. Ironically, the HIV virus itself has contributed a preeminent gene therapy tool, lentiviral vectors, to the treatment of HIV$^+$ patients, as well as those with numerous other genetic and infectious diseases. The advent of suppressive cART was unquestionably a turning point in the ability to limit HIV disease, improving the general health of patients and allowing focus on residual viral reservoir sites. The development of HSPC, T-cell, and DC-based approaches that inhibit HIV replication has progressed rapidly; countless strategies demonstrate substantial efficacy in vitro. The single greatest future challenge for HIV gene therapy, as for any HIV cure approach, is to translate success in vitro to the clinic. This includes efficient targeting of therapies to sites of viral persistence, and measurably reducing the size of viral reservoirs. The recent explosion of the CAR field should be more aggressively translated from patients with hematologic malignancies to patients with suppressed HIV$^+$ infection, building on the modest efficacy and favorable safety profiles of HIV CAR clinical trials undertaken almost two decades ago. Furthermore, it will be essential to combine strategies that protect HIV-susceptible cells while enabling them to specifically restrict infected cells. For example, CAR and DART approaches that "tether" virions or virus-producing cells would be well suited to co-express the mC46 inhibitor of HIV fusion, so that virus entry into these cells is minimized or inhibited. As our understanding of HIV persistence grows, so too does our recognition of this virus as a formidable foe. Likewise, we must continue to take the important steps forward, safely but deliberately, to test and optimize gene therapy-based strategies to cure HIV$^+$ patients.

## References

Aboulafia DM, Mitsuyasu RT, Miles SA (1991) Syngeneic bone-marrow transplantation and failure to eradicate HIV. AIDS 5:344

Aires Da Silva F, Santa-Marta M, Freitas-Vieira A et al (2004) Camelized rabbit-derived VH single-domain intrabodies against Vif strongly neutralize HIV-1 infectivity. J Mol Biol 340:525–542

Ali A, Kitchen SG, Chen IS et al (2016) HIV-1-specific chimeric antigen receptors based on broadly neutralizing antibodies. J Virol 90:6999–7006

Allers K, Hutter G, Hofmann J et al (2011) Evidence for the cure of HIV infection by CCR5 DELTA32/DELTA32 stem cell transplantation. Blood 117:2791–2799

Alvarnas JC, Le Rademacher J, Wang Y et al (2016) Autologous hematopoietic cell transplantation for HIV-related lymphoma: results of the BMT CTN 0803/AMC 071 trial. Blood 128:1050–1058

Amado RG, Mitsuyasu RT, Rosenblatt JD et al (2004) Anti-human immunodeficiency virus hematopoietic progenitor cell-delivered ribozyme in a phase I study: myeloid and lymphoid reconstitution in human immunodeficiency virus type-1-infected patients. Hum Gene Ther 15:251–262

An DS, Donahue RE, Kamata M et al (2007) Stable reduction of CCR5 by RNAi through hematopoietic stem cell transplant in non-human primates. Proc Natl Acad Sci 104: 13110–13115

Anderson J, Akkina R (2008) Human immunodeficiency virus type 1 restriction by human-rhesus chimeric tripartite motif 5alpha (TRIM 5alpha) in CD34(+) cell-derived macrophages in vitro and in T cells in vivo in severe combined immunodeficient (SCID-hu) mice transplanted with human fetal tissue. Hum Gene Ther 19:217–228

Anderson JS, Javien J, Nolta JA et al (2009) Preintegration HIV-1 inhibition by a combination lentiviral vector containing a chimeric TRIM5 alpha protein, a CCR5 shRNA, and a TAR decoy. Mol Ther 17:2103–2114

Andres C, Plana M, Guardo AC et al (2015) HIV-1 reservoir dynamics after vaccination and antiretroviral therapy interruption are associated with dendritic cell vaccine-induced T Cell Responses. J Virol 89:9189–9199

Angelucci E, Lucarelli G, Baronciani D et al (1990) Bone marrow transplantation in an HIV positive thalassemic child following therapy with azidothymidine. Haematologica 75:285–287

Ao Z, Wang X, Bello A et al (2011) Characterization of anti-HIV activity mediated by R88-APOBEC3G mutant fusion proteins in CD4+ T cells, peripheral blood mononuclear cells, and macrophages. Hum Gene Ther 22:1225–1237

Bahner I, Kearns K, Hao QL et al (1996) Transduction of human CD34+ hematopoietic progenitor cells by a retroviral vector expressing an RRE decoy inhibits human immunodeficiency virus type 1 replication in myelomonocytic cells produced in long-term culture. J Virol 70: 4352–4360

Bahner I, Sumiyoshi T, Kagoda M et al (2007) Lentiviral vector transduction of a dominant-negative Rev gene into human CD34+ hematopoietic progenitor cells potently inhibits human immunodeficiency virus-1 replication. Mol Ther 15:76–85

Banerjea A, Li MJ, Bauer G et al (2003) Inhibition of HIV-1 by lentiviral vector-transduced siRNAs in T lymphocytes differentiated in SCID-hu mice and CD34+ progenitor cell-derived macrophages. Mol Ther 8:62–71

Barker E, Planelles V (2003) Vectors derived from the human immunodeficiency virus, HIV-1. Front Biosci 8:d491–d510

Barouch DH, Picker LJ (2014) Novel vaccine vectors for HIV-1. Nat Rev Microbiol 12:765–771

Bauer G, Valdez P, Kearns K et al (1997) Inhibition of human immunodeficiency virus-1 (HIV-1) replication after transduction of granulocyte colony-stimulating factor-mobilized CD34+ cells from HIV-1-infected donors using retroviral vectors containing anti-HIV-1 genes. Blood 89:2259–2267

Beard BC, Trobridge GD, Ironside C et al (2010) Efficient and stable MGMT-mediated selection of long-term repopulating stem cells in nonhuman primates. J Clin Invest 120:2345–2354

Bednarik DP, Mosca JD, Raj NB et al (1989) Inhibition of human immunodeficiency virus (HIV) replication by HIV-trans-activated alpha 2-interferon. Proc Natl Acad Sci U S A 86:4958–4962

Benjamin R, Berges BK, Solis-Leal A et al (2016) TALEN gene editing takes aim on HIV. Hum Genet 135:1059–1070

Berger C, Flowers ME, Warren EH et al (2006) Analysis of transgene-specific immune responses that limit the in vivo persistence of adoptively transferred HSV-TK-modified donor T cells after allogeneic hematopoietic cell transplantation. Blood 107:2294–2302

Bevec D, Dobrovnik M, Hauber J et al (1992) Inhibition of human immunodeficiency virus type 1 replication in human T cells by retroviral-mediated gene transfer of a dominant-negative Rev trans-activator. Proc Natl Acad Sci U S A 89:9870–9874

Bex F, Hermans P, Sprecher S et al (1994) Syngeneic adoptive transfer of anti-human immunodeficiency virus (HIV-1)-primed lymphocytes from a vaccinated HIV-seronegative individual to his HIV-1-infected identical twin. Blood 84:3317–3326

Bibikova M, Beumer K, Trautman JK et al (2003) Enhancing gene targeting with designed zinc finger nucleases. Science 300:764

Blaese RM, Culver KW, Miller AD et al (1995) T lymphocyte-directed gene therapy for ADA-SCID; initial trial results after 4 years. Science 270:475–480

Bordignon C, Notarangelo LD, Nobili N et al (1995) Gene therapy in peripheral blood lymphocytes and bone marrow for ADA− immunodeficient patients. Science 270:470–475

Borducchi EN, Cabral C, Stephenson KE et al (2016) Ad26/MVA therapeutic vaccination with TLR7 stimulation in SIV-infected rhesus monkeys. Nature 540:284–287

Brauer F, Schmidt K, Zahn RC et al (2013) A rationally engineered anti-HIV peptide fusion inhibitor with greatly reduced immunogenicity. Antimicrob Agents Chemother 57:679–688

Braun SE, Taube R, Zhu Q et al (2012) In vivo selection of CD4(+) T cells transduced with a gamma-retroviral vector expressing a single-chain intrabody targeting HIV-1 tat. Hum Gene Ther 23:917–931

Breckpot K, Escors D, Arce F et al (2010) HIV-1 lentiviral vector immunogenicity is mediated by Toll-like receptor 3 (TLR3) and TLR7. J Virol 84:5627–5636

Briz V, Serramia MJ, Madrid R et al (2012) Validation of a generation 4 phosphorus-containing polycationic dendrimer for gene delivery against HIV-1. Curr Med Chem 19:5044–5051

Buchschacher GL Jr, Freed EO, Panganiban AT (1992) Cells induced to express a human immunodeficiency virus type 1 envelope gene mutant inhibit the spread of wild-type virus. Hum Gene Ther 3:391–397

Buehler DC, Marsden MD, Shen S et al (2014) Bioengineered vaults: self-assembling protein shell-lipophilic core nanoparticles for drug delivery. ACS Nano 8:7723–7732

Burns JC, Friedmann T, Driever W et al (1993) Vesicular stomatitis virus G glycoprotein pseudotyped retroviral vectors: concentration to very high titer and efficient gene transfer into mammalian and nonmammalian cells. Proc Natl Acad Sci USA 90:8033–8037

Capasso C, Garofalo M, Hirvinen M et al (2014) The evolution of adenoviral vectors through genetic and chemical surface modifications. Viruses 6:832–855

Caruso M, Klatzmann D (1992) Selective killing of CD4+ cells harboring a human immunodeficiency virus-inducible suicide gene prevents viral spread in an infected cell population. Proc Natl Acad Sci U S A 89:182–186

Chatterjee S, Johnson PR, Wong KK Jr (1992) Dual-target inhibition of HIV-1 in vitro by means of an adeno-associated virus antisense vector. Science 258:1485–1488

Chen SY, Bagley J, Marasco WA (1994) Intracellular antibodies as a new class of therapeutic molecules for gene therapy. Hum Gene Ther 5:595–601

Cheng L, Ma J, Li J et al (2017) Blocking type I interferon signaling enhances T cell recovery and reduces HIV-1 reservoirs. J Clin Invest 127:269–279

Coburn GA, Cullen BR (2002) Potent and specific inhibition of human immunodeficiency virus type 1 replication by RNA interference. J Virol 76:9225–9231

Contu L, La Nasa G, Arras M et al (1993) Allogeneic bone marrow transplantation combined with multiple anti-HIV-1 treatment in a case of AIDS. Bone Marrow Transplant 12:669–671

Das AT, Brummelkamp TR, Westerhout EM et al (2004) Human immunodeficiency virus type 1 escapes from RNA interference-mediated inhibition. J Virol 78:2601–2605

De Silva Feelixge HS, Stone D, Pietz HL et al (2016) Detection of treatment-resistant infectious HIV after genome-directed antiviral endonuclease therapy. Antiviral Res 126:90–98

Dean M, Carrington M, Winkler C et al (1996) Genetic restriction of HIV-1 infection and progression to AIDS by a deletion allele of the CKR5 structural gene. Hemophilia Growth And Development Study, multicenter AIDS Cohort Study, Multicenter Hemophilia Cohort Study,

San Francisco City Cohort, ALIVE Study [erratum appears in Science 1996 Nov 15;274 (5290):1069]. Science 273:1856–1862

Deeks SG, Wagner B, Anton PA et al (2002) A phase II randomized study of HIV-specific T-cell gene therapy in subjects with undetectable plasma viremia on combination antiretroviral therapy. Mol Ther 5:788–797

Deng J, Qu X, Lu P et al (2017) Specific and stable suppression of HIV provirus expression In vitro by chimeric zinc finger DNA methyltransferase 1. Mol Ther Nucleic acids 6:233–242

Digiusto DL, Krishnan A, Li L et al (2010) RNA-based gene therapy for HIV with lentiviral vector-modified CD34(+) cells in patients undergoing transplantation for AIDS-related lymphoma. Sci Transl Med 2:36ra43

Digiusto DL, Cannon PM, Holmes MC et al (2016) Preclinical development and qualification of ZFN-mediated CCR5 disruption in human hematopoietic stem/progenitor cells. Mol Ther Methods Clin Dev 3:16067

Dipaolo N, Ni S, Gaggar A et al (2006) Evaluation of adenovirus vectors containing serotype 35 fibers for vaccination. Mol Ther 13:756–765

Donahue RE, Bunnell BA, Zink MC et al (1998) Reduction in SIV replication in rhesus macaques infused with autologous lymphocytes engineered with antiviral genes. Nat Med 4:181–186

Dorman NM, Lever AM (2001) Investigation of RNA transcripts containing HIV-1 packaging signal sequences as HIV-1 antivirals: generation of cell lines resistant to HIV-1. Gene Ther 8:157–165

Douglas JL, Gustin JK, Viswanathan K et al (2010) The great escape: viral strategies to counter BST-2/tetherin. PLoS Pathog 6:e1000913

Durand CM, Ghiaur G, Siliciano JD et al (2012) HIV-1 DNA is detected in bone marrow populations containing CD4+ T cells but is not found in purified CD34+ hematopoietic progenitor cells in most patients on antiretroviral therapy. J Infect Dis 205:1014–1018

Edagwa B, Mcmillan J, Sillman B et al (2017) Long-acting slow effective release antiretroviral therapy. Expert Opin Drug Deliv:1–11

Eisele E, Siliciano RF (2012) Redefining the viral reservoirs that prevent HIV-1 eradication (Review). Immunity 37:377–388

Eleftheriadou I, Dieringer M, Poh XY et al (2017) Selective transduction of astrocytic and neuronal CNS subpopulations by lentiviral vectors pseudotyped with Chikungunya virus envelope. Biomaterials 123:1–14

Escarpe P, Zayek N, Chin P et al (2003) Development of a sensitive assay for detection of replication-competent recombinant lentivirus in large-scale HIV-based vector preparations. Mol Ther 8:332–341

Follenzi A, Ailles LE, Bakovic S et al (2000) Gene transfer by lentiviral vectors is limited by nuclear translocation and rescued by HIV-1 pol sequences. Nat Genet 25:217–222

Friedmann T (1992) A brief history of gene therapy. Nat Genet 2:93–98

Fuchs SP, Martinez-Navio JM, Piatak M Jr et al (2015) AAV-delivered antibody mediates significant protective effects against SIVmac239 challenge in the absence of neutralizing activity. PLoS Pathog 11:e1005090

Gabarre J, Leblond V, Sutton L et al (1996) Autologous bone marrow transplantation in relapsed HIV-related non-Hodgkin's lymphoma. Bone Marrow Transplant 18:1195–1197

Gaj T, Epstein BE, Schaffer DV (2016) Genome engineering using adeno-associated virus: basic and clinical research applications. Mol Ther 24:458–464

Garcia F, Climent N, Guardo AC et al (2013) A dendritic cell-based vaccine elicits T cell responses associated with control of HIV-1 replication. Sci Transl Med 5:166ra2

Gardner MR, Kattenhorn LM, Kondur HR et al (2015) AAV-expressed eCD4-Ig provides durable protection from multiple SHIV challenges. Nature 519:87–91

Garg H, Joshi A (2016) Conditional cytotoxic anti-HIV gene therapy for selectable cell modification. Hum Gene Ther 27:400–415

Gay W, Lauret E, Boson B et al (2004) Low autocrine interferon beta production as a gene therapy approach for AIDS: infusion of interferon beta-engineered lymphocytes in macaques chronically infected with SIVmac251. Retrovirology 1:29

Geyer MB, Brentjens RJ (2016) Review: current clinical applications of chimeric antigen receptor (CAR) modified T cells. Cytotherapy 18:1393–1409

Goldrick BA (2003) Bubonic plague and HIV. The delta 32 connection. Am J Nurs 103:26–27

Green M, Ishino M, Loewenstein PM (1989) Mutational analysis of HIV-1 Tat minimal domain peptides: identification of trans-dominant mutants that suppress HIV-LTR-driven gene expression. Cell 58:215–223

Hacein-Bey-Abina S, Garrigue A, Wang GP et al (2008) Insertional oncogenesis in 4 patients after retrovirus-mediated gene therapy of SCID-X1. J Clin Invest 118:3132–3142

Haim H, Si Z, Madani N et al (2009) Soluble CD4 and CD4-mimetic compounds inhibit HIV-1 infection by induction of a short-lived activated state. PLoS Pathog 5:e1000360

Hale M, Lee B, Honaker Y et al (2017a) Homology-directed recombination for enhanced engineering of chimeric antigen receptor T cells. Mol Ther Methods Clin Dev 4:192–203

Hale M, Mesojednik T, Romano Ibarra GS et al (2017b) Engineering HIV-resistant, anti-HIV chimeric antigen receptor T cells. Mol Ther 25:570–579

Hamm TE, Rekosh D, Hammarskjold ML (1999) Selection and characterization of human immunodeficiency virus type 1 mutants that are resistant to inhibition by the transdominant negative RevM10 protein. J Virol 73:5741–5747

Hanley PJ, Cruz CR, Savoldo B et al (2009) Functionally active virus-specific T cells that target CMV, adenovirus, and EBV can be expanded from naive T-cell populations in cord blood and will target a range of viral epitopes. Blood 114:1958–1967

Hanley PJ, Melenhorst JJ, Nikiforow S et al (2015) CMV-specific T cells generated from naive T cells recognize atypical epitopes and may be protective in vivo. Sci Transl Med 7:285ra63

Hansen SG, Ford JC, Lewis MS et al (2011) Profound early control of highly pathogenic SIV by an effector memory T-cell vaccine. Nature 473:523–527

Harrison GS, Long CJ, Maxwell F et al (1992) Inhibition of HIV production in cells containing an integrated, HIV-regulated diphtheria toxin A chain gene. AIDS Res Hum Retroviruses 8:39–45

Hassett JM, Zaroulis CG, Greenberg ML et al (1983) Bone marrow transplantation in AIDS. N Engl J Med 309:665

Hayakawa J, Washington K, Uchida N et al (2009) Long-term vector integration site analysis following retroviral mediated gene transfer to hematopoietic stem cells for the treatment of HIV infection. PLoS ONE 4:e4211

Henrich TJ, Hu Z, Li JZ et al (2013) Long-term reduction in peripheral blood HIV type 1 reservoirs following reduced-intensity conditioning allogeneic stem cell transplantation. J Infect Dis 207:1694–1702

Henrich TJ, Hanhauser E, Marty FM et al (2014) Antiretroviral-free HIV-1 remission and viral rebound after allogeneic stem cell transplantation: report of 2 cases. Ann Intern Med 161: 319–327

Henrich TJ, Hanhauser E, Hu Z et al (2015) Viremic control and viral coreceptor usage in two HIV-1-infected persons homozygous for CCR5 Delta32. AIDS 29:867–876

Herberman RB (1992) Adoptive therapy with purified CD8 cells in HIV infection. Semin Hematol 29:35–40

Hermann FG, Martinius H, Egelhofer M et al (2009) Protein scaffold and expression level determine antiviral activity of membrane-anchored antiviral peptides. Hum Gene Ther 20: 325–336

Herrera-Carrillo E, Liu YP, Berkhout B (2014) The impact of unprotected T cells in RNAi-based gene therapy for HIV-AIDS. Mol Ther 22:596–606

Hildinger M, Dittmar MT, Schult-Dietrich P et al (2001) Membrane-anchored peptide inhibits human immunodeficiency virus entry. J Virol 75:3038–3042

Hill AL, Rosenbloom DI, Goldstein E et al (2016) Real-time predictions of reservoir size and rebound time during antiretroviral therapy interruption trials for HIV. PLoS Pathog 12: e1005535

Ho M, Armstrong J, Mcmahon D et al (1993) A phase 1 study of adoptive transfer of autologous CD8+ T lymphocytes in patients with acquired immunodeficiency syndrome (AIDS)-related complex or AIDS. Blood 81:2093–2101

Holland HK, Saral R, Rossi JJ et al (1989) Allogeneic bone marrow transplantation, zidovudine, and human immunodeficiency virus type 1 (HIV-1) infection. Studies in a patient with non-Hodgkin lymphoma. Ann Intern Med 111:973–981

Horn PA, Keyser KA, Peterson LJ et al (2004) Efficient lentiviral gene transfer to canine repopulating cells using an overnight transduction protocol. Blood 103:3710–3716

Horster A, Teichmann B, Hormes R et al (1999) Recombinant AAV-2 harboring gfp-antisense/ ribozyme fusion sequences monitor transduction, gene expression, and show anti-HIV-1 efficacy. Gene Ther 6:1231–1238

Howe SJ, Mansour MR, Schwarzwaelder K et al (2008) Insertional mutagenesis combined with acquired somatic mutations causes leukemogenesis following gene therapy of SCID-X1 patients. J Clin Invest 118:3143–3150

Huang Y, Paxton WA, Wolinsky SM et al (1996) The role of a mutant CCR5 allele in HIV-1 transmission and disease progression. Nat Med 2:1240–1243

Hutter G, Ganepola S (2011) Eradication of HIV by transplantation of CCR5-deficient hematopoietic stem cells. Sci World J 11:1068–1076

Hutter G, Nowak D, Mossner M et al (2009) Long-term control of HIV by CCR5 Delta32/Delta32 stem-cell transplantation. N Engl J Med 360:692–698

Iwakuma T, Cui Y, Chang LJ (1999) Self-inactivating lentiviral vectors with U3 and U5 modifications. Virology 261:120–132

Jacobson SK, Calne RY, Wreghitt TG (1991) Outcome of HIV infection in transplant patient on cyclosporin. Lancet 337:794

Jaspers JE, Brentjens RJ (2017) Development of CAR T cells designed to improve antitumor efficacy and safety. Pharmacol Ther

Ji H, Jiang Z, Lu P et al (2016) Specific reactivation of latent HIV-1 by dCas9-SunTag-VP64-mediated Guide RNA targeting the HIV-1 promoter. Mol Ther 24:508–521

Jin Q, Marsh J, Cornetta K et al (2008) Resistance to human immunodeficiency virus type 1 (HIV-1) generated by lentivirus vector-mediated delivery of the CCR5Δ32 gene despite detectable expression of the HIV-1 co-receptors. J Gen Virol 89:2611–2621

Johnson PR, Schnepp BC, Zhang J et al (2009) Vector-mediated gene transfer engenders long-lived neutralizing activity and protection against SIV infection in monkeys. Nat Med 15:901–906

Johnston C, Harrington R, Jain R et al (2016) Safety and efficacy of combination antiretroviral therapy in human immunodeficiency virus-infected adults undergoing autologous or allogeneic hematopoietic cell transplantation for hematologic malignancies. Biol Blood Marrow Transpl 22:149–156

Jonnalagadda M, Brown CE, Chang WC et al (2013) Engineering human T cells for resistance to methotrexate and mycophenolate mofetil as an in vivo cell selection strategy. PLoS ONE 8: e65519

Kaminski R, Bella R, Yin C et al (2016) Excision of HIV-1 DNA by gene editing: a proof-of-concept in vivo study. Gene Ther 23:690–695

Kang EM, De Witte M, Malech H et al (2002a) Gene therapy-based treatment for HIV-positive patients with malignancies. J Hematother Stem Cell Res 11:809–816

Kang EM, De Witte M, Malech H et al (2002b) Nonmyeloablative conditioning followed by transplantation of genetically modified HLA-matched peripheral blood progenitor cells for hematologic malignancies in patients with acquired immunodeficiency syndrome. Blood 99:698–701

Katuwal M, Wang Y, Schmitt K et al (2014) Cellular HIV-1 inhibition by truncated old world primate APOBEC3A proteins lacking a complete deaminase domain. Virology 468–470: 532–544

Kaushal S, La Russa VF, Gartner S et al (1996) Exposure of human CD34+ cells to human immunodeficiency virus type 1 does not influence their expansion and proliferation of hematopoietic progenitors in vitro. Blood 88:130–137

Kearns K, Bahner I, Bauer G et al (1997) Suitability of bone marrow from HIV-1-infected donors for retrovirus-mediated gene transfer. Hum Gene Ther 8:301–311

Kiem HP, Wu RA, Sun G et al (2010) Foamy combinatorial anti-HIV vectors with MGMTP140 K potently inhibit HIV-1 and SHIV replication and mediate selection in vivo. Gene Ther 17: 37–49

Kim YG, Cha J, Chandrasegaran S (1996) Hybrid restriction enzymes: zinc finger fusions to Fok I cleavage domain. Proc Natl Acad Sci U S A 93:1156–1160

Kim SS, Peer D, Kumar P et al (2010) RNAi-mediated CCR5 silencing by LFA-1-targeted nanoparticles prevents HIV infection in BLT mice. Mol Ther 18:370–376

Kitchen SG, Levin BR, Bristol G et al (2012) In vivo suppression of HIV by antigen specific T cells derived from engineered hematopoietic stem cells. PLoS Pathog 8:e1002649

Klimas NG (1992) Clinical impact of adoptive therapy with purified CD8 cells in HIV infection. Semin Hematol 29:40–43; Discuss 43–44

Koenig S, Conley AJ, Brewah YA et al (1995) Transfer of HIV-1-specific cytotoxic T lymphocytes to an AIDS patient leads to selection for mutant HIV variants and subsequent disease progression. Nat Med 1:330–336

Kohn DB, Bauer G, Rice CR et al (1999) A clinical trial of retroviral-mediated transfer of a rev-responsive element decoy gene into CD34(+) cells from the bone marrow of human immunodeficiency virus-1-infected children. Blood 94:368–371

Kole R, Krainer AR, Altman S (2012) RNA therapeutics: beyond RNA interference and antisense oligonucleotides. Nat Rev Drug Discov 11:125–140

Kundu SK, Engleman E, Benike C et al (1998) A pilot clinical trial of HIV antigen-pulsed allogeneic and autologous dendritic cell therapy in HIV-infected patients. AIDS Res Hum Retroviruses 14:551–560

Lam S, Sung J, Cruz C et al (2015) Broadly-specific cytotoxic T cells targeting multiple HIV antigens are expanded from HIV+ patients: implications for immunotherapy. Mol Ther 23:387–395

Lane HC, Masur H, Longo DL et al (1984) Partial immune reconstitution in a patient with the acquired immunodeficiency syndrome. N Engl J Med 311:1099–1103

Lane HC, Zunich KM, Wilson W et al (1990) Syngeneic bone marrow transplantation and adoptive transfer of peripheral blood lymphocytes combined with zidovudine in human immunodeficiency virus (HIV) infection. Ann Intern Med 113:512–519

Lebbink RJ, De Jong DC, Wolters F et al (2017) A combinational CRISPR/Cas9 gene-editing approach can halt HIV replication and prevent viral escape. Sci Rep 7:41968

Lee TC, Sullenger BA, Gallardo HF et al (1992) Overexpression of RRE-derived sequences inhibits HIV-1 replication in CEM cells. New Biol 4:66–74

Lee SW, Gallardo HF, Gilboa E et al (1994) Inhibition of human immunodeficiency virus type 1 in human T cells by a potent Rev response element decoy consisting of the 13-nucleotide minimal Rev-binding domain. J Virol 68:8254–8264

Lee NS, Dohjima T, Bauer G et al (2002) Expression of small interfering RNAs targeted against HIV-1 rev transcripts in human cells. Nat Biotechnol 20:500–505

Legiewicz M, Badorrek CS, Turner KB et al (2008) Resistance to RevM10 inhibition reflects a conformational switch in the HIV-1 Rev response element. Proc Natl Acad Sci U S A 105:14365–14370

Leslie GJ, Wang J, Richardson MW et al (2016) Potent and broad inhibition of HIV-1 by a peptide from the gp41 heptad repeat-2 domain conjugated to the CXCR4 amino terminus. PLoS Pathog 12:e1005983

Levine BL, Mosca JD, Riley JL et al (1996) Antiviral effect and ex vivo CD4+ T cell proliferation in HIV-positive patients as a result of CD28 costimulation. Science 272:1939–1943

Li MJ, Bauer G, Michienzi A et al (2003) Inhibition of HIV-1 infection by lentiviral vectors expressing Pol III-promoted anti-HIV RNAs. Mol Ther 8:196–206

Lieberman J, Skolnik PR, Parkerson GR 3rd et al (1997) Safety of autologous, ex vivo-expanded human immunodeficiency virus (HIV)-specific cytotoxic T-lymphocyte infusion in HIV-infected patients. Blood 90:2196–2206

Limsirichai P, Gaj T, Schaffer DV (2016) CRISPR-mediated activation of latent HIV-1 expression. Mol Ther 24:499–507

Lisziewicz J, Sun D, Smythe J et al (1993) Inhibition of human immunodeficiency virus type 1 replication by regulated expression of a polymeric Tat activation response RNA decoy as a strategy for gene therapy in AIDS. Proc Natl Acad Sci U S A 90:8000–8004

Liu Q, Huang W, Zhang H et al (2014) Neutralizing antibodies against AAV2, AAV5 and AAV8 in healthy and HIV-1-infected subjects in China: implications for gene therapy using AAV vectors. Gene Ther 21:732–738

Liu L, Patel B, Ghanem MH et al (2015) Novel CD4-based bispecific chimeric antigen receptor designed for enhanced anti-HIV potency and absence of HIV entry receptor activity. J Virol 89:6685–6694

Liu B, Zou F, Lu L et al (2016) Chimeric antigen receptor T cells guided by the single-chain Fv of a broadly neutralizing antibody specifically and effectively eradicate virus reactivated from latency in CD4+ T lymphocytes isolated from HIV-1-infected individuals receiving suppressive combined antiretroviral therapy. J Virol 90:9712–9724

Lombardo A, Genovese P, Beausejour CM et al (2007) Gene editing in human stem cells using zinc finger nucleases and integrase-defective lentiviral vector delivery. Nat Biotechnol 25:1298–1306

Lori F, Lisziewicz J, Smythe J et al (1994) Rapid protection against human immunodeficiency virus type 1 (HIV-1) replication mediated by high efficiency non-retroviral delivery of genes interfering with HIV-1 tat and gag. Gene Ther 1:27–31

Maetzig T, Galla M, Baum C et al (2011) Gammaretroviral vectors: biology, technology and application. Viruses 3:677–713

Marschall AL, Dubel S, Boldicke T (2015) Specific in vivo knockdown of protein function by intrabodies. MAbs 7:1010–1035

Martinez-Navio JM, Fuchs SP, Pedreno-Lopez S et al (2016) Host anti-antibody responses following adeno-associated virus-mediated delivery of antibodies against HIV and SIV in rhesus monkeys. Mol Ther 24:76–86

Masiero S, Del Vecchio C, Gavioli R et al (2005) T-cell engineering by a chimeric T-cell receptor with antibody-type specificity for the HIV-1 gp120. Gene Ther 12:299–310

Matsuda Z, Yu X, Yu QC et al (1993) A virion-specific inhibitory molecule with therapeutic potential for human immunodeficiency virus type 1. Proc Natl Acad Sci U S A 90:3544–3548

Mavigner M, Watkins B, Lawson B et al (2014) Persistence of virus reservoirs in ART-treated SHIV-infected rhesus macaques after autologous hematopoietic stem cell transplant. PLoS Pathog 10:e1004406

Mcintyre GJ, Groneman JL, Yu YH et al (2009) 96 shRNAs designed for maximal coverage of HIV-1 variants. Retrovirology 6:55

Mcnamara LA, Onafuwa-Nuga A, Sebastian NT et al (2013) CD133+ hematopoietic progenitor cells harbor HIV genomes in a subset of optimally treated people with long-term viral suppression. J Infect Dis 207:1807–1816

Meyer J, Nick S, Stamminger T et al (1993) Inhibition of HIV-1 replication by a high-copy-number vector expressing antisense RNA for reverse transcriptase. Gene 129:263–268

Mitsuyasu RT, Anton PA, Deeks SG et al (2000) Prolonged survival and tissue trafficking following adoptive transfer of CD4zeta gene-modified autologous CD4(+) and CD8(+) T cells in human immunodeficiency virus-infected subjects. Blood 96:785–793

Mitsuyasu RT, Merigan TC, Carr A et al (2009) Phase 2 gene therapy trial of an anti-HIV ribozyme in autologous CD34+ cells. Nat Med 15:285–292

Miyoshi H, Blomer U, Takahashi M et al (1998) Development of a self-inactivating lentivirus vector. J Virol 72:8150–8157

Mock U, Machowicz R, Hauber I et al (2015) mRNA transfection of a novel TAL effector nuclease (TALEN) facilitates efficient knockout of HIV co-receptor CCR5. Nucleic Acids Res 43:5560–5571

Morton J, Davis MW, Jorgensen EM et al (2006) Induction and repair of zinc-finger nuclease-targeted double-strand breaks in Caenorhabditis elegans somatic cells. Proc Natl Acad Sci U S A 103:16370–16375

Nazari R, Joshi S (2008) CCR5 as target for HIV-1 gene therapy. Curr Gene Ther 8:264–272

Neagu MR, Ziegler P, Pertel T et al (2009) Potent inhibition of HIV-1 by TRIM5-cyclophilin fusion proteins engineered from human components. J Clin Invest 119:3035–3047

Neff CP, Zhou J, Remling L et al (2011) An aptamer-siRNA chimera suppresses HIV-1 viral loads and protects from helper CD4(+) T cell decline in humanized mice. Sci Transl Med 3:66ra6

Nishimura T, Kaneko S, Kawana-Tachikawa A et al (2013) Generation of rejuvenated antigen-specific T cells by reprogramming to pluripotency and redifferentiation. Cell Stem Cell 12:114–126

Norton TD, Miller EA, Bhardwaj N et al (2015) Vpx-containing dendritic cell vaccine induces CTLs and reactivates latent HIV-1 in vitro. Gene Ther 22:227–236

Novina CD, Murray MF, Dykxhoorn DM et al (2002) siRNA-directed inhibition of HIV-1 infection. Nat Med 8:681–686

Olszko ME, Trobridge GD (2013) Foamy virus vectors for HIV gene therapy. Viruses 5: 2585–2600

Onafuwa-Nuga A, Mcnamara LA, Collins KL (2010) Towards a cure for HIV: the identification and characterization of HIV reservoirs in optimally treated people. Cell Res 20:1185–1187

Page KA, Landau NR, Littman DR (1990) Construction and use of a human immunodeficiency virus vector for analysis of virus infectivity. J Virol 64:5270–5276

Park RJ, Wang T, Koundakjian D et al (2017) A genome-wide CRISPR screen identifies a restricted set of HIV host dependency factors. Nat Genet 49:193–203

Parolin C, Dorfman T, Palu G et al (1994) Analysis in human immunodeficiency virus type 1 vectors of cis-acting sequences that affect gene transfer into human lymphocytes. J Virol 68:3888–3895

Patel S, Jones RB, Nixon DF et al (2016a) T-cell therapies for HIV: preclinical successes and current clinical strategies. Cytotherapy 18:931–942

Patel S, Lam S, Cruz CR et al (2016b) Functionally active HIV-specific T cells that target gag and nef can be expanded from virus-naive donors and target a range of viral epitopes: implications for a cure strategy after allogeneic hematopoietic stem cell transplantation. Biol Blood Marrow Transpl: J Am Soc Blood Marrow Transplant 22:536–541

Perez EE, Riley JL, Carroll RG et al (2005) Suppression of HIV-1 infection in primary CD4 T cells transduced with a self-inactivating lentiviral vector encoding a membrane expressed gp41-derived fusion inhibitor. Clin Immunol 115:26–32

Peterson CW, Haworth KG, Burke BP et al (2016a) Multilineage polyclonal engraftment of Cal-1 gene-modified cells and in vivo selection after SHIV infection in a nonhuman primate model of AIDS. MolTherMethods ClinDev 3:16007

Peterson CW, Wang J, Norman KK et al (2016b) Long-term multilineage engraftment of autologous genome-edited hematopoietic stem cells in nonhuman primates. Blood 127: 2416–2426

Peterson CW, Benne C, Polacino P et al (2017) Loss of immune homeostasis dictates SHIV rebound after stem-cell transplantation. JCI Insight 2:e91230

Pham QT, Bouchard A, Grutter MG et al (2010) Generation of human TRIM5alpha mutants with high HIV-1 restriction activity. Gene Ther 17:859–871

Pichlmair A, Diebold SS, Gschmeissner S et al (2007) Tubulovesicular structures within vesicular stomatitis virus G protein-pseudotyped lentiviral vector preparations carry DNA and stimulate antiviral responses via Toll-like receptor 9. J Virol 81:539–547

Planelles V, Bachelerie F, Jowett JB et al (1995) Fate of the human immunodeficiency virus type 1 provirus in infected cells: a role for vpr. J Virol 69:5883–5889

Plavec I, Voytovich A, Moss K et al (1996) Sustained retroviral gene marking and expression in lymphoid and myeloid cells derived from transduced hematopoietic progenitor cells. Gene Ther 3:717–724

Qin XF, An DS, Chen IS et al (2003) Inhibiting HIV-1 infection in human T cells by lentiviral-mediated delivery of small interfering RNA against CCR5. Proc Natl Acad Sci 100:183–188

Ranga U, Woffendin C, Verma S et al (1998) Enhanced T cell engraftment after retroviral delivery of an antiviral gene in HIV-infected individuals. Proc Natl Acad Sci 95:1201–1206

Relander T, Johansson M, Olsson K et al (2005) Gene transfer to repopulating human CD34+ cells using amphotropic-, GALV-, or RD114-pseudotyped HIV-1-based vectors from stable producer cells. Mol Ther 11:452–459

Renneisen K, Leserman L, Matthes E et al (1990) Inhibition of expression of human immunodeficiency virus-1 in vitro by antibody-targeted liposomes containing antisense RNA to the env region. J Biol Chem 265:16337–16342

Richter M, Saydaminova K, Yumul R et al (2016) In vivo transduction of primitive mobilized hematopoietic stem cells after intravenous injection of integrating adenovirus vectors. Blood 128:2206–2217

Riddell SR, Greenberg PD, Overell RW et al (1992) Phase I study of cellular adoptive immunotherapy using genetically modified CD8+ HIV-specific T cells for HIV seropositive patients undergoing allogeneic bone marrow transplant. The Fred Hutchinson Cancer Research Center and the University of Washington School of Medicine, Department of Medicine. Div Oncol. Hum Gene Ther 3:319–338

Riddell SR, Elliott M, Lewinsohn DA et al (1996) T-cell mediated rejection of gene-modified HIV-specific cytotoxic T lymphocytes in HIV-infected patients. Nat Med 2:216–223

Riolobos L, Hirata RK, Turtle CJ et al (2013) HLA engineering of human pluripotent stem cells. Mol Ther 21:1232–1241

Roberts MR, Qin L, Zhang D et al (1994) Targeting of human immunodeficiency virus-infected cells by CD8+ T lymphocytes armed with universal T-cell receptors. Blood 84:2878–2889

Rossi JJ (2011) RNA nanoparticles come of age. Acta Biochim Biophys Sin (Shanghai) 43:245–247

Roy U, Rodriguez J, Barber P et al (2015) The potential of HIV-1 nanotherapeutics: from in vitro studies to clinical trials. Nanomedicine (Lond) 10:3597–3609

Rubin RH, Jenkins RL, Shaw BW Jr et al (1987) The acquired immunodeficiency syndrome and transplantation. Transplantation 44:1–4

Rustanti L, Jin H, Lor M et al (2017) A mutant Tat protein inhibits infection of human cells by strains from diverse HIV-1 subtypes. Virol J 14:52

Saayman SM, Lazar DC, Scott TA et al (2016) Potent and targeted activation of latent HIV-1 using the CRISPR/dCas9 activator complex. Mol Ther 24:488–498

Sanhadji K, Leissner P, Firouzi R et al (1997) Experimental gene therapy: the transfer of Tat-inducible interferon genes protects human cells against HIV-1 challenge in vitro and in vivo in severe combined immunodeficient mice. AIDS 11:977–986

Santoni De Sio FR, Cascio P, Zingale A et al (2006) Proteasome activity restricts lentiviral gene transfer into hematopoietic stem cells and is down-regulated by cytokines that enhance transduction. Blood 107:4257–4265

Sarver N, Cantin EM, Chang PS et al (1990) Ribozymes as potential anti-HIV-1 therapeutic agents. Science 247:1222–1225

Sather BD, Romano Ibarra GS, Sommer K et al (2015) Efficient modification of CCR5 in primary human hematopoietic cells using a megaTAL nuclease and AAV donor template. Sci Transl Med 7:307ra156

Sawyer SL, Wu LI, Emerman M et al (2005) Positive selection of primate TRIM5alpha identifies a critical species-specific retroviral restriction domain. Proc Natl Acad Sci 102:2832–2837

Saydaminova K, Ye X, Wang H et al (2015) Efficient genome editing in hematopoietic stem cells with helper-dependent Ad5/35 vectors expressing site-specific endonucleases under microRNA regulation. Mol Ther Methods Clin Dev 1:14057

Schaller T, Bauby H, Hue S et al (2014) New insights into an X-traordinary viral protein. Front Microbiol 5:126

Schleifman EB, Mcneer NA, Jackson A et al (2013) Site-specific genome editing in PBMCs with PLGA nanoparticle-delivered PNAs confers HIV-1 resistance in humanized mice. Mol Ther Nucleic Acids 2:e135

Schneider E, Lambermont M, Van Vooren JP et al (1997) Autologous stem cell infusion for acute myeloblastic leukemia in an HIV-1 carrier. Bone Marrow Transplant 20:611–612

Scholler J, Brady TL, Binder-Scholl G et al (2012) Decade-long safety and function of retroviral-modified chimeric antigen receptor T cells. Sci Transl Med 4:132ra53

Segura MM, Alba R, Bosch A et al (2008) Advances in helper-dependent adenoviral vector research. Curr Gene Ther 8:222–235

Setiawan LC, Kootstra NA (2015) Adaptation of HIV-1 to rhTrim5alpha-mediated restriction in vitro. Virology 486:239–247

Shen H, Cheng T, Preffer FI et al (1999) Intrinsic human immunodeficiency virus type 1 resistance of hematopoietic stem cells despite coreceptor expression. J Virol 73:728–737

Sheridan C (2011) Gene therapy finds its niche. Nat Biotechnol 29:121–128

Simonds RJ (1993) HIV transmission by organ and tissue transplantation. AIDS 7(Suppl 2): S35–S38

Singwi S, Joshi S (2000) Potential nuclease-based strategies for HIV gene therapy (Review). Front Biosci 5:D556–D579

Sloan DD, Lam CY, Irrinki A et al (2015) Targeting HIV reservoir in infected CD4 T Cells by dual-affinity re-targeting molecules (DARTs) that bind HIV envelope and recruit cytotoxic T Cells. PLoS Pathog 11:e1005233

Smith DH, Byrn RA, Marsters SA et al (1987) Blocking of HIV-1 infectivity by a soluble, secreted form of the CD4 antigen. Science 238:1704–1707

Stein S, Ott MG, Schultze-Strasser S et al (2010) Genomic instability and myelodysplasia with monosomy 7 consequent to EVI1 activation after gene therapy for chronic granulomatous disease. Nat Med 16:198–204

Steinberger P, Andris-Widhopf J, Buhler B et al (2000) Functional deletion of the CCR5 receptor by intracellular immunization produces cells that are refractory to CCR5-dependent HIV-1 infection and cell fusion. Proc Natl Acad Sci U S A 97:805–810

Sullenger BA, Gallardo HF, Ungers GE et al (1990) Overexpression of TAR sequences renders cells resistant to human immunodeficiency virus replication. Cell 63:601–608

Sullivan N, Sun Y, Binley J et al (1998) Determinants of human immunodeficiency virus type 1 envelope glycoprotein activation by soluble CD4 and monoclonal antibodies. J Virol 72: 6332–6338

Sung JA, Pickeral J, Liu L et al (2015) Dual-affinity re-targeting proteins direct T cell-mediated cytolysis of latently HIV-infected cells. J Clin Invest 125:4077–4090

Symons J, Vandekerckhove L, Hütter G et al (2014) Dependence on the CCR5 coreceptor for viral replication explains the lack of rebound of CXCR4-predicted HIV variants in the Berlin patient. Clin Infect Dis 59:596–600

Szewczyk M, Drzewinska J, Dzmitruk V et al (2012) Stability of dendriplexes formed by anti-HIV genetic material and poly(propylene imine) dendrimers in the presence of glucosaminoglycans. J Phys Chem B 116:14525–14532

Tan R, Xu X, Ogg GS et al (1999) Rapid death of adoptively transferred T cells in acquired immunodeficiency syndrome. Blood 93:1506–1510

Taylor JA, Vojtech L, Bahner I et al (2008) Foamy virus vectors expressing anti-HIV transgenes efficiently block HIV-1 replication. Mol Ther 16:46–51

Tebas P, Stein D, Binder-Scholl G et al (2013) Antiviral effects of autologous CD4 T cells genetically modified with a conditionally replicating lentiviral vector expressing long antisense to HIV.[Erratum appears in Blood. 2014 Jul 24;124(4):663]. Blood 121:1524–1533

Tebas P, Stein D, Tang WW et al (2014) Gene editing of CCR5 in autologous CD4 T cells of persons infected with HIV. N Engl J Med 370:901–910

Torlontano G, Di Bartolomeo P, Di Girolamo G et al (1992) AIDS-related complex treated by antiviral drugs and allogeneic bone marrow transplantation following conditioning protocol with busulphan, cyclophosphamide and cyclosporin. Haematologica 77:287–290

Trickett AE, Kelly M, Cameron BA et al (1998) A preliminary study to determine the effect of an infusion of cryopreserved autologous lymphocytes on immunocompetence and viral load in HIV-infected patients. J Acquir Immune Defic Syndr Hum Retrovirol 17:129–136

Trobridge GD, Beard BC, Gooch C et al (2008) Efficient transduction of pigtailed macaque hemtopoietic repopulating cells with HIV-based lentiviral vectors. Blood 111:5537–5543

Trobridge GD, Wu RA, Beard BC et al (2009) Protection of stem cell-derived lymphocytes in a primate AIDS gene therapy model after in vivo selection. PLoS ONE 4:e7693

Trobridge GD, Wu RA, Hansen M et al (2010) Cocal-pseudotyped lentiviral vectors resist inactivation by human serum and efficiently transduce primate hematopoietic repopulating cells. Mol Ther 18:725–733

Trono D, Feinberg MB, Baltimore D (1989) HIV-1 gag mutants can dominantly interfere with the replication of the wild-type virus. Cell 59:113–120

Turner ML, Watson HG, Russell L et al (1992) An HIV positive haemophiliac with acute lymphoblastic leukaemia successfully treated with intensive chemotherapy and syngeneic bone marrow transplantation. Bone Marrow Transplant 9:387–389

Urnov FD, Miller JC, Lee YL et al (2005) Highly efficient endogenous human gene correction using designed zinc-finger nucleases. Nature 435:646–651

Varela-Rohena A, Molloy PE, Dunn SM et al (2008) Control of HIV-1 immune escape by CD8 T cells expressing enhanced T-cell receptor. Nat Med 14:1390–1395

Venkatesh LK, Arens MQ, Subramanian T et al (1990) Selective induction of toxicity to human cells expressing human immunodeficiency virus type 1 Tat by a conditionally cytotoxic adenovirus vector. Proc Natl Acad Sci U S A 87:8746–8750

Vercruysse T, Pardon E, Vanstreels E et al (2010) An intrabody based on a llama single-domain antibody targeting the N-terminal alpha-helical multimerization domain of HIV-1 rev prevents viral production. J Biol Chem 285:21768–21780

Veres G, Escaich S, Baker J et al (1996) Intracellular expression of RNA transcripts complementary to the human immunodeficiency virus type 1 gag gene inhibits viral replication in human CD4+ lymphocytes. J Virol 70:8792–8800

Walker JE, Chen RX, Mcgee J et al (2012) Generation of an HIV-1-resistant immune system with CD34(+) hematopoietic stem cells transduced with a triple-combination anti-HIV lentiviral vector. J Virol 86:5719–5729

Wang H, Liu Y, Li Z et al (2008) In vitro and in vivo properties of adenovirus vectors with increased affinity to CD46. J Virol 82:10567–10579

Wang Z, Tang Z, Zheng Y et al (2010) Development of a nonintegrating Rev-dependent lentiviral vector carrying diphtheria toxin A chain and human TRAF6 to target HIV reservoirs. Gene Ther 17:1063–1076

Wang CX, Sather BD, Wang X et al (2014) Rapamycin relieves lentiviral vector transduction resistance in human and mouse hematopoietic stem cells. Blood 124:913–923

Wang J, Exline CM, Declercq JJ et al (2015) Homology-driven genome editing in hematopoietic stem and progenitor cells using ZFN mRNA and AAV6 donors. Nat Biotechnol 33:1256–1263

Wang Z, Guo Y, Han W (2017) Current status and perspectives of chimeric antigen receptor modified T cells for cancer treatment. Protein & cell: [Epub ahead of print 2017 May 2]

Weerasinghe M, Liem SE, Asad S et al (1991) Resistance to human immunodeficiency virus type 1 (HIV-1) infection in human CD4+ lymphocyte-derived cell lines conferred by using retroviral vectors expressing an HIV-1 RNA-specific ribozyme. J Virol 65:5531–5534

Wheeler LA, Trifonova R, Vrbanac V et al (2011) Inhibition of HIV transmission in human cervicovaginal explants and humanized mice using CD4 aptamer-siRNA chimeras. J Clin Invest 121:2401–2412

Whiteside TL, Elder EM, Moody D et al (1993) Generation and characterization of ex vivo propagated autologous CD8+ cells used for adoptive immunotherapy of patients infected with human immunodeficiency virus. Blood 81:2085–2092

Woffendin C, Ranga U, Yang Z et al (1996) Expression of a protective gene-prolongs survival of T cells in human immunodeficiency virus-infected patients. Proc Natl Acad Sci 93:2889–2894

Wu X, Liu H, Xiao H et al (1996) Inhibition of human and simian immunodeficiency virus protease function by targeting Vpx-protease-mutant fusion protein into viral particles. J Virol 70:3378–3384

Xu H, Svarovskaia ES, Barr R et al (2004) A single amino acid substitution in human APOBEC3G antiretroviral enzyme confers resistance to HIV-1 virion infectivity factor-induced depletion. Proc Natl Acad Sci U S A 101:5652–5657

Yam P, Jensen M, Akkina R et al (2006) Ex vivo selection and expansion of cells based on expression of a mutated inosine monophosphate dehydrogenase 2 after HIV vector transduction: effects on lymphocytes, monocytes, and CD34+ stem cells. Mol Ther 14:236–244

Yanez-Munoz RJ, Balaggan KS, Macneil A et al (2006) Effective gene therapy with nonintegrating lentiviral vectors. Nat Med 12:348–353

Younan PM, Polacino P, Kowalski JP et al (2013) Positive selection of mC46-expressing CD4+ T cells and maintenance of virus specific immunity in a primate AIDS model. Blood 122: 179–187

Younan PM, Peterson CW, Polacino P et al (2015) Lentivirus-mediated gene transfer in hematopoietic stem cells is impaired in SHIV-infected, ART-treated nonhuman primates. Mol Ther 23:943–951

Yuan J, Wang J, Crain K et al (2012) Zinc-finger nuclease editing of human cxcr4 promotes HIV-1 CD4+ T cell resistance and enrichment. Mol Ther 20:849–859

Yukl SA, Boritz E, Busch M et al (2013) Challenges in detecting HIV persistence during potentially curative interventions: a study of the Berlin patient. PLoS Pathog 9:e1003347

Zaia JA, Rossi JJ, Murakawa GJ et al (1988) Inhibition of human immunodeficiency virus by using an oligonucleoside methylphosphonate targeted to the tat-3 gene. J Virol 62:3914–3917

Zennou V, Petit C, Guetard D et al (2000) HIV-1 genome nuclear import is mediated by a central DNA flap. Cell 101:173–185

Zhen A, Kamata M, Rezek V et al (2015) HIV-specific immunity derived from chimeric antigen receptor-engineered stem cells. Mol Ther: J Am Soc Gene Ther 23:1358–1367

Zhen A, Rezek V, Youn C et al (2017) Targeting type I interferon-mediated activation restores immune function in chronic HIV infection. J Clin Invest 127:260–268

Zhou J, Rossi J (2014) Cell-type-specific aptamer and aptamer-small interfering RNA conjugates for targeted human immunodeficiency virus type 1 therapy. J Investig Med 62:914–919

Zhou J, Li H, Li S et al (2008) Novel dual inhibitory function aptamer-siRNA delivery system for HIV-1 therapy. Mol Ther 16:1481–1489

Zhou J, Swiderski P, Li H et al (2009) Selection, characterization and application of new RNA HIV gp 120 aptamers for facile delivery of Dicer substrate siRNAs into HIV infected cells. Nucleic Acids Res 37:3094–3109

Zhou J, Satheesan S, Li H et al (2015) Cell-specific RNA aptamer against human CCR5 specifically targets HIV-1 susceptible cells and inhibits HIV-1 infectivity. Chem Biol 22:379–390

Zufferey R, Dull T, Mandel RJ et al (1998) Self-inactivating lentivirus vector for safe and efficient in vivo gene delivery. J Virol 72:9873–9880

Zufferey R, Donello JE, Trono D et al (1999) Woodchuck hepatitis virus posttranscriptional regulatory element enhances expression of transgenes delivered by retroviral vectors. J Virol 73:2886–2892

Printed in the United States
By Bookmasters